'*Doing Spatial History* provides a rich [...] sources, spaces and concepts useful fo[r...] cal research. More than that, it showcases vibrant and accessible conversations across historical disciplines, different periods and an array of places, underlining the enormous potential of the emerging and energetic field of spatial history.'

Diarmid A. Finnegan, *Queen's University Belfast, UK*

'With an ecumenical spirit, *Doing Spatial History* offers a concrete guide to researching the lived experience of the past. Authors from various fields show us how sources ranging from police records and maritime contracts to novels and maps can enrich and complicate the humanities. Researchers will especially appreciate the introduction, which offers both a helpful entry point to the field and an overview of its literature.'

Susan Schulten, *University of Denver, USA*

'For a spatial historian who has been working for some 20 years to introduce space as an analytical category into the historical and cultural sciences, it is a particular pleasure to see this "guide to using historical sources" published. It is long overdue. For real and imagined places – from bars to borders – the volume shows what difference "doing spatial history" can make.'

Susanne Rau, *University of Erfurt, Germany*

Doing Spatial History

This volume provides a practical introduction to spatial history through the lens of the different primary sources that historians use. It is informed by a range of analytical perspectives and conveys a sense of the various facets of spatial history in a tangible, case-study based manner.

The chapter authors hail from a variety of fields, including early modern and modern history, architectural history, historical anthropology, economic and social history, as well as historical and human geography, highlighting the way in which spatial history provides a common forum that facilitates discussion across disciplines. The geographical scope of the volume takes readers on a journey through central, western, and east central Europe, to Russia, the Mediterranean, the Ottoman Empire, and East Asia, as well as North and South America, and New Zealand. Divided into three parts, the book covers particular types of sources, different kinds of space, and specific concepts, tools and approaches, offering the reader a thorough understanding of how sources can be used within spatial history specifically but also the different ways of looking at history more broadly.

Very much focussing on *doing* spatial history, this is an accessible guide for both undergraduate and postgraduate students within modern history and its related fields.

Riccardo Bavaj is Professor of Modern History, University of St Andrews, UK, and Co-Director of the Institute for Transnational and Spatial History. His research focusses on the intellectual and spatial history of 20th-century Germany. He has co-edited *Germany and 'the West'* and *Zivilisatorische Verortungen* (with Martina Steber).

Konrad Lawson is Lecturer in Modern History, University of St Andrews, UK, and Co-Director of the Institute for Transnational and Spatial History. His research focusses on modern East Asian history and the aftermaths of Japanese empire.

Bernhard Struck is Reader of Modern History, University of St Andrews, UK, and Founding Director of the Institute for Transnational and Spatial History. His research focusses on continental European History, c. 1750 to early 20th century, comparative and transnational history. He has co-edited *Shaping the Transnational Sphere* (with Davide Rodogno and Jakob Vogel).

The Routledge Guides to Using Historical Sources

How does the historian approach primary sources? How do interpretations differ? How can such sources be used to write history?

The *Routledge Guides to Using Historical Sources* series introduces students to different sources and illustrates how historians use them. Titles in the series offer a broad spectrum of primary sources and, using specific examples, examine the historical context of these sources and the different approaches that can be used to interpret them.

Reading Russian Sources
A Student's Guide to Text and Visual Sources from Russian History
Edited by George Gilbert

History and Economic Life
A Student's Guide to approaching Economic and Social History sources
Edited by Georg Christ and Philipp R. Roessner

Approaching Historical Sources in their Contexts
Space, Time and Performance
Edited by Sarah Barber and Corinna M. Peniston-Bird

Reading Primary Sources
The Interpretation of Texts from Nineteenth and Twentieth Century History, 2nd edition
Edited by Miriam Dobson and Benjamin Ziemann

Sources for the History of Emotions
A Guide
Edited by Katie Barclay, Sharon Crozier-De Rosa and Peter N. Stearns

Games of History
Games and Gaming as Historical Sources
Apostolos Spanos

Doing Spatial History
Edited by Riccardo Bavaj, Konrad Lawson and Bernhard Struck

Sources in the History of Psychiatry, from 1800 to the Present
Edited by Chris Millard and Jennifer Wallis

For more information about this series, please visit: https://www.routledge.com/Routledge-Guides-to-Using-Historical-Sources/book-series/RGHS

Doing Spatial History

Edited by Riccardo Bavaj, Konrad Lawson, and Bernhard Struck

LONDON AND NEW YORK

Cover image: Students Occupying Odeon Theatre in Paris, May 16, 1968 (b/w photo). Photo © AGIP / Bridgeman Images

First published 2022
by Routledge
2 Park Square, Milton Park, Abingdon, Oxon OX14 4RN

and by Routledge
605 Third Avenue, New York, NY 10158

Routledge is an imprint of the Taylor & Francis Group, an informa business

© 2022 selection and editorial matter, Riccardo Bavaj, Konrad Lawson and Bernhard Struck; individual chapters, the contributors

The right of Riccardo Bavaj, Konrad Lawson and Bernhard Struck to be identified as the authors of the editorial material, and of the authors for their individual chapters, has been asserted in accordance with sections 77 and 78 of the Copyright, Designs and Patents Act 1988.

All rights reserved. No part of this book may be reprinted or reproduced or utilised in any form or by any electronic, mechanical, or other means, now known or hereafter invented, including photocopying and recording, or in any information storage or retrieval system, without permission in writing from the publishers.

Trademark notice: Product or corporate names may be trademarks or registered trademarks, and are used only for identification and explanation without intent to infringe.

British Library Cataloguing-in-Publication Data
A catalogue record for this book is available from the British Library

Library of Congress Cataloging-in-Publication Data
A catalog record has been requested for this book

ISBN: 978-0-367-26154-2 (hbk)
ISBN: 978-0-367-26156-6 (pbk)
ISBN: 978-0-429-29173-9 (ebk)

DOI: 10.4324/9780429291739

Typeset in Times New Roman
by SPi Technologies India Pvt Ltd (Straive)

Contents

List of figures ix
List of contributors xi

Introduction: Spatial history – an expansive field 1
RICCARDO BAVAJ

PART I
Working with sources 37

1 **Maps** 39
  BERNHARD STRUCK AND RICCARDO BAVAJ

2 **Travel guides** 56
  JAMES KORANYI

3 **Novels, autobiographies, and memoirs** 73
  SARAH DEUTSCH

4 **Newspaper archives** 87
  SHERRY OLSON AND PETER HOLLAND

5 **Architectural drawings** 102
  DESPINA STRATIGAKOS

PART II
Exploring spaces 119

6 **Ships** 121
  MATTHEW YLITALO AND SARAH EASTERBY-SMITH

viii *Contents*

7	**Bars** KATE FERRIS	139
8	**Rivers** MARK HARRIS	154
9	**Infrastructures** FRITHJOF BENJAMIN SCHENK	171
10	**Border zones** LISA HELLMAN	188

PART III
Reflecting on concepts, tools, and approaches — 205

11	**Lefebvrean landscapes** DAWN HOLLIS	207
12	**Maritoriality** MICHAEL TALBOT	222
13	**Regional imaginaries** KONRAD LAWSON	237
14	**Economic geographies** ANTONIS HADJIKYRIACOU	252
15	**Digital mapping** TIM COLE AND ALBERTO GIORDANO	274

Selected bibliography — 288
Index — 295

Figures

1.1	Cover sheet, *Mappa Polski* [1827]	40
1.2	Map of 'administrative partition and key places', *Mappa Polski* [1827]	44
1.3	Map of 'manufactories and factories', *Mappa Polski* [1827]	46
1.4	Map of 'religious rites', *Mappa Polski* [1827]	48
1.5	Map of 'languages', *Mappa Polski* [1827]	48
2.1	Photograph of woman from rural Caraș-Severin in the Banat	64
2.2	Map of the Burzenland	65
2.3	Map of the church fortress at Viscri / Deutschweisskirch	66
2.4	Sketch of detail of door	67
4.1	'Dorothy Fitzgerald reading to her siblings', c. 1905: Older children supervised much of the play and learning of younger children, and organized their adventures on the hills, at the shore or in the bush	91
5.1	Heinrich Hoffmann, postcard of the Berghof, c. 1936	104
5.2	Alois Degano, ground-floor plan for the expansion of Haus Wachenfeld, dated 16 November 1935 and approved by local building authorities on 22 January 1936	106
5.3	Heinrich Hoffmann, postcard of the Great Hall, c. 1936. The marble fireplace is visible at the far end of the room. To the right, through the opening, is the small living room. At the far right, a tapestry hides a movie screen	107
5.4	Heinrich Hoffmann, view of the window in the Great Hall, c. 1936. Note the large map table in front of the window, and to the left, the globe satirized by Charlie Chaplin in his 1940 film, *The Great Dictator*	107
5.5	Photograph of Neville Chamberlain (left of Hitler) and other guests having tea with Hitler in the Great Hall of the Berghof, as seen on the cover of the 24 September 1938 issue of *Anglo-German Review*. The visit occurred on 15 September 1938, when the British prime minister travelled to the Obersalzberg to discuss the international crisis brewing over Hitler's insistence on invading Czechoslovakia	111

5.6	An unsigned ground-floor plan for the expanded Berghof, not built, n.d.	114
5.7	An unsigned second-floor plan for the expanded Berghof showing the library (*Bibliothek*), not built, n.d. Immediately to the right of the library staircase was the entrance to Hitler's study	115
6.1	Whaling barque *SS Esquimaux* at ice edge, 1890s, with her reflection visible in calm water in the foreground. Note the steam funnel as well as the 'crow's nest' (lookout platform) fastened to the main mast, with man inside	123
6.2	Map of the North Atlantic, showing Newfoundland, Baffin, and Greenland. Dundee, which is on the east coast of central Scotland, is indicated on the map but is not named. The city is due west from Bell Rock, on the north bank of the inlet (Firth of Tay)	124
8.1	Map of the Amazon area of Guiana. It belongs to the work "Relation de la Riviere des Amazones", Paris, 1680	159
8.2	German map of the Guianas based on Ralegh's account, 1599 (John Carter Brown, J590 B915v GVL8.1 / 2-SIZE, reproduced with the kind permission of the John Carter Brown Library)	161
8.3	Part of a map of the Amazon showing riverbank Jesuit missions and other centres of the population, 1753 (Biblioteca Pública de Evora, GAV 4 N25)	162
8.4	Map by Manuel Ferreira of the lower and middle parts of Tapajós River as far as Jesuits have worked	165
9.1	Friedrich List's *Das deutsche Eisenbahn-System*, 1833	174
9.2	Map of the Russian Empire, 1900	176
9.3	Building of the Moscow-Vindavo-Rybinsk Railway Company in Saint Petersburg (today *Vitebskii vokzal*), erected 1900–1904	178
10.1	Isaac Titsingh's (c.1740–1812) depictions of the Chinese (above) and the Dutch (below) trading quarters	194
14.1	Digital Elevation Model (DEM) and distribution of extant villages	260
14.2	Heatmap showing spatial distribution of wheat production	262
14.3	Heatmap showing spatial distribution of wine production on the basis of the grape juice tax	264
14.4	Heatmap showing spatial distribution of wine consumption of the basis of the tavern tax	264
14.5	Heatmap showing spatial distribution of olive production	265
14.6	Heatmap showing spatial distribution of olive oil production	265
14.7	Heatmap showing spatial distribution of cotton production	266
14.8	Heatmap showing spatial distribution of linen production	267
14.9	Heatmap showing spatial distribution of silk production	267
14.10	Heatmap showing spatial distribution of honey production	268
14.11	Heatmap showing spatial distribution of carob production	268

Contributors

Riccardo Bavaj is Professor of Modern History, University of St Andrews, and Co-Director of the Institute for Transnational and Spatial History. His research focusses on the intellectual and spatial history of 20th-century Germany. His publications include *Zivilisatorische Verortungen: Der 'Westen' an der Jahrhundertwende (1880-1930) [Civilisational Mappings]* (ed. with Martina Steber) (2018), *Germany and 'the West': The History of a Modern Concept* (ed. with Martina Steber) (2015), *Der Nationalsozialismus: Entstehung, Aufstieg und Herrschaft [A Spatial History of Nazism]* (2016).

Tim Cole is Professor of Social History and Director of the Bristow Institute, University of Bristol. His research interests cover social and environmental history, historical geography, and digital humanities, Holocaust and memory landscapes. His is a founding member (together with Alberto Giordano) of the 'Holocaust Geographies Collaborative'. His publications include *The Holocaust in the Twenty-First Century: Relevance and Challenges in the Digital Age* (ed. with Simone Gigliotti) (2021), *Holocaust Landscapes* (2016), *Traces of the Holocaust: Journeying In and Out of the Ghettos* (2013).

Sarah Deutsch is Professor of History, Duke University. Her research areas focus on racial, gender, and class differences in relation to spatial formations and urban history in the 19th and 20th centuries. Her publications include *Making a Modern U.S. West: The Contested Terrain of a Region and Its Borders, 1898–1940* (2022), *Women and the City: Gender, Space and Power in Boston, 1870–1940* (2000), and *From Ballots to Breadlines: American Women, 1920–1940* (1994).

Sarah Easterby-Smith is Senior Lecturer of Modern History, University of St Andrews. Her research revolves around the history of science, cultural and social history of science in Britain and France, the production and movement of natural knowledge in the 18th century, networks of natural historians and botanists around 1800. Her publications include *Cultivating Commerce: Cultures of Botany in Britain and France, 1760–1815* (2017) and 'Recalcitrant Seeds: Material Culture and the Global History of Science', *Past & Present* (2019).

Kate Ferris is Reader of Modern History, University of St Andrews, and Principal Investigator of the ERC-funded research project 'Dictatorship as Experience: A Comparative History of Everyday Life and "Lived Experience" of Dictatorship in Mediterranean Europe (1922–1975)'. Her publications include *Everyday Life in Fascist Venice, 1929–1940* (2012), *Imagining 'America' in Late Nineteenth Century Spain* (2016), and *The Politics of Everyday Life in Fascist Italy: Outside the State?* (ed. with Joshua Arthurs and Michael Ebner) (2017).

Alberto Giordano is Professor of Geography, Texas State University. He is past President of the University Consortium for Geographic Information Science (UCGIS) and a Fulbright recipient. His research interests include holocaust and genocide geography (Italy, Budapest, Armenia) and historical GIS, as well as policy applications of GIScience. He is a founding member of the 'Holocaust Geographies Collaborative' (www.holocaust-geographies.org). His publications include *Geographies of the Holocaust* (ed. with Tim Cole and Anne Kelly Knowles) (2014).

Antonis Hadjikyriacou is Adjunct Lecturer of Ottoman and Turkish history at Panteion University, Athens, and Affiliated Scholar at the Center for Spatial and Textual Analysis (CESTA) at Stanford University. His research focusses on social and economic history, geographic information systems (GIS), environmental and climate history, and the spatial history of the Mediterranean and Ottoman world. His publications include *Insularity in the Ottoman World*, special issue of *Princeton Papers* (ed.) (2017), and 'The Respatialization of Cypriot Insularity during the Age of Revolutions', in Matthias Middell and Megan Maruschke (eds.), *The French Revolution as a Moment of Respatialization* (2019).

Mark Harris is Professor of Historical Anthropology, University of St Andrews. His research focusses on peasantries, peasant social life, and their history along the waterways of the Amazon river and its tributaries. His publications include *Life on the Amazon: The Anthropology of a Brazilian Peasant Village* (2000), *Rebellion on the Amazon: The Cabanagem, Race, and Popular Culture in North Brazil, 1798–1840* (2010), 'Rethinking Amerindian Spaces in Brazilian History', *Ethnohistory* (2018) (with Silvia Espelt-Bombin).

Lisa Hellman is Research Leader for the group 'Coerced Circulation of Knowledge' at the Bonn Center for Dependency and Slavery Studies, and a Pro Futura fellow at the Swedish Collegium for Advanced Study. She works at the intersection between social, cultural, maritime, and global history in East and Central Asia. Her publications include 'Enslaved in Dzungaria: What an Eighteenth-Century Crocheting Instructor Can Teach Us about Overland Globalisation', *Journal of Global History* (2021), and *This House Is Not a Home: European Everyday Life in Canton and Macao 1730–1830* (2018).

Peter Holland (†) was Professor of Geography, University of Otago. His research was focused on colonists' experiences and environmental knowledge in New Zealand as well as the transformation of native ecosystems into productive farm land and the perception of landscapes. His publications include *Home in the Howling Wilderness: Settlers and the Environment in Southern New Zealand* (2013) and 'Learning about the Environment in Early Colonial New Zealand', in Tom Brooking and Eric Pawson (eds.), *Seeds of Empire: The Environmental Transformation of New Zealand* (2010).

Dawn Hollis is Leverhulme Postdoctoral Research Fellow, University of St Andrews. Her research focusses on the human experience of mountains and mountainous landscapes during the early modern and modern period as well as mountaineering and its construction as a modern sport. She currently works on the Leverhulme-funded project 'Mountains in Ancient Literature and Culture and Their Postclassical Reception'. Her publications include 'Mountain Gloom and Mountain Glory', *ISLE* (2019) and *Mountain Dialogues from Antiquity to Modernity* (ed. with Jason König) (2021).

James Koranyi is Associate Professor of Modern History, University of Durham. He is a historian of east-central Europe. His research focus revolves around topics including memory culture, minorities and diasporas, nationalism, and landscape mainly on Romania, Serbia, Hungary, and Germany in the 19th and 20th century. He is the author of 'Space: Empires, Nations, and Borders', in Irina Livezeanu and Árpád von Klimó (eds.), *The Routledge History of East Central Europe since 1700* (2015) (with Bernhard Struck) and *Migrating Memories: Romanian Germans in Modern Europe* (2022).

Konrad Lawson is Lecturer in East Asian History, University of St Andrews, and Co-Director of the Institute for Transnational and Spatial History. He is a transnational historian focussing on modern East Asia, especially the early aftermaths of Japanese empire throughout the region. His publications include 'Universal Crime, Particular Punishment: Trying the Atrocities of the Japanese Occupation as Treason in the Philippines, 1947–1953', *Comparativ* (2013) and 'Reimagining the Postwar International Order', in Simon Jackson and Alanna O'Malley (eds.), *The Institution of International Order* (2018).

Sherry Olson is Professor Emerita of Geography, McGill University. A historical geographer, her research focusses on the intersection between economy, culture, and environment. She has worked on forest history, railways, and the pricing of water. More recent work has focussed on urban history and urban environments in relation to social and economic questions as well as public health issues and disease (especially in Montreal). Her publications include *Peopling the North American City: Montreal, 1840–1900* (2011) (with Patricia A. Thornton) and 'Appetite for Grass:

Re-Engineering Landscapes of Otago and Southland, 1864–1914', *New Zealand Geographer* (2020) (with Peter Holland).

Frithjof Benjamin Schenk is Professor of Eastern European History, University of Basel. His research focusses on Tsarist Russia from the 18th to the 20th century, the history of memory, imperial biographies, infrastructure planning, railways and mobility, and urban history. His publications include *Russlands Fahrt in die Moderne: Mobilität und sozialer Raum im Eisenbahnzeitalter [Russia's Journey into Modernity]* (2014) and *Alexander Nevskij: Heiliger – Fürst – Nationalheld: Eine Erinnerungsfigur im russischen kulturellen Gedächtnis, 1263–2000* (2004).

Despina Stratigakos is Professor of Architectural History, University of Buffalo. Her research revolves around themes that connect architecture and power, urban history, and the role of gender in architecture in the 20th century. Her publications include *Hitler's Northern Utopia: Building the New Order in Occupied Norway* (2020), *Where Are the Women Architects* (2016), *Hitler at Home* (2015), *A Women's Berlin: Building the Modern City* (2008).

Bernhard Struck is Reader of Modern History, University of St Andrews, and founding Director of the Institute for Transnational and Spatial History. His research interests include the history of Germany, Poland, and France in the 18th and 19th centuries, transnational and comparative history, the history of travel and cartography. His publications include *Nicht West – nicht Ost. Frankreich und Polen in der Wahrnehmung deutscher Reisender zwischen 1750 und 1850 [Neither West nor East]* (2006) and *Revolution, Krieg und Verflechtungen: Deutschland und Frankreich, 1789–1815* (2008).

Michael Talbot is Senior Lecturer in the History of the Ottoman Empire and Modern Middle East, University of Greenwich. His research focusses on maritime law and practices, urban history, and Ottoman interactions with the wider world. In 2018 he was selected as one of the BBC and AHRC 'New Generation Thinkers'. His publications include *British-Ottoman Relations, 1661–1807: Commerce and Diplomatic Practice in Eighteenth-Century Istanbul* (2017) and 'Separating the Waters from the Sea: The Place of Islands in Ottoman Maritime Territoriality during the Eighteenth Century', *Princeton Papers* (2017).

Matthew Ylitalo is a PhD researcher at the University of St Andrews and the University of Dundee. His training background is in history, classics, and geography. His research, funded by the Scottish Graduate School for the Arts & Humanities (SGSAH), focusses on Dundee whaling, its social, economic, and scientific dimensions as well as its global and trans-maritime connections between the 1820s and 1920s. His publications include 'Maritime Labour and Economic Opportunity: Shetlanders and the Dundee Arctic Whaling Trade during the Late Nineteenth Century', *International Journal of Maritime History* (2019).

Introduction: Spatial history
An expansive field

Riccardo Bavaj

History is a matter of time. Historians, it might be argued, are primarily experts in temporal matters: They are concerned with chronology, continuity, and change; they explore origins and historical processes; they explain what happened, when, and why; they situate matters in their historical context and treat them as historically conditioned. Anachronism and presentism are among the historian's worst enemies, as are assumptions of transhistorical timelessness. History is made through time, and historians historicize. Of course, this description of the historian's profession is highly stylized, to the point of caricature.[1] Nonetheless, it captures important elements of the scholarly self-understanding, methodological approach, and disciplinary skill set of historians. This volume aims to show that historical scholarship benefits from a heightened sensitivity to the *spatial* dimensions of history. It illuminates ways of doing *spatial history*.

Varieties of spatial history

Spatial history comes in a wide variety of forms. For example, it can involve a greater attentiveness to the physicality and materiality of spaces – to the fact that history quite literally 'takes place'. It can also comprise an analysis of spaces as imagined and discursively constructed. Here we might think of mental maps produced by travel guides and landscape paintings, as well as imaginaries evoked by spatial concepts ranging from 'the West', 'the Balkans', and 'the Global South' to 'city', 'countryside', and 'home'.

Spatial history also engages with practices of territorialization, the drawing of borders and creation of infrastructure. It dissects the processes of knowledge production, which are generated in and shaped by distinct sites such as laboratories, salons, and universities. It explores spaces both as constituted by and as constituting social relations and human interaction. Space no longer appears as a mere stage, unaffected by (nor itself affecting) the social interaction unfolding on it. Instead, it is produced by this very social interaction, and it is also itself an agent in the constitution of society. To be sure, spatial history also draws on maps. But it uses maps less as an illustration of seemingly objective geographical realities, and more as primary sources in their own right. Maps appear as forms of symbolic appropriation

DOI: 10.4324/9780429291739-1

and spatial knowledge; they can be read as visual modes of persuasion. Finally, spatial history can involve the *creation* of maps, among other forms of data visualization. It can entail the use of computational methods embraced by historical geographic information systems (GIS) for the purpose of generating new insights as well as new questions.

Needless to say, spatial history also comes in other forms. Some scholars have a fairly expansive, multi-faceted sense of this emerging field, while others see it as more or less synonymous with historical GIS. This might seem confusing. However, it is hardly surprising. We need only consider spatial history's manifold origins and the divergent trajectories across both national and disciplinary academic communities that have shaped it.

In Anglo-American academia, for instance, 'spatial history' is now frequently associated with GIS-driven, data-intensive research. This is often clustered around specific institutional centres and collaborative networks. Over the years, this field has been populated by a variety of practitioners from the discipline of history.[2] Initially, however, it was shaped largely by historical geographers and environmental historians – the former particularly prominent in Britain, and the latter in North America.[3] To give two examples: (1) The Stanford-based 'Spatial History Project' was founded in 2007 by environmental historian Richard White. In an influential, agenda-setting paper – appropriately titled 'What Is Spatial History?' – White carefully defined the purpose of this project. According to this paper, spatial history comprises the creation of visual representations 'of the interrelation of time and space' through digital tools. The underlying rationale is to use such visualizations as 'a means of doing research', rather than for illustration and effective communication.[4] (2) The 'Digital Humanities Hub' at the University of Lancaster emerged from various GIS-based initiatives. Since 2006, these initiatives have been led by Ian Gregory. Though formally a member of staff in a history department, Gregory is a geographer by training, and he has long been at the forefront of historical GIS.[5] He has also clearly been inspired by White's conceptualization of spatial history. Over the past ten years, Gregory has invoked this term in order to describe a 'more applied', research question–driven approach that transcends a mainly technology-centred, database-focused remit – an approach, to be sure, that remains very much rooted in data visualization and analysis.[6] This understanding has an earlier precursor in Loren Siebert's 'GIS spatial history of Tokyo' (1997).[7] Only recently, however, has it gained traction. Importantly, this approach to data visualization informs the over 600-page long *Routledge Companion to Spatial History* (2018), a milestone in the field, as well as the Indiana University Press series 'Spatial Humanities' (2010–). As the name suggests, the latter publishing outlet has a wider thematic and interdisciplinary scope. Nonetheless, it remains equally committed to GIS and to other cognate tools of digital mapping.[8]

From the viewpoint of historical geography and environmental history, this narrower definition of 'spatial history' certainly makes sense. For historical geographers, a more expansive understanding of the term may threaten to duplicate the remit of their own subdiscipline. This is broadly concerned with 'the geographies of the past', as well as 'the influence of the past in shaping the geographies of the present and the future'.[9] A more expansive understanding may also place too much emphasis on 'history'. This might be unpalatable to representatives of a field that, with some notable exceptions, has long been institutionally integrated into geography departments.[10] To some geographers, moreover, the term 'spatial' might still smack of a predominately quantitative, abstract, and pattern-fixated spatial science once in fashion during the 1950s and 1960s (more on this further below).[11] Likewise, the eyebrows of environmental historians might be raised by a notion of 'spatial history' that is fundamentally – and not just digitally – concerned with the interaction between humans and the physical environment, and even with 'nature'. To these historians, such a formulation of spatial history may appear as a superstructure that is at best superfluous and at worst an unwanted redesign of scholarly identities.[12]

And yet, as this volume suggests, spatial history is more than simply '*digital* spatial history'.[13] Indeed, the term 'spatial history' has been invoked in a variety of scholarly contexts, well beyond the realm of digital mapping.[14] It was also used well *before* historical GIS began to take off. The classic point of reference is *The Road to Botany Bay*, Paul Carter's study of the exploration and colonial settlement of Australia in the late 18th and 19th centuries. This book was first published in 1987, just in time for the following year's bicentenary of white settlement in Australia. *The Road to Botany Bay* dissected acts of place-naming, surveying, mapping, and other practices of symbolic appropriation as ways in which Australia was 'brought into being'.[15] Carter drew on a range of primary sources, including explorers' journals, travel accounts, memoirs, letters, poems, paintings, maps, and photographs. His narrative homed in on the very process of exploring, journeying, and settling, as well as the act of writing about these activities.

The Road to Botany Bay was Carter's first book. It was clearly marked by his initial preoccupations as a writer, literary critic, and cultural theorist. However, it also foreshadowed some of the future traits of his transdisciplinary profile as a scholar of creative research, place-making, and urban design. Carter had emigrated from Britain to Australia in the early 1980s. He understood his book both as a way of dealing with his own disorienting impressions of his new home country and as a new mode of writing history.[16]

Carter's historiographical agenda was fittingly captured by the subtitle of the book's first British edition: *An Essay in Spatial History*. Certainly, subsequent American publishers found more marketable variations on this subtitle, which was soon rendered as *An Exploration of Landscape and History*. However, this did nothing to detract from the prominence ascribed to the term 'spatial history' throughout the text. It is frequently invoked, and it comes to serve as the reader's steady companion. Very self-consciously, Carter pitched his analysis of the linguistic and cultural construction of landscape as a means to counter a particular approach to historical writing. This approach had reduced 'space to a stage' – a stage which had somehow always been there, naturally and objectively, and on which events were 'unfolding in time alone', in accordance with the 'framework of European chronology'. Carter also identified this approach among 'historical geographies', and he referred to it as 'cause-and-effect narrative history'.

For Carter, this kind of history displayed a tendency to 'confuse routes with roots'. It was written from the point of view of an 'all-seeing spectator', seemingly equipped with a 'panoramic eye'. Spatial history, instead, entailed a zooming in on 'spatial experiences', 'fantasies', 'horizons', and practices such as 'choosing directions, applying names, [and] inhabiting the country'. It explored spaces and places in the making, with a focus on the attitudes, desires, and beliefs of those involved in this process. This is why Carter's book was 'a history of roads, footprints, trails of dust and foaming wakes' *as described and delineated* by explorers, observers, surveyors, settlers, town planners, and other place-makers. These actors drew on metaphors and other literary devices. Moreover, Carter's approach refused to submit to a simple chronological structure. It teased out the incongruities, confusions, ironies, tentative moves, back-and-forth reflections, and meandering associations involved in the process of place-making. Carter embraced the idea of 'spatiality as a form of non-linear writing'.[17]

The Road to Botany Bay has been outlined here at length because it brought the term 'spatial history' into being. In addition to this, however, Carter's book still illuminates some of the key facets of what has become a wide-ranging field. Of particular importance is an acute sensitivity to the historical contingency and cultural constructedness of space. At times, of course, Carter oscillates uneasily between the declared emphasis on authorial intentions behind historical sources and an acknowledgement of rhetorical convention. Moreover, the perspectives of scholar and historical actor are often elided. Many such figures, including James Cook and Matthew Flinders, are portrayed as intimately involved in the endeavour of writing and 'making spatial history'.[18] Nonetheless, and despite some stylistic idiosyncrasies and far-fetched allusions, *The Road to Botany Bay* is still worth reading. It certainly deserves more than a mere passing reference in the historiography of spatial history.

Helpfully, the continued relevance of Carter's work was recently acknowledged in an insightful piece on 'spatial history as scholarly practice'. This was published in 2015 by Zephyr Frank, the current director of the Stanford Spatial History project. Frank has long worked at the intersection of history and literature, as shown by his study *Reading Rio de Janeiro*. Accordingly, Frank sees Carter's book as an 'exemplary work of spatial history'. More than this, however, Frank also makes the case for a broader understanding of spatial history as 'a general sensibility about the role of space in explaining historical structure and change as well as the importance of the historical construction of certain kinds of space'. As a 'tool-based technical method', historical GIS certainly offers a fruitful angle for spatial history, but it does not measure the whole terrain.[19]

This conceptualization of spatial history was echoed by historian of science and cartographer William Rankin in a conversation on 'walls, borders and boundaries', published by the *American Historical Review* in 2017. In this discussion, Rankin asked whether digitally focused notions of spatial history might be subject to 'a recency illusion driven by digital enthusiasm'.[20] Rankin is well-versed in GIS and other computational methods. Due to his initial training as an architect, however, he has been equally at home in the analogue spheres of spatial theory.[21]

In this vein, it may also be worth recalling that, in 1995, social historian Raphael Samuel and historical geographer Felix Driver devoted a special issue of the *History Workshop Journal* to the topic of 'Spatial History: Re-Thinking the Idea of Place'. This volume gathered articles from a range of geographers, including Doreen Massey, Jane M. Jacobs, and Charles Withers. The goal was to facilitate cross-disciplinary dialogue, 15 years before any invocations of GIS-focused 'spatial history'.[22] We might also consider here the work of geographer and political philosopher Stuart Elden. Around 2000, Elden was busily searching for a suitable term to describe the centrality of space and place in the thought of Martin Heidegger and Michel Foucault, as well as the (related) analytical work these terms performed for both philosophers. Elden consciously decided on the term 'spatial history', as opposed to a 'history of space', or spaces.[23] Finally, we might profitably cast our gaze beyond the Anglosphere. This quickly reveals that relevant cognates of 'spatial history' typically come with ecumenical meanings. Here, a strictly digital methodological approach has only very recently moved into the semantic orbit of 'spatial history'.[24]

Rankin aptly describes spatial history as 'an emerging subfield at the intersection of history, geography, and the digital humanities'.[25] And yet locating it more precisely remains a challenge. There have been fruitful reflections, for instance, on the relationship between historical geography and environmental history.[26] However, little attempt has been made to situate spatial history in relation to either of these more established fields – beyond terminological discussions between historical geographers such as Alan Baker or Chris Philo over the potential differences between 'historical geography' and 'geographical history'.[27]

The value of spatial history

It will probably come as a relief to many readers that this volume makes no attempt to tease out disciplinary distinctions in minute detail. Nor will there be any exercise in academic boundary-policing, let alone corporate-style 'takeover bids'.[28] Academic fields tend to be notorious shape shifters – ever-changing clusters of knowledge production. The field of historical geography has changed enormously over the past 30 to 40 years. The landscape is very different to the one which faced Paul Carter in the mid-1980s. Today, one would be hard pressed to advance an idea of 'spatial history' as somehow superior to a subdiscipline supposedly still anchored in naïvely empiricist epistemologies. In fact, as I suggest here, the value of 'spatial history' lies elsewhere.

Lure, signpost, and translation aid

On the most basic level, the term 'spatial history' helps to raise awareness among historians that, as stated at the beginning of this introduction, spatial matters are an essential part of historical scholarship. Space, in other words, is not something 'to be left to geographers', to paraphrase David Harvey, one of the most influential geographers of recent decades.[29] More importantly, it serves as a signpost for historians to adapt knowledge on spatial matters from other disciplines. The term 'spatial history' is used here to direct historians to find inspiration in relevant cognate fields. These include historical geography, cultural and human geography, cartography, anthropology, and literary studies, to name but a few.

A heightened sensitivity to the spatial dimensions of history entails possible forays into the pages of the *Journal of Historical Geography*; *Progress in Human Geography*; *Dialogues in Human Geography*; *Environment and Planning*; *Geografiska Annaler*; *Transactions of the Institute of British Geographers*; *Annals of the American Association of Geographers*; *Geoforum*; *GeoHumanities*; *cultural geographies*; *Journal of Cultural Geography*; *Space and Culture*; *Landscape Research*; *Political Geography*; *Geopolitics*; *Journal of Borderlands Studies*; *Borderlands*; *Antipode*; *Emotion, Space and Society*; *Gender, Place & Culture*; *Cartographica*; and *Imago Mundi*.

Pertinent monograph series are also relevant here. These include *Cambridge Studies in Historical Geography* (–2007); *Studies in Historical Geography*; *Tauris Historical Geography Series*; *Routledge Research in Historical Geography*; *Historical Geography and Geosciences*; *Mappings: Society/Theory/Space* (–1998); *Critical Geographies* (–2005); *Culture, Politics, and the Built Environment*; *Space, Place and Society* (–2012); *Place, Memory, Affect*; *Routledge Research in Culture, Space and Identity*; *Penn Studies in Landscape Architecture*; *Geocriticism and Spatial Literary Studies*; *SpatioTemporality/RaumZeitlichkeit*; *Knowledge and Space*; and *Dialectics of the Global*.

Particularly, 'spatial history' provides a useful translation aid that helps mediate between the disciplines of history and geography. Certainly, there have been areas of more intensive exchange in this respect, especially in environmental history and the history of science. Typically, however, historians have only infrequently turned to colleagues from geography departments for advice on conceptual or methodological matters. In fact, a cursory look at introductions and companions to historiography reveals that historical geography has often been considered beyond the remit of historians.[30] Geographers, for their part, have time and again identified common ground between historians and geographers. They have frequently endeavoured to 'bridge the divide'.[31] And yet, most students of history will never have encountered 'historical geography' during their studies.

There have been exceptions. H.P.R. Finberg's *Approaches to History: A Symposium* was published in 1962. It comprised chapters on 'universal history', 'local history', 'archaeology and place-names', the 'history of science', and 'historical geography'. Finberg was Head of the Department of English

Local History at the University of Leicester. This had been established by his predecessor W.G. Hoskins, who was most well-known for his 1955 book *The Making of the English Landscape*. Finberg emphasized the local historian's 'lively topographical sense' and the need for 'a pair of legs not easily tired' when exploring the geology, architecture, and other features of the local landscape. For someone cut from this cloth, it was a natural choice to commission a chapter on historical geography for a 'symposium' on the riches of current historiography.[32] The scholar who contributed this chapter was H.C. Darby, Professor of Geography at University College London and, at the time, Britain's foremost historical geographer.[33]

Finberg's *Approaches to History* has remained a rare example of an introduction to the study of history which includes a contribution from a historical geographer. It was only 50 years later, in 2012, that there appeared another overview of 'research methods for history' with input from historical geography. This contribution focused on 'GIS, spatial technologies and digital mapping'.[34] The choice of topic, as well as the fact that the volume was edited by urban historians (as opposed to a rural landscape-oriented local historian), demonstrates the shifting ground of collaboration between geographers and historians.[35]

But we are dealing here with notable exceptions to a clear trend. On the whole, examples of collaboration have been few and far between. Most surveys of historical writing have dispensed with expertise from the neighbouring field of historical geography. They have been largely bereft of analytically refined geographical perspectives.[36]

'Spatial history' may help to address this blind spot. It may serve to communicate some of the theoretical, conceptual, and thematic preoccupations of geography to history students. Indeed, such a translation aid is, in part, required due to disciplinary differences in citation technique, publishing habits, and writing style, as well as in thinking. On an international level, this translation work becomes even more important. The subdiscipline of historical geography tends to be less prominent beyond the Anglosphere. Here, historians may be even less likely to turn to their counterparts in geography. In Germany, for example, historical geography has been much depleted over the years, with only two chairs left in the field.[37] Place matters.

Boundary spanner, ecumene, and GeoHumanities

As mentioned, 'spatial history' draws historians' attention to the spatial dimensions of history, and to space-related knowledge production beyond the confines of history departments. It also, however, helps to facilitate conversations among historians of different hues and specializations. Part of spatial history's appeal is the potential it holds for the creation of a common forum across, especially, environmental history, landscape history, local and regional history, transnational and global history, urban history, architectural history, the history of cartography, and the history of science. Needless to say, historical GIS is also an aspect of this forum. In this integrative sense, spatial historians act as 'boundary spanners', to use a term from collaborative public management. They build and maintain connections between these and other cognate fields, working at their interstices and mediating knowledge in 'mutually enriching' ways.[38]

Tied in with this notion of 'boundary spanner' is the idea of spatial history as 'ecumene'. The adoption of this word, with its ancient Greek origins, is inspired by polymath and cultural geographer Denis Cosgrove. In 1994, Cosgrove founded a journal under this name (later changed to the more search engine–friendly *cultural geographies*). Originally referring to the 'dwelling place of humankind', for Cosgrove the word was to encapsulate a scholarly 'attachment ... to geographical knowledge and the geographical imagination ... well beyond disciplinary frontiers'.[39] It was conceived both as 'subdisciplinary intervention' and 'transdisciplinary engagement'. As the journal's subsequent editors, Philip Crang and Don Mitchell, would later put it, this enterprise was guided by a commitment to 'forging an ecumenical intellectual space'. This was to be a 'heterogenous space of engagements and transactions' that would allow for a 'variety of intellectual styles and voices', rather than an 'intellectual territory' to be marked and defended.[40]

Moreover, as highlighted in Crang and Mitchell's 2000 editorial, 'genuinely worldly and ecumenical' also meant extending the 'intellectual landscape' beyond the 'usual anglophone North Atlantic axes'.[41] Indeed, it seems clear that this brief must ultimately translate into a global and less Eurocentric 'research space'. Such a space must confront and interrogate 'both its national confinements and international hierarchies'. Precisely this point is forcefully argued by historian Dominic Sachsenmaier in his outline of 'world history as ecumenical history'.[42]

'Spatial history' in this ecumenical sense is one aspect of a much larger conversation which, for the past ten years, has evolved under the name of 'GeoHumanities'. This discussion has been driven in particular by human and cultural geographers such as Tim Cresswell, Deborah P. Dixon, and Harriet Hawkins, with much support from the American Association of Geographers. The goal is to (re-)connect the discipline of geography to the realm of the humanities. This is redolent of the deeper roots of geography, which lie in both classical and Renaissance humanism prior to the discipline's shift towards a quantifying, measuring, calculating spatial science. This process reached its heights in the 1960s, but it remains dominant in some areas of the discipline.

GeoHumanities aims to extend, in a cross-disciplinary vein, the work of 'humanistic geographers' such as Yi-Fu Tuan, Edward Relph, David Lowenthal, and Anne Buttimer.[43] From the 1970s, these scholars advanced a more interpretive, text- and image-based approach centred on place and human experience. They explored how 'meanings arise from human-environment interactions' and 'how humans make the Earth into home'. Likewise, GeoHumanities recognizes – and was to some extent initiated by – a growing attention to space and place within the arts and humanities. This has been particularly marked in literary studies, philosophy, and history, with 'spatial history' expressly counted among its interlocutors.

Like 'spatial history', 'GeoHumanities' comes in both expansive and more narrowly-defined GIS-based forms. It has oscillated between 'geo' in the sense of 'Earth as dwelling place' and 'geo' in the sense of 'location'. However, most practitioners have been leaning towards an understanding that foregrounds 'the concepts of space and place rather than the techniques and epistemologies of GIS'.[44] An awareness of this oscillation, as well as a preference for a more encompassing engagement with space and place, is the basis for an engagement among spatial historians in this most congenial intellectual endeavour.

Analytical focus and conceptual frameworks

For all its many facets, cross-disciplinary connections, and boundary-spanning work, it is precisely the focus on space and place that lends 'spatial history' coherence and a shared perspective.[45] Space and place constitute key organizing principles. They provide analytical focus and inform conceptual frameworks. Spatial history is not so much defined by *what* scholars choose to investigate (mountains, rivers, the sea, and so on), or what kinds of sources they use (maps, maps, and more maps). More importantly, it is (or ought to be) defined by the *way* in which sources are analyzed and by *how* scholars go about writing history. It is a matter of perspective, methodology, and analytical approach – a way of seeing, thinking, and 'knowing'.[46] The next section considers the origins of this specific intellectual approach in more detail.

'The spatial turn'

Spatial means of knowledge production have been promoted ever more widely across the humanities and the social sciences over the past 30 to 40 years. This process of dissemination, as well as the growing uptake of such approaches, is frequently described as 'the spatial turn'. The term was first evoked by US literary critic and theorist Fredric Jameson in 1988. Jameson intended the term to distinguish postmodernism from modernism as a different 'form of interrelationship between time and space' – one in which space, simply put, was given greater weight and significance.[47]

One year after Jameson first invoked the idea of a 'spatial turn', a more important intervention was made by Edward Soja, a US geographer and urban theorist. Soja would become the most enthusiastic proponent of the idea of a 'spatial turn', and he would remain so until his death in 2015. Soja initially used the concept rather tentatively to describe an evolving spatial perspective among the scholarly trajectories of various social critics and cultural theorists, ranging from Henri Lefebvre and Michel Foucault to John Berger and Anthony Giddens.[48] Increasingly, though, Soja developed 'the spatial turn' into a shorthand and rallying cry for what he called – in echo of Jameson – a 'reassertion of space' after the 'triumph of historicism'.

According to both Soja and Jameson, this 'triumph' had gripped large segments of social thought from the late 19th century onwards. For instance, as

Soja wryly observed, some economic theorists conceptualized the economy 'as if it were packed solidly on to the head of a pin, in a fantasy world with virtually no spatial dimensions'.[49] An avowed Marxist, Soja took particular issue with the near-complete absence of a 'geographical imagination' from the universalist, development-oriented, linear progressivist models of Marxist thought. Like its liberal antagonists, Marxism tended to classify the world into backwards/forwards, traditional/modern, thereby not only temporalizing space but outright 'annihilating' it. As early as 1980, Soja was calling for the Marxist focus on the 'social relations of class' to be complemented with a perspective on 'spatial relations of uneven development'. A decade later, he began to explicitly frame his agenda in terms of a 'spatial turn'.[50]

Since the 1990s, the spatial turn has predominantly gained traction in Soja's sense of 'spatializing the historical narrative',[51] rather than in a more narrowly defined conduit of GIS-driven research.[52] It is certainly the case that the increased availability and commercialization of technologies such as GPS (Global Positioning System) and GIS, coupled with the invention of the World Wide Web and the rise of the global internet, have fostered 'geo'-related spatial semantics ('cyberspace', 'web portal', 'network', and so on). However, these semantics were part of a much wider conversation; they were not generally tied to the proliferation of a single technology. The rising interest in space was fuelled by a range of trends, some paradoxical, others self-evident.

These trends included, *first*, the transformation of communication technologies more broadly. This transformation gave several commentators cause to conclude that they were witnessing the eventual formation of a 'global village', as described by media theorist Marshall McLuhan in the 1960s. What David Harvey dubbed 'time-space compression' in 1989,[53] French cultural theorist Paul Virilio a few years later ominously framed as an imminent 'abolition' of space: 'little will remain of the expanse of this planet which will be ... reduced to nothing by the teletechnologies of generalised interactivity'.[54] Others were more cheerful and anticipated the advent of a 'borderless world'. They pointed in particular to the global integration of financial markets and the acceleration of global economic transactions. This would amount to nothing less than 'the end of geography'.[55]

This emerging discourse on a supposed 'annihilation of space' is reminiscent of 19th-century reflections on both capitalism (Karl Marx) and railways (Heinrich Heine).[56] It has had two principal effects. Firstly, 'space' became a talking point precisely because commentators repeatedly forecast its imminent disappearance. Secondly, however, suspicions began to grow about just how 'borderless' or 'spaceless' the world of global capitalism and the internet really was. Critics pointed to the 'limits of despatialization',[57] and to the continued importance of borders and border control. In an age of increased mobility and mobile communication, the question 'Where on Earth are you?' arguably became more rather than less pertinent. Indeed, it is no coincidence that this question also provided the title for a witty and thought-provoking piece on borders, passports, and visas by historian and journalist Frances

Stonor Saunders.[58] Predictions of an 'end of the nation-state' proved premature, or altogether misplaced, while interest in 'space' surged.[59]

The *second* key trend behind the rising interest in space is closely linked to the end of the Cold War. This gave rise to a renewed sense of geo-historical contingency. It re-sensitized scholars to the historical conditionality of geopolitical constellations. Of course, interrogating and thinking beyond Cold War bipolarity had been possible before the wall came down in 1989.[60] Nonetheless, what the leading human geographer Derek Gregory called the 'disclosure of ... taken-for-granted geographical imaginaries' certainly became easier without the discursive straight-jackets of the superpower conflict.[61] To be sure, one of Gregory's intellectual mentors, the literary critic Edward Said, had already engaged with 'imaginative geographies' as early as 1977/78, in his studies on 'Orientalism'.[62] And yet, it was not until the 1990s that a critical mass of scholars, hailing from various disciplines, turned their eyes to spatial imaginaries such as 'the West', 'Eastern Europe', and 'the Third World'.[63] In much the same way, the collapse of the Soviet Union not only triggered a process of deterritorialization but also one of reterritorialization. This saw borders redrawn and supranational bonds rebuilt within 'post-wall Europe'.[64] As with spatial imaginaries, scholars now also began to unpack the history of evolving 'regimes of territorialization'.[65]

In addition to technological and historical changes, the growing interest in space has also been underpinned by a *third* trend. In short, 'the spatial turn' is a 'child of postmodernism'. It is closely informed by a pattern of thought that originated from a time *preceding* the digital revolution and the end of the Cold War.[66] I have already mentioned Fredric Jameson, a major proponent of this architectural and cultural movement. In 1984, Jameson published a foundational article in the *New Left Review* called 'Postmodernism, or The Cultural Logic of Late Capitalism'.[67] This is not the place here to rehash Jameson's arguments, let alone the multiple facets of the hotly debated intellectual impulse of postmodernism.[68] Suffice to say that Jameson perceived the emergence of a certain type of eclectic, playful, and ambiguous architecture in the mid- to late-1970s. He considered this indicative of a fundamentally new orientation of society.

Jameson identified Los Angeles as a key site in this regard. Tellingly, it was also the city where some of the most prominent postmodern geographers and urban theorists were based, not least Michael Dear, Michael Storper, Jennifer Wolch, and Edward Soja. They came to be known as the 'Los Angeles School (of Urbanism)' – a name apparently first given to them in Mike Davis' magnificent history of LA, *City of Quartz* (1990).[69] Jameson himself was not based in Los Angeles. Academically, however, he placed considerable emphasis on an analysis of the city, and specifically of John Portman's Westin Bonaventure Hotel (1974–6). For Jameson, this hotel – which remains the largest in LA – represented a 'postmodern hyperspace'. The complex ensemble of entryways, escalators, elevators, lobby, and 'glass skin', he argued, made for an experience of disorientation and 'dissociation', 'transcending the capacities of the individual human body to locate itself'. According to

Jameson, this spatial experience symbolized the 'incapacity of our minds ... to map the great global multinational and decentred communicational network in which we find ourselves caught as individuals'.[70]

Such architectural analysis, with its far-reaching claims about a 'postmodern condition' of society as a whole, was strongly influenced by poststructuralist philosophy. For obvious reasons, this extensive and expansive scholarly and intellectual phenomenon can only be captured here with the broadest of brushstrokes. The key site of poststructuralism was Paris. In 1979, the French philosopher Jean-François Lyotard published *The Postmodern Condition*. This book declared the end of all 'master narratives', with their emphasis on universal progress and Enlightenment notions of reason and truth. Such grand explanatory frameworks and legitimizing sense-making tools, he stated, had 'lost their credibility'. This 'incredulity towards metanarratives' nurtured reservations about linear chronologies more generally and instead fostered a tendency to think in terms of simultaneities and juxtapositions.[71]

Intimately linked to Lyotard's contentions was Jacques Derrida's theory of deconstruction. This was most comprehensively laid out in a trilogy of works published in 1967. Truth, Derrida argued, was always relative to certain conceptual frameworks, with their embedded hierarchies and oppositions ('reason'/'emotion', 'culture'/'nature', etc.). We can only 'know' what such frameworks allow us to know; direct access to the world is denied to us. We live 'not inside reality, but inside our *representations* of it', as literary scholar Christopher Butler has aptly put it. To quote Derrida's most famous statement, there is 'nothing outside the text'. Derrida aimed to 'deconstruct' those textual representations, and the hidden assumptions informing them, thereby subverting any confidence in categories of 'common sense', as well as the persuasive power of rhetorical commonplaces.[72]

These formulations, together with cognate interventions from Richard Rorty, Roland Barthes, Hayden White, and many others, became subsumed under the catchword of the 'linguistic turn'. Without the linguistic turn, the 'spatial turn' would have looked rather different – indeed, it may never have occurred.[73] For example, it is not for nothing that cartographer and historical geographer Brian Harley, whose work became one of the most frequently cited within the 'spatial turn' literature, called one of his most influential essays 'deconstructing the map' (1989). This contribution made a case for reading maps like texts, deciphering their visual rhetoric (the thickness of a line, the choice of colour, shading and hatching, etc.), identifying symbolic hierarchies, and detecting silences.[74]

Significantly, another of Harley's essays had the title 'maps, knowledge, and power' (1988).[75] This points to another important source of the spatial turn's theoretical inspiration. It was philosopher and historian Michel Foucault who explored the relationship between language, knowledge, and power, especially in his works from the 1960s. Foucault focused particularly on the production of knowledge through law, criminology, medicine, and other disciplines. He was especially concerned with the ways in which science defined and delineated seemingly objective categories such as 'normality' and 'deviance', 'sanity' and 'madness', 'reason' and 'folly'. Foucault's goal was to tease out the discursive construction of power relations, and the ensuing dynamics of inclusion and exclusion. Moreover, he explored the spaces in which such 'normalizing' delimitations became manifest, including workhouses, asylums, hospitals, and prisons. Little wonder that geographers have found much food for thought in Foucault's work, which was to become a key reference point in discussions on 'the spatial turn'.[76]

This was also because of a lecture that Foucault had given in 1967 to a group of architects. Shortly before Foucault's death in 1984, this lecture was published under the title 'Of Other Spaces'; it was based on a 12-minute radio broadcast from 1966 entitled 'Utopia and Literature'. Foucault's lecture contained the now famous and rather sweeping statement that, while 'the great obsession of the nineteenth century was ... history, with its themes of development and ... the ever-accumulating past', 'the present epoch will perhaps be above all the epoch of space' – 'the epoch of juxtaposition, the epoch of the near and far, of the side-by-side, of the dispersed'.

Such relational dynamics lie at the heart of the lecture's central term – 'heterotopia'. This concept has inspired much scholarship, but it is notoriously ambiguous. It was originally a medical term denoting the dislocation of a certain tissue type co-existing with tissue at its usual anatomical location. In this context, however, 'heterotopia' refers broadly to the – real and imagined – disrupting, suspending, and reverting of spatial routines in certain sites in relation to other sites and practices of place-making. These range from children transforming their parents' bed into a playground, to the different and, in part, contradictory symbolic meanings ascribed to gardens, from antiquity to modern times, both as 'the smallest parcel of the world' and its 'totality'. 'Heterotopia', Foucault stated, is a space 'which draws us out of ourselves'.[77]

'Heterotopy', in the sense of places *within* but not *of* the city, inhabited by 'strangers' and marginalized 'others', also surfaces in the work of Henri Lefebvre. This French philosopher, urban theorist, and polymathic thinker has become something of a patron saint of the broad church of spatial scholars. In Lefebvre's work, however, the term 'heterotopy' is embedded in a theory of *Urban Revolution* (1970). It is turned, productively and dialectically, into utopian spaces of resistance, difference, and liberation.[78] This dialectical dynamic, one of the Marxist hallmarks of Lefebvre's writings, also informed the book that was to become the most frequently cited reference in spatially minded scholarship: *The Production of Space* (1974). The 1991 translation of this book was undertaken by the one-time English Situationist Donald Nicholson-Smith.[79] It made Lefebvre a household name among Anglo-American geographers, and later among scholars of other disciplines, many of them beyond the Anglosphere. Geographers and urbanists such as David Harvey, Mark Gottdiener, and, especially, Edward Soja had already begun to spread the Lefebvrian gospel before the translation of *The Production of Space*. However, the book's perfectly timed publication made it part and parcel of 'the spatial turn' (even though 1991 was also, sadly, the year of the 90-year-old Lefebvre's death).[80]

The title of this important – if anything but straightforward – book encapsulated Lefebvre's main message: Space cannot be reduced to Euclidean geometry, the Cartesian coordinate system, or the Newtonian notion of absolute space, bucket- and container-like. Instead, space needs to be approached as something socially produced. In an attempt to overcome the long-standing dualism between mind and matter, and to mediate between subjective perception and objective physicality, Lefebvre advanced his now oft-quoted and diversely interpreted spatial triad of conceived, perceived, and lived space, or, put differently, of 'representations of space', spatial practice, and 'representational space'.

More is said about this in our *Guide to Spatial History* under 'Social Space and Political Protest'.[81] Here it must suffice to point out that a key facet of Lefebvre's triad was the utopian potential located within its dialectic tensions. This might be situated, for example, between the 'conceived space' of urban planners, architects, and engineers on the one hand, and, on the other, the counterintuitive, creative, 'happening'- and performance art-like uses of this space – fleetingly and momentarily mediated through 'spatial practice'. Lefebvre referred to such uses as 'lived space'. Crucially, this dialectic interplay entailed the utopian possibility of turning 'abstract space' – homogenous, unitary, capitalist – into 'differential space', which enabled difference, experiential particularity, and liberation.[82]

In this regard, it is important to note the intellectual lineage and historical setting of Lefebvre's thought. These aspects have been elucidated with great subtlety by a range of scholars, including Stuart Elden, Christian Schmid, and Łukasz Stanek.[83] Two are worth mentioning here. First, Lefebvre was indebted not only to Marx, but also to Nietzsche and Heidegger. These figures were notoriously sensitive to the ambivalence of modernity (alienation, atomization, dissociation). They also expressed criticism with respect to the idea of geometric space, and embraced the notion of 'poetic dwelling'.[84] Second, Lefebvre was involved in the avant-garde movement of the Situationist International alongside artist, filmmaker, and anti-capitalist thinker Guy Debord, and was engaged in political activism around the events of May 1968. This began to unfold at the University of Nanterre, Lefebvre's workplace and the epitome of 1960s functionalist architecture. Constructed in a bleak Paris suburb, the university became a hotspot of student radicalism: 'Beneath the pavement, the beach'.[85]

In this sense, 'the spatial turn' can clearly be traced back to the 1960s. That said, it is important to note that the spatial turn manifested itself differently, and at different times, across both national and disciplinary boundaries. Moreover, in some respects it was as much a 'turn' as it was a 're-turn'.[86] For example, in post-1945 Germany, historical writing had been marked by a virtual absence of cartographic material and geographical perspectives (with the possible exception of local and regional history).[87] This lacuna was largely due to the aftershock of the prominence of 'space' and spatial sciences in the Third Reich.[88] Even as late as 1996, British historian Richard J. Evans was able to write in a review article about 'the strange aversion of modern German historians to maps'.[89]

Introduction: Spatial history 15

This situation only began to change when 'the spatial turn' set in around the turn of the millennium.⁹⁰ Two years after the publication of Evans' article, German historian Jürgen Osterhammel titled a now oft-cited literature review 'the *return* of space'.⁹¹ Certainly, the work of Lefebvre and other spatial theorists gained much attention among German scholars. And yet, 'the spatial turn' was also a reappropriation and renewed adaptation of intellectual traditions that had been 'lost'.⁹²

For instance, Karl Schlögel, a historian of Eastern Europe, very consciously chose a paraphrased quote from 19th-century political and human geographer Friedrich Ratzel as the title for what would become Germany's landmark volume on 'the spatial turn': *In Space We Read Time*.⁹³ Schlögel's book received some criticism, especially from German geographers.⁹⁴ And yet it was anything but an uncritical reactivation of national traditions. The book effectively introduced German audiences to the work of Anglo-American geographers such as Matthew H. Edney, Derek Gregory, and Edward Soja. Moreover, Schlögel also reminded readers of the longstanding French tradition of thinking about history and geography in tandem. This was manifest in, for example, the geo-historical work of Paul Vidal de la Blache, Lucien Febvre, Fernand Braudel, and other Annales historians.⁹⁵

Further national and disciplinary variation has been in evidence here. The 'linguistic turn' mentioned above has typically segued into a more wide-ranging 'cultural turn'. The analytical focus there has been on 'meaning', 'interpretation', and 'imaginaries'. Such an orientation took hold in Anglo-American geography around 15 years earlier than was the case in Germany.⁹⁶ To some extent, certain manifestations of this tendency have harked back to even earlier advances in geography, which aimed to foreground 'experience' and 'imagination' instead of, say, settlement patterns and geographic dispersion.⁹⁷

In the 1980s, geographers such as Denis Cosgrove, Stephen Daniels, and James and Nancy Duncan combined 'humanistic' approaches (mentioned above under 'GeoHumanities') with both Marxist and poststructuralist perspectives. They adopted the vision-centred analytical notion of landscape as a 'way of seeing' and ordering the world, and married these ideas to the concept of 'social formation', which captured the historical context of economic domination and cultural hegemony.⁹⁸

Despite the specifically Marxist anchoring of this scholarship, the 'cultural turn' in Anglo-American geography was predominantly centred on 'meaning', 'vision', and 'text'. It reinforced the representation-oriented facet of 'the spatial turn'. Around 1990, there began to appear publications with titles such as *Writing Worlds* and *New Words, New Worlds*, all of which pointed in the same direction.⁹⁹ A decade later, however, this trend was partly challenged and partly complemented by tendencies toward 're-materialization', which drew attention to the agency of the 'non-human' or 'more-than-human'.¹⁰⁰ These intellectual tendencies have since given rise to a vast literature, including discussions on affect-, materiality- and performance-centred, so-called non-representational theory, which cannot be summarized here.¹⁰¹

Space and place

Suffice to say that the discursive force field of 'the spatial turn' has in many respects been constituted by the tensions between 'mind and matter' as they were first dialectically mediated in Lefebvre's work.[102] Many interventions in this field have grappled with the relationship between the materiality and constructedness of space, and have often reacted to a perceived imbalance in this dialectic in one or the other direction.[103] A range of sociologists, including Pierre Bourdieu and Markus Schroer, have offered an important insight here, that space, though socially constituted, shapes social interaction and confines the range of possible uses, behaviours, and routines. It does so both through its physicality and, crucially, the material qualities *ascribed* to it. 'Social space' and 'physical space' are mutually constitutive.[104] William Rankin recently made a particularly welcome intervention in this regard. He suggested that we should call off the fight against the old ghosts of environmental determinism and the postmodern bugbears of epistemic relativism. Instead, we might usefully replace the terms 'determine' and 'construct' with 'interact' or 'co-produce'.[105]

As already mentioned, it is primarily the focus on space and place that lends spatial history coherence and a shared perspective. The preceding depiction of 'the spatial turn', and especially of Lefebvre's multi-perspectival triad, suggests that this is more easily said than translated into scholarly practice. Both 'space' and 'place' are contested terms. Sometimes, they are defined in diametrically opposed ways.[106] And yet many scholars tend to agree with geographer John Agnew's suggestion that 'place' can be unpacked according to a threefold dimensionality which comprises location, locale, and sense of place. According to this formulation, 'place' may take on three meanings: (1) it may refer to a specific point in space, locatable in terms of longitude and latitude, and usually conceived of in relation to other locations; (2) it may point to a locale as a setting of social interaction which is historically and culturally conditioned, such as a seminar room, workplace, or restaurant; (3) it may be associated with memories, emotions, symbolic meanings, and a sense of belonging.[107]

Place, then, may be intimately connected with 'spatial imaginaries', a term already invoked earlier in this introduction. Indeed, place and space are closely intertwined. To be sure, some scholars prefer one term over the other as a focal point for their conceptual framework. Nonetheless, it is difficult to talk about the one without also talking about the other.[108] Places are, after all, constitutive of space – and vice versa. As sociologist Martina Löw puts it: 'Space is a relational arrangement of living beings and social goods *at places*'.[109] Löw distinguishes here between two ways in which space is constituted. First, this is achieved through 'spacing'. This can take the form of surveying and marking borders, the erection of buildings, the displaying of goods in the supermarket, or the self-positioning of people vis-à-vis other people. Second, space is constituted through 'synthesis' – that is, the thinking together of goods and/or people. This is based on 'perception, imagination, and memory'.[110]

Löw's relational approach dovetails nicely with Doreen Massey's understanding of space. This three-point conceptualisation entails, first, space as the 'product of interrelations' and as 'constituted through interactions, from the immensity of the global to the intimately tiny'; second, 'the sphere of the possibility of the existence of multiplicity … in which distinct trajectories coexist'; and third, 'as always under construction [and] in the process of being made': 'The spatial is social relations "stretched out"'.[111] Massey's conceptualization of space, with its emphasis on contingency and becoming, may prove particularly appealing to historians, given their disciplinary reservations about assumptions of transhistorical timelessness alluded to at the beginning of this introduction. Her idea of 'space-time', encapsulating the interwovenness of history and space, as well as time and geography, has proven enormously fruitful for geographers, and may no doubt also speak to spatially attuned historians.[112]

The aim of this volume

This volume aims to provide a practical introduction to the emerging field of spatial history. It does so on the basis of a variety of primary sources, and it is informed by a range of analytical perspectives. It seeks to convey a sense of the multiple facets of spatial history in a tangible, case-study based manner. The emphasis of this volume is very much on *doing* spatial history.

To date, there exists a sizeable literature that provides overviews as well as detailed analyses of spatial theories, methodologies, and themes. Its authors are often human and cultural geographers, but also scholars with a background in sociology, anthropology, philosophy, literature, and architecture.[113] Histories of geography and geographical thought abound.[114] So too do studies on specific spatial theorists such as Henri Lefebvre.[115] There also exists an equally sizeable German-language literature focusing on individual thinkers, disciplinary discourses, and methodological approaches.[116] Two of the most-cited German titles on spatial theory have recently been translated: Doris Bachmann-Medick's *Cultural Turns* and Martina Löw's *The Sociology of Space*.[117]

Now available in English is also Susanne Rau's *History, Space, and Place*, which provides an introduction to the historical 'analysis of the spatial dimensions of society'. This book neatly complements an overview of spatial concepts, analytical frameworks, and disciplinary approaches to spatial studies with a more hands-on, step-by-step guide to the historical investigation of the formations, dynamics, perceptions, and uses of space. The main emphasis lies, however, on providing a thematic, methodological, and historiographical survey of the field. The geographical focus is principally centred on Germany and France.[118]

Also relevant here is the recent publication of several edited volumes and special journal issues which have been inspired by 'the spatial turn' and revolve around the topic of space and history. For example, Paul Stock's *The Uses of Space in Early Modern History*, which originated from a seminar series held at the London School of Economics, brings together specialists from various disciplines, such as archaeology, historical geography, and literary studies. Contributors explore a variety of spaces, including vernacular houses, parish churches, scientific academies, cities, imperial border zones, as well as coastal and insular 'microregions'. A majority of chapters focus on British history.[119]

Many of the contributions to Stock's volume throw into sharp relief how 'space' as an analytical category can be usefully deployed, both to advance our knowledge of a particular subject matter and to engage wider historiographical discussions. For instance, archaeologist Matthew Johnson analyzes English vernacular houses and highlights the importance of materiality in shaping the meanings of 'living space'. As he observes, 'things, including spaces, are not passive or secondary. They materialize practices and meanings in a way that is not reducible simply or only to the written or spoken word, or to a pre-existing set of values.'[120] Historical geographer and intellectual historian Robert J. Mayhew, meanwhile, unpacks various layers of Thomas Malthus' *Essay on the Principle of Population* (1798) through a scalar analysis of global, national, and local 'ways of ... worldmaking'.[121] Other authors shed light on the historicity of space itself. Notable here is a chapter by historical geographer Michael Heffernan, who describes 18th-century endeavours in measuring and re-conceptualizing space at the Paris Academy of Sciences. Heffernan's contribution exemplifies the ways in which space was produced within a specific scientific setting.[122]

In a similar way, other case study collections concentrate on specific areas of spatial history, both thematically and geographically. Most of these publications are primarily targeted at specialists working in the field.[123] The oft-cited volume by Barney Warf and Santa Arias, *The Spatial Turn*, exhibits considerable geographical and disciplinary diversity, and it offers much theoretical inspiration – especially in the chapters by literary scholars Arias, Pamela K. Gilbert, and Mariselle Meléndez, which are also based on a sustained engagement with primary sources.[124]

Finally, Karl Schlögel's *In Space We Read Time*, mentioned above, is another regularly cited book. It provides a collage of vignettes that are wonderfully stimulating. The book also has a certain polemical dimension, and makes no bones about its scepticism towards approaches that are tightly focussed on discourse, language, and representation. Schlögel makes the case – more or less metaphorically – for getting out of one's armchair and putting on some hiking boots. Historians should wander around, hunt for material traces of their object of study, and read hard surfaces (of buildings etc.). After all, 'history takes place', and historians should engage much more with the physical sites and material environments of the past. Schlögel's aim to (re-)sensitize historians to the materiality of history is laudable, but it runs the risk of conveying a rather one-sided view of spatial history.[125]

Doing Spatial History aims to showcase the many approaches to this endeavour. Authors have been selected from a variety of different fields, including early modern and modern history, architectural history, historical anthropology, economic and social history, as well as historical and human geography. After all, part of the beauty of spatial history is that it provides a common forum which facilitates discussion across disciplines. It is able to encompass and connect a range of more established fields, such as historical geography, environmental history, and urban history, to name but a few. Moreover, the geographical scope of this volume is such that it takes readers on a journey through central, western, and east central Europe, to Russia, the Mediterranean, the Ottoman Empire, and East Asia, as well as North and South America, and New Zealand.

The logic behind the three-tier structure of this volume is to offer readers different entry points to the field of spatial history. Chapters of Part I revolve around a particular type of primary source (maps, travel guides, novels, newspapers, and architectural drawings); chapters of Part II zoom in on a particular kind of space (ships, bars, rivers, infrastructures, and border zones), while chapters of Part III engage a particular concept, tool or approach ('Lefebvrean landscapes', 'maritoriality', regional imaginaries, and historical GIS). Of course, all chapters in this volume use sources, explore spaces, and engage with concepts, tools or approaches. And yet, the individual sections of this volume signal to the reader that, in the chapters which ensue, a particular aspect of spatial history is foregrounded and takes centre stage.

Notes

1 There is, for instance, a multi-layered discussion among historians on 'presentism'. For a recent overview, see Marcus Colla, 'The Spectre of the Present: Time, Presentism and the Writing of Contemporary History', *Contemporary European History* 30, 2021, 124–35; see also David Armitage, 'In Defense of Presentism', in Darrin M. McMahon (ed.), *History and Human Flourishing*, Oxford: Oxford University Press, forthcoming. Available HTTP: https://scholar.harvard.edu/files/armitage/files/in_defence_of_presentism.pdf (accessed 31 March 2021); Marek Tamm and Laurent Olivier (eds.), *Rethinking Historical Time: New Approaches to Presentism*, London: Bloomsbury, 2019. I would like to thank Fiona Banham, Alex Burkhardt, John Clark, Tim Cresswell, Diarmid Finnegan, Lauren Holmes, James Koranyi, and Jacqueline Rose for help, comments, and insights. Special thanks go to my co-editors Konrad Lawson and Bernhard Struck.
2 See, for instance, the projects led by David R. Ambaras and Kate McDonald, *Bodies and Structures: Deep-Mapping Modern East Asian History*. Available HTTP: https://scalar.chass.ncsu.edu/bodies-and-structures/index (accessed 31 March 2021), and Kelly O'Neill, *Beautiful Spaces: An Exercise in Narrative Mapping & Crimean History*. Available HTTP: https://beautifulspaces.omeka.fas.harvard.edu/ (accessed 31 March 2021).
3 These are, of course, only relative degrees of prominence in the spatial distribution of academic subdisciplines. Anne Kelly Knowles, a foundational figure in the field of historical GIS, is a historical geographer academically socialized and based in the United States (bar an earlier teaching stint at Aberystwyth).

4 Richard White, 'What Is Spatial History?', *The Spatial History Project*, February 2010, pp. 2, 6. Available HTTP: https://web.stanford.edu/group/spatialhistory/cgi-bin/site/pub.php?id=29 (accessed 31 March 2021).
5 In 1994, Ian Gregory completed an MSc in Geographical Information Science at the University of Edinburgh – a programme that had been established as early as 1985 and was the first of its kind. See Bruce M. Gittings, 'Reflections on Forty Years of Geographical Information in Scotland: Standardisation, Integration and Representation', *Scottish Geographical Journal* 125, 2009, 78–94, here 81.
6 Ian N. Gregory and Alistair Geddes, 'Introduction: From Historical GIS to Spatial Humanities: Deepening Scholarship and Broadening Technology', in id. (eds.), *Toward Spatial Humanities: Historical GIS & Spatial History*, Bloomington: Indiana University Press, 2014, pp. ix–xix, here p. xv; see also Ian N. Gregory et al., *Troubled Geographies: A Spatial History of Religion and Society in Ireland*, Bloomington: Indiana University Press, 2014, p. 7; for an earlier example see Robert Schwartz, Ian Gregory, and Thomas Thévenin, 'Spatial History: Railways, Uneven Development, and Population Change in France and Great Britain, 1850–1914', *Journal of Interdisciplinary History* 42, 2011, 53–88; see also, in a similar vein, Jennifer Bonnell and Marcel Fortin, 'Introduction', in id. (eds.), *Historical GIS Research in Canada*, Calgary: University of Calgary Press, 2014, pp. ix–xix, here p. xii.
7 Loren Siebert, *Creating a GIS Spatial History of Tokyo*, PhD thesis, Seattle: University of Washington, 1997; id., 'Using GIS to Document, Visualize, and Interpret Tokyo's Spatial History', *Social Science History* 24, 2000, 537–74. It is not without irony that Siebert chose the term 'spatial history' because of its expansive connotations: 'The term *spatial* is abstract enough to encompass geography, planning, demography, urban design, and architecture as well as other, less visible phenomena, yet it is concrete enough to evoke images of actual places. And the term *history* is generic enough to include changes in all sorts of temporal phenomena, from natural and built environments to changing social, economic, and political conditions and events, virtually all of which have some connection to place.' Ibid., 540 (original emphasis).
8 Ian Gregory, Don DeBats, and Don Lafreniere (eds.), *The Routledge Companion to Spatial History*, London: Routledge, 2018. The book series 'Spatial Humanities' is edited by David J. Bodenhamer, John Corrigan, and Trevor M. Harris, who have also edited particular volumes in this series, e.g. *The Spatial Humanities: GIS and the Future of Humanities Scholarship*, Bloomington: Indiana University Press, 2010; and *Deep Maps and Spatial Narratives*, Bloomington: Indiana University Press, 2015.
9 Michael Heffernan, 'Historical Geography', in Derek Gregory et al. (eds.), *The Dictionary of Human Geography*, 5th ed., Oxford: Wiley-Blackwell, 2009. Available HTTP: https://search-credoreference-com.ezproxy.st-andrews.ac.uk/content/entry/bkhumgeo/historical_geography/0?institutionId=2454 (accessed 31 March 2021). The following volumes offer a sense of the enormous breadth of the field: Mona Domosh, Michael Heffernan, and Charles W.J. Withers (eds.), *The SAGE Handbook of Historical Geography*, vols. 1–2, London: Sage, 2020; John Morrissey et al., *Key Concepts in Historical Geography*, London: Sage, 2014; Brian Graham and Catherine Nash (eds.), *Modern Historical Geographies*, Harlow: Longman, 2000.
10 See Charles W.J. Withers, Mona Domosh and Michael Heffernan, 'Introduction', in id. (eds.), *Handbook of Historical Geography*, vol. 1, pp. xxvii–l, here p. xxx.
11 I am grateful to Diarmid Finnegan for pointing this out. For the 'quantitative (re)turn' in geography during the 1950s and 1960s, see David Livingstone, *The*

Geographical Tradition, Oxford: Blackwell, 1993, pp. 316–28; see also Tim Cresswell, *Geographic Thought: A Critical Introduction*, Malden, Mass.: Wiley-Blackwell, 2013, pp. 79–102.

12 In 1999, however, when pondering the 'relationship between environmental history and national history' – and without giving any thought to computational techniques – Richard White had still been entertaining the idea that 'somewhat haltingly a history of space is taking shape that draws inspiration from, complements, and will certainly challenge some of the efforts in geography'. Richard White, 'The Nationalization of Nature', *The Journal of American History* 86, 1999, 976–86, here 976–7.

13 This term has been chosen wisely by Harriet Hawkins, 'History, Geography and the GeoHumanities', in Domosh, Heffernan and Withers (eds.), *Handbook of Historical Geography*, vol. 2, pp. 1019–41, here p. 1025 (own emphasis).

14 See, for instance, Mark Bassin, Christopher Ely and Melissa K. Stockdale (eds.), *Space, Place, and Power in Modern Russia: Essays in the New Spatial History*, DeKalb: Northern Illinois University Press, 2010; Courtney J. Campbell, 'Space, Place, and Scale: Human Geography and Spatial History in *Past and Present*', *Past & Present* 239, 2018, e23–e45; Ruby Ekkel, 'Woman's Sphere Remodelled: A Spatial History of the Victorian Woman's Christian Temperance Union, 1887–1914', *Victorian Historical Journal* 91/1, 2020, 90–111; Laura C. Forster, 'The Paris Commune in London and the Spatial History of Ideas, 1871–1900', *The Historical Journal* 62, 2019, 1021–44; Matthew Boyd Goldie, 'Spatial History: Estres, Edges, and Contents', *Studies in the Age of Chaucer* 40, 2018, 379–87; Allen M. Howard and Richard M. Shain (eds.), *The Spatial Factor in African History*, Leiden: Brill, 2005 (where 'spatial history' is frequently invoked without any reference to GIS); Pascal Schillings and Alexander van Wickeren, 'Towards a Material and Spatial History of Knowledge', *Historical Social Research* 40, 2015, 203–18.

15 Paul Carter, *The Road to Botany Bay: An Essay in Spatial History*, Minneapolis: University of Minnesota, 1987, p. 60.

16 See id., *Amplifications: Poetic Migration, Auditory Memory*, London: Bloomsbury, 2019, pp. 1–2.

17 Id., *Road to Botany Bay*, pp. xvi–xix, xxi–xxii.

18 Ibid., pp. xxi, 33, 179. Carter was, of course, acutely aware of some of the potential difficulties a book like his, 'which speculates about history *as writing*', might present to historians. Paul Carter and David Malouf, 'Spatial History', *Textual Practice* 3/2, 1989, 173–83, here 173 (original emphasis). This broadcast conversation from 27 March 1988 was also published as 'Space, Writing and Historical Identity', *Thesis Eleven* 22/1, 1989, 92–105. Connections may be drawn here to the semi-fictional work by Bruce Chatwin, *The Songlines*, New York: Penguin, 1987, and to the study by historian and anthropologist Greg Dening, *Islands and Beaches: Discourse on a Silent Land: Marquesas, 1774–1880*, Honolulu: University of Hawaii Press, 1980 (see esp. ch. 1: 'Names and Places', and ch. 2: 'Space and Time').

19 Zephyr Frank, 'Spatial History as Scholarly Practice', in Patrik Svensson and David Theo Goldberg (eds.), *Between Humanities and the Digital*, Cambridge, Mass.: MIT Press, 2015, pp. 411–28, here pp. 420–1; see also id., *Reading Rio de Janeiro: Literature and Society in the Nineteenth Century*, Stanford: Stanford University Press, 2016; for an early example of a study deeply indebted to Carter,

see J.K. Noyes, *Colonial Space: Spatiality in the Discourse of German South West Africa, 1884–1915*, Chur: Harwood, 1992.
20 William Rankin, in 'AHR Conversation: Walls, Borders, and Boundaries in World History', *American Historical Review* 122, 2017, 1501–53, here 1511.
21 Of course, it would be a gross misrepresentation to suggest that practitioners of historical GIS tend to be oblivious of spatial theory, or of its history. Far from it. See, especially, Bodenhamer, Corrigan and Harris (eds.), *Deep Maps and Spatial Narratives*; Alexander von Lünen and Charles Travis (eds.), *History and GIS: Epistemologies, Considerations and Reflections*, New York: Springer, 2013.
22 Felix Driver and Raphael Samuel, 'Spatial History: Rethinking the Idea of Place', *History Workshop Journal* 39, Spring 1995, v–vii, 136–92.
23 Stuart Elden, *Mapping the Present: Heidegger, Foucault and the Project of a Spatial History*, London: Continuum, 2001.
24 Historical GIS is, for instance, absent from the two main German works in the field: Susanne Rau, *History, Space, and Place*, trans. Michael Thomas Taylor, London: Routledge, 2019, based on 2nd German ed. 2017, first published 2013; Karl Schlögel, *In Space We Read Time: On the History of Civilization and Geopolitics*, trans. Gerrit Jackson, Chicago and London: University of Chicago Press, 2016, German 2003.
25 William Rankin, 'How the Visual Is Spatial: Contemporary Spatial History, Neo-Marxism, and the Ghost of Braudel', *History and Theory* 59, 2020, 311–42, here 311.
26 See, especially, David Demeritt, 'The Nature of Metaphors in Cultural Geography and Environmental History', *Progress in Human Geography* 18, 1994, 163–85; Michael Williams, 'The Relations of Environmental History and Historical Geography', *Journal of Historical Geography* 20, 1994, 3–21; and the review article by Simon Naylor, 'Historical Geography: Natures, Landscapes, Environments', *Progress in Human Geography* 30, 2006, 792–802.
27 See, above all, Alan R.H. Baker, *Geography and History: Bridging the Divide*, Cambridge: Cambridge University Press, 2003, especially pp. 36, 62–71, 209–27; id., 'Classifying Geographical History', *The Professional Geographer* 59, 2007, 344–56; Chris Philo, 'History, Geography, and the "Still Greater Mystery" of Historical Geography', in Derek Gregory, Ron Martin and Graham Smith (eds.), *Human Geography: Society, Space and Social Science*, Basingstoke: Palgrave Macmillan, 1994, pp. 252–81; see also the overview by Silvia Elena Piovan, *The Geohistorical Approach: Methods and Applications*, New York: Springer, 2020, pp. 5–17. Baker also included the term 'spatial history' in his considerations on the relationship between history and geography, defining it as the historical analysis of the 'cultural organisation of space' and the 'cultural construction of location and distance', i.e., the 'changing spatial relations among and within places', especially in the context of 'macrospaces' such as countries and continents. By contrast, he viewed Carter's *Road to Botany Bay*, as well as Driver and Samuel's special issue of the *History Workshop Journal*, as examples of 'place history'. Baker, *Geography and History*, pp. 65, 67.
28 See, in a similar vein, the astute comments by Naylor, 'Historical Geography', and ibid., 793, for the ironic expression 'takeover bid'.
29 David Harvey, 'Geographical Knowledge in the Eye of Power: Reflections on Derek Gregory's *Geographical Imaginations*', *Annals of the Association of American Geographers* 85, 1995, 160–4, here 161: 'the geographical imagination is far too pervasive and important a facet of intellectual life to be left alone to geographers.'

30 See, for instance, Michael Bentley (ed.), *Companion to Historiography*, London: Routledge, 1997; Stefan Berger, Heiko Feldner and Kevin Passmore (eds.), *Writing History: Theory and Practice*, London: Bloomsbury, 2003, 3rd ed. 2020; Jeremy Black and Donald MacRaild, *Studying History*, London: Palgrave Macmillan, 1997, 4th ed. 2017; David Cannadine (ed.), *What Is History Now?*, Basingstoke: Palgrave Macmillan, 2002; Anna Green and Kathleen Troup (eds.), *The Houses of History: A Critical Reader in Twentieth-Century History and Theory*, Manchester: Manchester University Press, 1999, 2nd ed. 2016; Ludmilla Jordanova, *History in Practice*, London: Bloomsbury, 2000, 3rd ed. 2019; Jorma Kalela, *Making History*, Basingstoke: Palgrave Macmillan, 2012; Peter Lambert and Phillipp Schofield (eds.), *Making History: An Introduction to the History and Practices of a Discipline*, London: Routledge, 2004; Ulinka Rublack (ed.), *A Concise Companion to History*, Oxford: Oxford University Press, 2012; Axel Schneider and Daniel Woolf (eds.), *The Oxford History of Historical Writing*, vol. 5, Oxford: Oxford University Press, 2011; Marek Tamm and Peter Burke (eds.), *Debating New Approaches to History*, London: Bloomsbury, 2019; John Tosh, *The Pursuit of History*, London: Routledge, 1984, 6th ed. 2015.

31 Baker, *Geography and History*. A more optimistic note is struck by Miles Ogborn, 'The Relations between Geography and History: Work in Historical Geography in 1997', *Progress in Human Geography* 23, 1999, 97–108, here 103–4, stating that historians and historical geographers 'certainly talk, read each other's work, and contribute to the same journals and conferences' – increasingly so 'on the basis that they are addressing the same sorts of questions with the same sorts of tools'.

32 H.P.R. Finberg, 'Local History', in id. (ed.), *Approaches to History: A Symposium*, London: Routledge, 1962, pp. 111–25, here pp. 123–4.

33 H.C. Darby, 'Historical Geography', in Finberg (ed.), *Approaches to History*, pp. 127–56. See *The Common Lands of England & Wales*, London: Collins, 1963, a report of the Royal Commission on Common Lands by Hoskins and geographer Dudley Stamp, for another example from that time for geo-historical cooperation.

34 Keith Lilley, 'GIS, Spatial Technologies and Digital Mapping', in Simon Gunn and Lucy Faire (eds.), *Research Methods for History*, Edinburgh: Edinburgh University Press, 2012, pp. 121–40; for a revised and expanded version see Keith Lilley and Catherine Porter, 'GIS, Spatial Technologies and Digital Mapping', in Gunn and Faire (eds.), *Research Methods for History*, 2nd ed., 2016, pp. 125–46.

35 Like Finberg and Hoskins in their times, however, both editors are based in Leicester. The volume emerged from the institutional contexts of the Centre for Urban History at Leicester, established in 1985, and the Centre for Urban Culture at Nottingham, founded in 2000 (renamed as 'Urban Culture Network' in 2011).

36 Further exceptions that prove the rule are Peter Claus and John Marriott, *History: An Introduction to Theory, Method and Practice*, New York: Pearson, 2012, which includes a chapter on 'geography'; Sarah Barber and Corinna M. Peniston-Bird (eds.), *History beyond the Text*, London and New York: Routledge, 2009, which introduces students, among other things, to the study of landscape and architecture; see also, with a focus on landscape history, Nicola Whyte, 'Spatial History', in Sasha Handley, Rohan McWilliam and Lucy Noakes (eds.), *New Directions in Social and Cultural History*, London: Bloomsbury, 2018, pp. 233–51. Beyond the genre of surveys and companions, of course, collaborations between historians and geographers have been more frequent – especially in the fields of environmental history and the history of science, as mentioned above. See, for example, Diarmid

A. Finnegan and Jonathan Jeffrey Wright (eds.), *Spaces of Global Knowledge: Exhibition, Encounter and Exchange in an Age of Empire*, Farnham: Ashgate, 2015.

37 See, in this context, the downbeat statement by one of Germany's 'veterans' in the field, Klaus Fehn, 'Historische Geographie', in Hans-Jürgen Goertz (ed.), *Geschichte: Ein Grundkurs*, Hamburg: Rowohlt, 1998, pp. 394–407, here esp. pp. 394–5, 404; see also the more recent assessment by Hans Gebhardt, 'Historische Geographie und kritische Humangeographie: Einige vorläufige Überlegungen', in Jan-Erik Steinkrüger and Winfried Schenk (eds.), *Zwischen Geschichte und Geographie, zwischen Raum und Zeit*, Berlin: Lit Verlag, 2015, pp. 1–6, here pp. 2–3.

38 Heike Jöns and Tim Freytag, 'Boundary Spanning in Social and Cultural Geography', *Social & Cultural Geography* 17, 2016, 1–22, here 4 (with a particular focus on transnational knowledge transfer). The term is borrowed loosely from Paul Williams, 'The Competent Boundary Spanner', *Public Administration* 80, 2002, 103–24.

39 Denis Cosgrove, 'Editorial', *Ecumene* 1, 1994, 1–5, here 1–2. The word also acted as an acronym here of the journal's initial subtitle: 'environment, **cu**lture, **mean**ing'.

40 Philip Crang and Don Mitchell, '*cultural geographies*: an ecumenical journal', *cultural geographies* 9/1, 2002, 1–2.

41 Philip Crang and Don Mitchell, 'Editorial', *Ecumene* 7, 2000, 1–6, here 2, 4.

42 Dominic Sachsenmaier, 'World History as Ecumenical History?', *Journal of World History* 18, 2007, 465–89, here 469, 473.

43 On 'humanistic geography' see the wide-ranging volume, dedicated to Yi-Fu Tuan, by Paul C. Adams, Steven Hoelscher and Karen E. Till (eds.), *Textures of Place: Exploring Humanist Geographies*, Minneapolis: University of Minnesota Press, 2001; see also Cresswell, *Geographic Thought*, pp. 103–21. David Lowenthal, especially, produced much 'boundary-spanning' work. He was trained in both history and geography, and became instrumental in the founding of heritage studies. A prominent point of reference among historians is his magisterial *The Past Is a Foreign Country*, Cambridge: Cambridge University Press, 1985; see also *The Past Is a Foreign Country Revisited*, Cambridge: Cambridge University Press, 2015.

44 Tim Cresswell and Deborah P. Dixon, 'GeoHumanities', in Douglas Richardson et al. (eds.), *The International Encyclopedia of Geography*, Hoboken: Wiley & Sons, 2017, pp. 1–9, here pp. 3–4, 6–7. Available HTTP: https://doi.org/10.1002/9781118786352.wbieg1169 (accessed 31 March 2021). See also Tim Cresswell et al., 'Imagining and Practicing the Geohumanities: Past, Present, Future', *GeoHumanities* 1, 2015, 1–19; as well as the two foundational volumes by Stephen Daniels et al. (eds.), *Envisioning Landscapes, Making Worlds: Geography and the Humanities*, London and New York: Routledge, 2011; Michael Dear et al. (eds.), *GeoHumanities: Art, History, Text at the Edge of Place*, London and New York: Routledge, 2011.

45 That 'space' has by no means always been a central analytical term for the field of historical geography has been shown, in relation to Darby's work, by Hugh Prince, 'H.C. Darby and the Historical Geography of England', in H.C. Darby, *The Relations of History and Geography: Studies in England, France and the United States*, Exeter: University of Exeter Press, 2002, pp. 63–88, here p. 71.

46 The term 'spatial ways of knowing' is the key concept of Manasvini Narayana, *The Spatial Turn in History: Implications for Curriculum in Higher Education*, PhD thesis, Montreal: Concordia University, 2019, esp. pp. 4–12.

47 Fredric Jameson, 'Postmodernism and Utopia', in id. et al., *Utopia Post Utopia: Configurations of Nature and Culture in Recent Sculpture and Photography*, Cambridge, Mass.: MIT Press, 1988, pp. 11–32, here p. 11. This chapter was later included, in a slightly revised form, in Fredric Jameson, *Postmodernism, or, The Cultural Logic of Late Capitalism*, London: Verso, 1991, pp. 154–80.

48 Edward W. Soja, *Postmodern Geographies: The Reassertion of Space in Critical Social Theory*, 8th ed., London and New York: Verso, 2003, first published 1989, pp. 16, 30–2, 39, 154. The term was used in a similar vein, with respect to the evolving discipline of geography during the late 19th century, by Paul Rabinow, *French Modern: Norms and Forms of the Social Environment*, Cambridge, Mass.: MIT Press, 1989, pp. 139, 195. That said, there was still no mention of it in a foundational article by Soja, which was completed by 1988, and which formed the basis of two key chapters in *Postmodern Geographies*: Edward W. Soja, 'Modern Geography, Western Marxism, and the Restructuring of Critical Social Theory', in Richard Peet and Nigel Thrift (eds.), *New Models in Geography: The Political-Economy Perspective*, vol. 2, London: Unwin Hyman, 1989, pp. 318–47. On the origins of 'the spatial turn' see also Nikolai Roskamm, 'Das Reden vom Raum: Zur Aktualität des *Spatial Turn*: Programmatik, Determinismus und "sozial konstruierter Raum"', *Peripherie* 32, 2012, 171–89, here 173-5.

49 Soja, *Postmodern Geographies*, pp. 30–2. This turn of phrase has come to enjoy great popularity among spatially minded scholars. Playing on the old anti-scholastic saying 'How many angels can dance on the head of a pin?', Doreen Massey famously used the phrase in her *Spatial Divisions of Labour: Social Structures and the Geography of Production*, Basingstoke: Macmillan, 1984, 2nd ed. 1995, p. 51: 'Nothing much happens, bar angels dancing, on the head of a pin.' The turn of phrase reappears, with a similar thrust, in White, 'Spatial History', 1.

50 Soja, *Postmodern Geographies*, pp. 31, 33; id., 'Taking Space Personally', in Barney Warf and Santa Arias (eds.), *The Spatial Turn: Interdisciplinary Perspectives*, London and New York: Routledge, 2009, pp. 11–35, here p. 21; see also id., 'The Socio-Spatial Dialectic', *Annals of the Association of American Geographers* 70, 1980, 207–25; id., *My Los Angeles: From Urban Restructuring to Regional Urbanization*, Berkeley: University of California Press, 2014, pp. 172–80, 188–90; as well as the important book by Neil Smith, *Uneven Development: Nature, Capital, and the Production of Space*, 3rd ed., with a new afterword by the author and a foreword by David Harvey, London: Verso, 2010, 2nd ed. with a new preface 1990, first published 1984.

51 Soja, *Postmodern Geographies*, p. 13.

52 For the view that 'GIS lies at the heart of this so-called spatial turn, see David J. Bodenhamer, John Corrigan and Trevor M. Harris, 'Introduction', in id. (eds.), *The Spatial Humanities*, pp. vii–xv, here p. vii; see also Gregory et al. (eds.), *Routledge Companion to Spatial History*, pp. 1–2. William Rankin rightly calls this 'surprisingly straightforward'. Rankin, 'How the Visual Is Spatial', 313. References to 'the spatial turn' became more frequent, especially, from the turn of the millennium, mostly without any reference to GIS: see, for example, Denis Cosgrove, 'Introduction: Mapping Meaning', in id. (ed.), *Mappings*, London: Reaktion, 1999, pp. 1–23, here p. 7; Simon Gunn, 'The Spatial Turn: Changing Histories of Space and Place', in id. and Robert J. Morris (eds.), *Identities in Space: Contested Terrains in the Western City since 1850*, Aldershot: Ashgate, 2001, pp. 1–14; John Pickles, 'Social and Cultural Cartographies and the Spatial Turn in Social Theory', *Journal of Historical Geography* 25, 1999, 93–8. The

exception here was the special issue edited by Anne Kelly Knowles, 'Historical GIS: The Spatial Turn in Social Science History', *Social Science History* 24/3, 2000.
53 David Harvey, *The Condition of Postmodernity: An Enquiry into the Origins of Cultural Change*, Malden, Mass.: Blackwell, 1989; id., 'Between Space and Time: Reflections on the Geographical Imagination', *Annals of the Association of American Geographers* 80, 1990, 418–34, here esp. 425–33.
54 Paul Virilio, 'The Third Interval: A Critical Transition', *Art & Design* 7/1–2, January/February 1992, 78–85, here 85.
55 Richard O'Brien, *Global Financial Integration: The End of Geography*, London: Pinter for Royal Institute of International Affairs, 1992; Kenichi Ohmae, *The Borderless World: Power and Strategy in the Global Marketplace*, New York: Harper, 1990.
56 In 1857, Marx spoke of an 'annihilation of space through time' in relation to the capital-induced need to overcome distance in ever shorter timespans. On this, see Roland Wenzlhuemer, 'Globalization, Communication and the Concept of Space in Global History', *Historical Social Research* 35/1, 2010, 19–47, here 24. On the railways, see the classic account by Wolfgang Schivelbusch, *The Railway Journey*, Berkeley: University of California Press, 1986, German 1977, ch. 3: 'Railroad Space and Railroad Time', and for Heinrich Heine, ibid., p. 37. Connections can also be drawn to new conceptualizations of time and space at the turn of the century, imaginatively explored in relation to new communication and transportation technologies by Stephen Kern, *The Culture of Time and Space, 1880–1918*, with a new preface, Cambridge, Mass.: Harvard University Press, 2003, first published 1983, esp. chs. 6–9: 'The Nature of Space', 'Form', 'Distance', and 'Direction'.
57 Daniela Ahrens, *Grenzen der Enträumlichung: Weltstädte, Cyberspace und transnationale Räume in der globalisierten Moderne*, Wiesbaden: VS Verlag für Sozialwissenschaften, 2001.
58 Frances Stonor Saunders, 'Where on Earth Are You?', *London Review of Books* 38/5, March 2016, 7–12.
59 Jean-Marie Guéhenno, *The End of the Nation-State*, Minneapolis: University of Minnesota Press, 1995, French 1993.
60 See, for instance, Jürgen Dinkel, *The Non-Aligned Movement: Genesis, Organization and Politics (1927–1992)*, Leiden: Brill, 2018, German 2015; id., Steffen Fiebrig and Frank Reichherzer (eds.), *Nord/Süd: Perspektiven auf eine globale Konstellation*, Berlin and Boston: De Gruyter, 2020; Simo Mikkonen and Pia Koivunen (eds.), *Beyond the Divide: Entangled Histories of Cold War Europe*, New York: Berghahn, 2018; Frank Reichherzer, Emmanuel Droit and Jan Hansen (eds.), *Den Kalten Krieg vermessen: Über Reichweite und Alternativen*, Berlin and Boston: De Gruyter, 2018.
61 Derek Gregory, 'Geographical Imaginary', in id. et al. (eds.), *Dictionary of Human Geography*. Available HTTP: https://search-credoreference-com.ezproxy.st-andrews.ac.uk/content/entry/bkhumgeo/geographical_imaginary/0?institutionId=2454 (accessed 31 March 2021).
62 Edward W. Said, *Orientalism*, new ed., London: Penguin, 2003, first published 1978; id., 'Orientalism', *The Georgia Review* 31, 1977, 162–206, part 1: 'Imaginative Geography and Its Representations'.
63 See, especially, Arturo Escobar, *Encountering Development: The Making and Unmaking of the Third World*, Princeton: Princeton University Press, 1995; Christopher GoGwilt, *The Invention of the West: Joseph Conrad and the*

Double-Mapping of Europe and Empire, Stanford: Stanford University Press, 1995; Walter D. Mignolo, *The Darker Side of the Renaissance: Literacy, Territoriality, and Colonization*, Ann Arbor: University of Michigan Press, 1995; Larry Wolff, *Inventing Eastern Europe: The Map of Civilization on the Mind of the Enlightenment*, Stanford: Stanford University Press, 1994; see also Derek Gregory, 'Imaginative Geographies', *Progress in Human Geography* 19, 1995, 447–85; and Michael C. Frank, 'Imaginative Geography as a Travelling Concept: Foucault, Said and the Spatial Turn', *European Journal of English Studies* 13, 2009, 61–77; more on this literature in Konrad Lawson, Riccardo Bavaj and Bernhard Struck, *A Guide to Spatial History: Areas, Aspects, and Avenues of Research* (2021), Available HTTPS: spatialhistory.net/guide; see also the chapter by Konrad Lawson in this volume.

64 Kristina Spohr, *Post Wall, Post Square: Rebuilding the World after 1989*, London: William Collins, 2019.
65 See, above all, the influential article by Charles Maier, 'Consigning the Twentieth Century to History: Alternative Narratives for the Modern Era', *American Historical Review* 105, 2000, 807–31. As Maier mentions in a footnote, he 'first developed the concept of territoriality as a key to periodization' in 1996. Ibid., 808, fn. 2. More on this in Lawson, Bavaj and Struck, *Guide to Spatial History*.
66 Doris Bachmann-Medick, *Cultural Turns: New Orientations in the Study of Culture*, Berlin and Boston: De Gruyter, 2016, German 2006, p. 211.
67 Fredric Jameson, 'Postmodernism, or The Cultural Logic of Late Capitalism', *New Left Review* I/146, July–August 1984, 53–92.
68 See, instead, the useful volumes by geographers Paul Cloke, Chris Philo and David Sadler, *Approaching Human Geography: An Introduction to Contemporary Theoretical Debates*, London: Paul Chapman, 1991, pp. 170–201; Michael J. Dear and Steven Flusty (eds.), *The Spaces of Postmodernity: Readings in Human Geography*, Oxford: Blackwell, 2002; Claudio Minca (ed.), *Postmodern Geography: Theory and Praxis*, Oxford: Blackwell, 2001; the landmark article by Michael J. Dear, 'The Postmodern Challenge: Reconstructing Human Geography', *Transactions of the Institute of British Geographers* 13, 1988, 262–74; the recent overview by Claudio Minca, 'Postmodern Turn in Geography', in Audrey Kobayashi (ed.), *International Encyclopedia of Human Geography*, vol. 10, 2nd ed., Amsterdam: Elsevier, 2020, pp. 323–31; and more generally Brian McHale (ed.), *The Cambridge Introduction to Postmodernism*, Cambridge: Cambridge University Press, 2015; see also, as an early assessment from a historian's point of view, Jane Caplan, 'Postmodernism, Poststructuralism, and Deconstruction: Notes for Historians', *Central European History* 22, 1989, 260–78.
69 Mike Davis, *City of Quartz: Excavating the Future in Los Angeles*, London and New York: Verso, 1990; see also Michael J. Dear, 'The Los Angeles School of Urbanism: An Intellectual History', *Urban Geography* 24, 2003, 493–509.
70 Jameson, 'Postmodernism', 83–4.
71 Jean-François Lyotard, *The Postmodern Condition: A Report on Knowledge*, trans. Geoff Bennington and Brian Massumi, with a foreword by Fredric Jameson, Minneapolis: University of Minnesota Press, 1984, French 1979.
72 Christopher Butler, *Postmodernism: A Very Short Introduction*, Oxford: Oxford University Press, 2002, p. 21 (own emphasis).
73 The interrelation of 'turns' is brought out well in Bachmann-Medick, *Cultural Turns*.
74 J.B. Harley, 'Deconstructing the Map', *Cartographica* 26/2, 1989, 1–20.

75 Id., 'Maps, Knowledge, and Power', in Denis Cosgrove and Stephen Daniels (eds.), *The Iconography of Landscape: Essays on the Symbolic Representation, Design and Use of Past Environments*, Cambridge: Cambridge University Press, 1988, pp. 277–312. Both essays, 'Deconstructing the Map' and 'Maps, Knowledge, and Power', are included in the posthumously published collection of articles: J.B. Harley, *The New Nature of Maps: Essays in the History of Cartography*, ed. Paul Laxton, Baltimore: Johns Hopkins University Press, 2001, which also includes a critical introduction by J.H. Andrews.

76 See, especially, Felix Driver, 'Power, Space and the Body: A Critical Assessment of Foucault's *Discipline and Punish*', *Environment and Planning D: Society and Space* 3, 1985, 425–46; Thomas R. Flynn, *Sartre, Foucault, and Historical Reason*, vol. 2: *Poststructuralist Mapping of History*, Chicago: University of Chicago Press, 2005; Robert J. Mayhew, 'Historical Geography 2007–2008: Foucault's Avatars – Still in (the) Driver's Seat', *Progress in Human Geography* 33, 2009, 387–97; Chris Philo, 'Foucault's Geography', in Mike Crang and Nigel Thrift (eds.), *Thinking Space*, London and New York: Routledge, 2000, pp. 205–38; and the reader by Jeremy W. Crampton and Stuart Elden (eds.), *Space, Knowledge and Power: Foucault and Geography*, Aldershot: Ashgate, 2007. Elden has authored numerous books on Foucault, including *Mapping the Present*, and, more recently, *Foucault's Last Decade*, Cambridge: Polity, 2016; *Foucault: The Birth of Power*, Cambridge: Polity, 2017; and *The Early Foucault*, Cambridge: Polity, 2021. Needless to say, Foucault's work gained much attention among historians, too. See, pars pro toto, Allan Megill, 'The Reception of Foucault by Historians', *Journal of the History of Ideas* 48, 1987, 117–41.

77 Michel Foucault, 'Of Other Spaces', *Diacritics* 16/1, 1986, 22–7, here 22–3; see also the later and, in part, more accurate translation: 'Different Spaces', in Michel Foucault, *Aesthetics: Essential Works, 1954–1984*, London: Allen Lane, 1998, pp. 175–85; for a helpful commentary see Peter Johnson, 'Unravelling Foucault's "Different Spaces"', *History of the Human Sciences* 19/4, 2006, 75–90; id., 'The Geographies of Heterotopia', *Geography Compass* 7, 2013, 790–803; see also Kevin Hetherington, *The Badlands of Modernity: Heterotopia and Social Ordering*, London: Routledge, 1997; Mariangela Palladino and John Miller (eds.), *The Globalization of Space: Foucault and Heterotopia*, London: Pickering & Chatto, 2015; Robert J. Topinka, 'Foucault, Borges, Heterotopia: Producing Knowledge in Other Spaces', *Foucault Studies* 9, 2010, 54–70; see also the chapter by Matthew Ylitalo and Sarah Easterby-Smith in this volume.

78 Henri Lefebvre, *The Urban Revolution*, Minneapolis: University of Minnesota Press, 2003, French 1970, pp. 9–11. This point is convincingly made by Johnson, 'Foucault's "Different Spaces"', 83–4.

79 Henri Lefebvre, *The Production of Space*, trans. Donald Nicholson-Smith, Oxford: Blackwell, 1991, French 1974; on the revolutionary avant-garde movement of the Situationist International, founded in 1957, see Sam Cooper, *The Situationist International in Britain: Modernism, Surrealism, and the Avant-Gardes*, London and New York: Routledge, 2017; James Trier, *Guy Debord, the Situationist International, and the Revolutionary Spirit*, Leiden: Brill, 2019.

80 A selection of the vast scholarship on Lefebvre is provided below in fn. 115. See, here, David Harvey, *Social Justice and the City*, London: Edward Arnold, 1973; Mark Gottdiener, *Social Production of Urban Space*, Austin: University of Texas Press, 1985; Edward W. Soja, 'The Spatiality of Social Life: Towards a Transformative Retheorisation', in Derek Gregory and John Urry (eds.), *Social Relations and*

Spatial Structures, Basingstoke: Palgrave Macmillan, 1985, pp. 90–127; see also Peter Arnade, Martha Howell and Walter Simons, 'Fertile Spaces: The Productivity of Urban Space in Northern Europe', *The Journal of Interdisciplinary History* 32, 2002, 515–48; Stuart Elden, 'Politics, Philosophy, Geography: Henri Lefebvre in Recent Anglo-American Scholarship', *Antipode* 33, 2001, 809–25; Eleonore Kofman and Elizabeth Lebas, 'Lost in Transposition: Time, Space and the City', in Henri Lefebvre, *Writings on Cities*, trans. and ed. Eleonore Kofman and Elizabeth Lebas, Oxford: Blackwell, 1996, pp. 3–60; as well as Derek Gregory, 'Lefebvre, Lacan and the Production of Space', in Georges B. Benko and Ulf Strohmayer (eds.), *Geography, History and Social Sciences*, Dordrecht: Kluwer, 1995, pp. 15–44; Andy Merrifield, 'Henri Lefebvre: A Socialist in Space', in Crang and Thrift (eds.), *Thinking Space*, pp. 167–82; Rob Shields, *Lefebvre, Love and Struggle: Spatial Dialectics*, London and New York: Routledge, 1998.
81 Lawson, Bavaj and Struck, *Guide to Spatial History*.
82 A helpful analysis is provided by Andy Merrifield, *Henri Lefebvre: A Critical Introduction*, London and New York: Routledge, 2006, pp. 99–120; see also the chapter by Dawn Hollis in this volume.
83 Stuart Elden, *Understanding Henri Lefebvre*, London: Continuum, 2004; Christian Schmid, *Stadt, Raum und Gesellschaft: Henri Lefebvre und die Theorie der Produktion des Raumes*, Stuttgart: Steiner, 2005, 2nd ed. 2010; Łukasz Stanek, *Henri Lefebvre on Space: Architecture, Urban Research, and the Production of Theory*, Minneapolis: University of Minnesota Press, 2011.
84 Stuart Elden, 'There Is a Politics of Space Because Space Is Political: Henri Lefebvre and the Production of Space', *Radical Philosophy Review* 10, 2007, 101–16, here 111; see also, in this context, Martin Heidegger, 'Building Dwelling Thinking' (1951), in Sharon M. Meagher (ed.), *Philosophy and the City: Classic to Contemporary Writings*, Albany: SUNY Press, 2008, pp. 119–25; see most recently Ryan L. Allen, 'Resurrecting the Archaic: Symbols and Recurrence in Henri Lefebvre's Revolutionary Romanticism', *Modern Intellectual History* 18, 2021, 474–96.
85 On Nanterre, see Victor Colette, 'From American Dream to Nightmare on the Left: Student Revolts, the "Crèche Sauvage" and the Slums: The University of Nanterre, 1962–71', in Jill Pellew and Miles Taylor (eds.), *Utopian Universities: A Global History of the New Campuses of the 1960s*, London: Bloomsbury, 2020, pp. 351–67. The Paris 1968 slogan 'Beneath the pavement, the beach' is picked up by McKenzie Wark, *The Beach Beneath the Street: The Everyday Life and Glorious Times of the Situationist International*, London: Verso, 2011.
86 This Janus-faced directionality is also emphasized in Doris Bachmann-Medick, 'Cultural Turns', version: 2.0, *Docupedia-Zeitgeschichte*, 17 June 2019. Available HTTP: <http://docupedia.de/zg/Bachmann-Medick_cultural_turns_v2_de_2019?oldid=132484> (accessed 31 March 2021).
87 See, especially, Karl-Georg Faber, 'Was ist eine Geschichtslandschaft?' (1968), in Pankraz Fried (ed.), *Probleme und Methoden der Landesgeschichte*, Darmstadt: WBG, 1978, pp. 390–424; id., 'Geschichtslandschaft – Région historique – Section in History: Ein Beitrag zur vergleichenden Wissenschaftsgeschichte', *Saeculum* 30, 1979, 4–21; see also Riccardo Bavaj, 'Was bringt der "Spatial Turn" der Regionalgeschichte? Ein Beitrag zur Methodendiskussion', *Westfälische Forschungen* 56, 2006, 457–84; Sigrid Hirbodian, Christian Jörg and Sabine Klapp (eds.), *Methoden und Wege der Landesgeschichte*, Ostfildern: Jan Thorbecke, 2015;

Nina Lohmann, 'Der "Raum" in der deutschen Geschichtswissenschaft', *Acta Universitatis Carolinae Studia Territorialia* 3/4, 2010, 47–93, here esp. 83–6.

88 See Riccardo Bavaj, *Der Nationalsozialismus: Entstehung, Aufstieg und Herrschaft*, Berlin: be.bra, 2016, pp. 54–60, 135–45; Paolo Giaccaria and Claudio Minca (eds.), *Hitler's Geographies: The Spatialities of the Third Reich*, Chicago and London: University of Chicago Press, 2016; Guntram Henrik Herb, *Under the Map of Germany: Nationalism and Propaganda, 1918–1945*, London: Routledge, 1997; Claus-Christian W. Szejnmann and Maiken Umbach (eds.), *Heimat, Region, and Empire: Spatial Identities under National Socialism*, Basingstoke: Palgrave Macmillan, 2012.

89 Richard J. Evans, 'From Unification to World War' (review of Wolfgang J. Mommsen, *Propyläen Geschichte Deutschlands*, vols. 7/1 and 7/2), *German Historical Institute London Bulletin*, 1996, 15–26, here 23.

90 The main reference points are David Blackbourn, *A Sense of Place: New Directions in German History*, German Historical Institute London: The 1998 Annual Lecture, London: German Historical Institute, 1999; Jürgen Osterhammel, 'Die Wiederkehr des Raumes: Geopolitik, Geohistorie und historische Geographie', *Neue Politische Literatur* 43, 1998, 374–97; Karl Schlögel, *Im Raume lesen wir die Zeit: Über Zivilisationsgeschichte und Geopolitik*, Munich: Hanser, 2003, English 2016: *In Space We Read Time*.

91 Osterhammel, 'Die Wiederkehr des Raumes' (own emphasis); see also id., 'Raumerfassung und Universalgeschichte', in Gangolf Hübinger, Jürgen Osterhammel and Erich Pelzer (eds.), *Universalgeschichte und Nationalgeschichten: Ernst Schulin zum 65. Geburtstag*, Freiburg/Breisgau: Rombach, 1994, pp. 51–72; id., 'Raumbeziehungen: Internationale Geschichte, Geopolitik und historische Geographie', in Wilfried Loth and Jürgen Osterhammel (eds.), *Internationale Geschichte: Themen, Ergebnisse, Aussichten*, Munich: Oldenbourg, 2000, pp. 287–308; and, of course, his magnum opus *The Transformation of the World: A Global History of the Nineteenth Century*, Princeton: Princeton University Press, 2014, German 2009.

92 See also Christoph Nübel, 'Was ist neu am "spatial turn"? Potentiale und Grenzen deutscher geschichtswissenschaftlicher Raumkonzepte vom 19. Jahrhundert bis heute', *Historische Mitteilungen der Ranke-Gesellschaft* 27, 2015, 160–85.

93 Schlögel, *In Space We Read Time*. The original quote can be found in Friedrich Ratzel, 'Geschichte, Völkerkunde und historische Perspektive', *Historische Zeitschrift* 93, 1904, 1–46, here 28; on this, see also Jörg Döring and Tristan Thielmann, 'Was lesen wir im Raume? Der *Spatial Turn* und das geheime Wissen der Geographen', in id. (eds.), *Spatial Turn: Das Raumparadigma in den Kultur- und Sozialwissenschaften*, Bielefeld: transcript, 2008, pp. 7–45, here p. 22; on both Schlögel and 'the spatial turn' in Eastern European history more generally, see the review articles by Nick Baron, 'New Spatial Histories of Twentieth Century Russia and the Soviet Union: Surveying the Landscape', *Jahrbücher für Geschichte Osteuropas* 55, 2007, 374–400; id., 'New Spatial Histories of 20th-Century Russia and the Soviet Union: Exploring the Terrain', *Kritika. Explorations in Russian and Eurasian History* 9, 2008, 433–48.

94 See, for instance, Gerhard Hard, 'Der *Spatial Turn*, von der Geographie her beobachtet', in Döring and Thielmann (eds.), *Spatial Turn*, pp. 263–315; Julia Lossau and Roland Lippuner, 'Geographie und *Spatial Turn*', *Erdkunde* 58, 2004, 201–11.

95 See Paul Vidal de la Blache, *Principles of Human Geography*, London: Constable & Co., 1926, French 1922; Lucien Febvre, *A Geographical Introduction to History*, London: Kegan Paul & Co., 1925, French 1922; and Fernand Braudel, *The Mediterranean and the Mediterranean World in the Age of Philip II*, 2 vols., Berkeley: University of California Press, 1972, French 1949; see also Susan W. Friedman, *Marc Bloch, Sociology and Geography: Encountering Changing Disciplines*, Cambridge: Cambridge University Press, 1996; Michael Heffernan and Karen M. Morin, 'Between History and Geography', in Domosh, Heffernan and Withers (eds.), *Handbook of Historical Geography*, vol. 1, 25–46; Eric Piltz, 'Unbestimmte Oberflächen: Rezeptionen und Konvergenzen von Geographie und Geschichtswissenschaft im *cultural* und *spatial turn*', in Elisabeth Tiller and Christoph O. Mayer (eds.), *RaumErkundungen: Einblicke und Ausblicke*, Heidelberg: Universitätsverlag Winter, 2011, pp. 213–34; Rankin, 'How the Visual Is Spatial'; Charles W.J. Withers, 'Place and the "Spatial Turn" in Geography and in History', *Journal of the History of Ideas* 70, 2009, 637-58, here 645–6, 658.

96 For an attempt of transnational knowledge transfer see, for instance, Wolfgang Natter and Ute Wardenga, 'Die "neue" und "alte" *Cultural Geography* in der anglo-amerikanischen Geographie', *Berichte zur deutschen Landeskunde* 77/1, 2003, 71–90.

97 See, especially, David Lowenthal, 'Geography, Experience, and Imagination: Towards a Geographical Epistemology', *Annals of the Association of American Geographers* 51, 1961, 241–60; David Lowenthal and Hugh C. Prince, 'The English Landscape', *Geographical Review* 54, 1964, 309–46; Hugh C. Prince, 'Real, Imagined and Abstract Worlds of the Past', *Progress in Geography* 3/1, 1971, 1–86.

98 Denis Cosgrove, *Social Formation and Symbolic Landscape*, Madison: University of Wisconsin Press, 1984, new ed. with an 'introductory essay', 1998; Cosgrove and Daniels (eds.), *Iconography of Landscape*; James and Nancy Duncan, '(Re)reading the Landscape', *Environment and Planning D: Society and Space* 6, 1988, 117–26; on this trend see also the perceptive remarks by Tim Cresswell, 'Landscape and the Obliteration of Practice', in Kay Anderson et al. (eds.), *Handbook of Cultural Geography*, London: Sage, 2003, pp. 269–81, here esp. pp. 272–3; more on this literature in Lawson, Bavaj and Struck, *Guide to Spatial History*.

99 Trevor J. Barnes and James S. Duncan (eds.), *Writing Worlds: Discourse, Texts and Metaphor in the Representation of Landscape*, London: Routledge, 1992; Chris Philo (ed.), *New Words, New Worlds: Reconceptualising Social and Cultural Geography*, Aberystwyth: Cambrian Printers, 1991; see also James Duncan and David Ley (eds.), *Place/Culture/Representation*, London: Routledge, 1993; Peter Jackson, *Maps of Meaning: An Introduction to Cultural Geography*, London: Unwin Hyman, 1989; Allan Pred, *Lost Words and Lost Worlds: Modernity and the Language of Everyday Life in Late Nineteenth-Century Stockholm*, Cambridge: Cambridge University Press, 1990; as well as the influential text by political theorist Timothy Mitchell, 'The World as Exhibition', *Comparative Studies in Society and History* 31, 1989, 217–36.

100 See, especially, Peter Jackson, 'Rematerializing Social and Cultural Geography', *Social & Cultural Geography* 1, 2000, 9–14; Chris Philo, 'More Words, More Worlds: Reflections on the "Cultural Turn" and Human Geography', in Ian Cook et al. (eds.), *Cultural Turns/Geographical Turns: Perspectives on Cultural Geography*, Harlow: Prentice Hall, 2000, pp. 26–53.

101 See now Paul Simpson, *Non-Representational Theory*, London and New York: Routledge, 2021.
102 See, for instance, Ralph Kingston, 'Mind over Matter? History and the Spatial Turn', *Cultural and Social History* 7, 2010, 111–21; Paul Stock, 'History and the Uses of Space', in id. (ed.), *The Uses of Space in Early Modern History*, Basingstoke: Palgrave Macmillan, 2015, pp. 1–18.
103 See, for example, Leif Jerram, 'Space: A Useless Category for Historical Analysis?', *History and Theory* 52, 2013, 400–19; Katrina Navickas, 'Why I am Tired of Turning: A Theoretical Interlude', 2011. Available HTTP: http://www.historyworkingpapers. org/?page_id=225 (accessed 31 March 2021); id., 'Thirdspace? Historians and the Spatial Turn, with a Case Study of Political Graffiti in Late Eighteenth- and Early Nineteenth-Century England', in Sam Griffiths and Alexander von Lünen (eds.), *Spatial Cultures: Towards a New Social Morphology of Cities Past and Present*, London and New York: Routledge, 2016, pp. 67–75; and Julia Lossau, 'Spatial Turn', in Frank Eckardt (ed.), *Handbuch Stadtsoziologie*, Wiesbaden: VS Verlag für Sozialwissenschaften, 2012, pp. 185–98.
104 See, especially, Pierre Bourdieu, 'Social Space and the Genesis of Appropriated Physical Space' (1991), *International Journal of Urban and Regional Research* 42/1, 2018, 106–14; Markus Schroer, 'Spatial Theories/Social Construction of Spaces', in Anthony M. Orum (ed.), *The Wiley Blackwell Encyclopedia of Urban and Regional Studies*, Hoboken: Wiley-Blackwell, 2019; see also id., *Räume, Orte, Grenzen: Auf dem Weg zu einer Soziologie des Raums*, Frankfurt/Main: Suhrkamp, 2006.
105 Rankin, 'How the Visual Is Spatial', 342; see also Beat Kümin, 'Turns and Perspectives', in C. Scott Dixon and Beat Kümin (eds.), *Interpreting Early Modern Europe*, London and New York: Routledge, 2019, pp. 471–88, here p. 477.
106 Compare, for instance, the definitions offered by French historian and philosopher Michel de Certeau, *The Practice of Everyday Life*, Berkeley: University of California Press, 1984, French 1980, with those advanced by Chinese-American geographer Yi-Fu Tuan, *Space and Place: The Perspective of Experience*, Minneapolis: University of Minnesota Press, 1977 – both of which have proven influential across various disciplines.
107 See John A. Agnew, *Place and Politics: The Geographical Mediation of State and Society*, London: Routledge, 1987; on the latter point see also Gaston Bachelard, *The Poetics of Space*, Boston: Beacon, 1994, French 1957; see more generally Tim Cresswell, *Place: An Introduction*, Malden, Mass.: Wiley-Blackwell, 2004, 2nd ed. 2015; id., *In Place/Out of Place: Geography, Ideology, and Transgression*, Minneapolis: University of Minnesota Press, 1996; id., 'Place', in Kobayashi, *International Encyclopedia of Human Geography*, pp. 117–24.
108 See, for instance, the revealing remarks by Tim Cresswell, 'Landscape and the Obliteration of Practice', p. 269: 'I like place because of its everyday nature. Place, unlike space and landscape, permeates our everyday life and provides meaning in people's lives. Places are quite clearly "lived". At the other extreme space has an analytical quality about it. It rubs shoulders easily with social theory and allows conversations to occur with other social science disciplines.' See also Edward S. Casey, 'How to Get from Space to Place in a Fairly Short Stretch of Time', in Steven Feld and Keith H. Basso (eds.), *Senses of Place*, Santa Fe: School of American Research Press, 1996, pp. 13–52; id., *The Fate of Place: A Philosophical History*, Berkeley: University of California Press, 1997; David

Harvey, 'From Space to Place and Back Again: Reflections on the Condition of Postmodernity', in John Bird et al. (eds.), *Mapping the Futures: Local Cultures, Global Change*, London: Routledge, 1993, pp. 3–29.

109 Martina Löw, *The Sociology of Space: Materiality, Social Structures, and Action*, Basingstoke: Palgrave Macmillan, 2016, German 2001, p. 232 (own emphasis).

110 Ibid., pp. 134–5. Löw draws here on Anthony Giddens ('spacing') and Norbert Elias ('synthesis'). See Anthony Giddens, *The Constitution of Society: Outline of the Theory of Structuration*, Cambridge: Polity, 1992; and Norbert Elias, *The Court Society*, trans. Edmund Jephcott, New York: Pantheon, 1983, German 1969.

111 Doreen Massey, *For Space*, London: Sage, 2005, p. 9; id., *Space, Place and Gender*, Cambridge: Polity, 1994, p. 2. This is, to an extent, reminiscent of the famous statement by sociologist Georg Simmel, for whom the boundary was 'not a spatial fact with sociological consequences, but a sociological fact that forms itself spatially'. Georg Simmel, 'The Sociology of Space' (1908), in David Frisby and Mike Featherstone (eds.), *Simmel on Culture: Selected Writings*, London: Sage, 1997, pp. 137–69, here p. 142.

112 See, especially, Felicity Callard, 'Doreen Massey', in Phil Hubbard and Rob Kitchin (eds.), *Key Thinkers on Space and Place*, 2nd ed., London: Sage, 2011, first published 2004, pp. 299-306; see also David Featherstone and Joe Painter (eds.), *Spatial Politics: Essays for Doreen Massey*, Malden, Mass.: John Wiley & Sons, 2013; Marion Werner et al. (eds.), *Doreen Massey: Critical Dialogues*, Newcastle upon Tyne: Agenda Publishing, 2018; for a different conceptualization of the relation between space and time, see the proposal by Philip Ethington, Professor of 'History, Political Science and Spatial Sciences', to write history, in a very fundamental sense, as spatial representation of the past: 'Placing the Past: "Groundwork" for a Spatial Theory of History', *Rethinking History* 11, 2007, 465–93.

113 Titles that fall into this category include Crang and Thrift (eds.), *Thinking Space*; Cresswell, *Place*; Tim Edensor, Ares Kalandides and Uma Kothari (eds.), *The Routledge Handbook of Place*, London and New York: Routledge, 2020; Jen Jack Gieseking et al. (eds.), *The People, Place, and Space Reader*, London and New York: Routledge, 2014; Hubbard and Kitchin (eds.), *Key Thinkers on Space and Place*; Setha Low, *Spatializing Culture: The Ethnography of Space and Place*, London and New York: Routledge, 2017, id. and Denise Lawrence-Zúñiga (eds.), *The Anthropology of Space and Place: Locating Culture*, Oxford: Blackwell, 2003; Les Roberts, *Spatial Anthropology: Excursions in Liminal Space*, London: Rowman & Littlefield, 2018; Rob Shields, *Spatial Questions: Cultural Topologies and Social Spatialisations*, London: Sage, 2013; Robert T. Tally Jr., *Spatiality*, London and New York: Routledge, 2013; id. (ed.), *Teaching Space, Place, and Literature*, London and New York: Routledge, 2018; Fran Tonkiss, *Space, the City and Social Theory: Social Relations and Urban Forms*, Cambridge: Polity, 2005; and Verena Andermatt Conley, *Spatial Ecologies: Urban Sites, State and World-Space in French Cultural Theory*, Liverpool: Liverpool University Press, 2012; as well as the philosophical treatments by Edward S. Casey, *Fate of Place*; id., *Getting Back Into Place: Toward a Renewed Understanding of the Place-World*, Bloomington: Indiana University Press, 1993, 2nd ed. 2009; J. Nicholas Entrikin, *The Betweenness of Place: Towards a Geography of Modernity*, Basingstoke: Macmillan, 1991; Jeff Malpas, *Place and Experience: A Philosophical Topography*, London and New York: Routledge,

1999, 2nd ed. 2018. Handbooks of human and cultural geography are part of this strand of literature as well, e.g. Paul Cloke, Philip Crang and Mark Goodwin (eds.), *Introducing Human Geographies*, London and New York: Routledge, 1999, 3rd ed. 2014; Anderson et al.(eds.), *Handbook of Cultural Geography*; Georges B. Benko and Ulf Strohmayer (eds.), *Human Geography*, London and New York: Routledge, 2004; John A. Agnew and David N. Livingstone (eds.), *The SAGE Handbook of Geographical Knowledge*, California: Sage, 2011; John A. Agnew and James S. Duncan (eds.), *The Wiley-Blackwell Companion to Human Geography*, Malden: Wiley Blackwell, 2011; Roger Lee et al. (eds.), *The SAGE Handbook of Human Geography*, Thousand Oaks: Sage, 2014; Nuala C. Johnson, Richard H. Schein and Jamie Winders (eds.), *The Wiley-Blackwell Companion to Cultural Geography*, Malden, Mass.: Wiley-Blackwell, 2013.

114 Cresswell, *Geographic Thought*; Ron Johnston and J.D. Sidaway, *Geography and Geographers*, London: Arnold, 1979, 7th ed. 2016; Ron Johnston and Michael Williams (eds.), *A Century of British Geography*, Oxford: Oxford University Press, 2003; Livingstone, *The Geographical Tradition*; Richard Peet, *Modern Geographical Thought*, Malden, Mass.: Blackwell, 1998; see also, with a more confined scope, Trevor J. Barnes and Eric Sheppard (eds.), *Spatial Histories of Radical Geography: North America and Beyond*, Hoboken: John Wiley & Sons, 2019; Gavin Bowd and Daniel Clayton, *Impure and Worldly Geography: Pierre Gourou and Tropicality*, London and New York: Routledge, 2019; Federico Ferretti, *Anarchy and Geography: Reclus and Kropotkin in the UK*, London and New York: Routledge, 2019; Anne Marie Claire Godlewska, *Geography Unbound: French Geographic Science from Cassini to Humboldt*, Chicago and London: University of Chicago Press, 1999; Susan Schulten, *The Geographical Imagination in America, 1880–1950*, Chicago and London: University of Chicago Press, 2001; Neil Smith, *American Empire: Roosevelt's Geographer and the Prelude to Globalization*, Berkeley: University of California Press, 2003; Michael Williams with David Lowenthal and William M. Denevan, *To Pass on a Good Earth: The Life and Work of Carl O. Sauer*, Charlottesville: University of Virginia Press, 2014.

115 See, for instance, Chris Butler, *Henri Lefebvre: Spatial Politics, Everyday Life and the Right to the City*, London and New York: Routledge, 2012; Elden, *Understanding Lefebvre*; Kanishka Goonewardena et al. (eds.), *Space, Difference, Everyday Life: Reading Henri Lefebvre*, London and New York: Routledge, 2008; Merrifield, *Henri Lefebvre*; Schmid, *Stadt, Raum und Gesellschaft*; Stanek, *Lefebvre on Space*; Łukasz Stanek et al. (eds.), *Urban Revolution Now: Henri Lefebvre in Social Research and Architecture*, London and New York: Routledge, 2014; Sue Middleton, *Henri Lefebvre and Education*, London and New York: Routledge, 2014; see also Jenny Bauer and Robert Fischer (eds.), *Perspectives on Henri Lefebvre: Theory, Practices and (Re)Readings*, Berlin and Boston: De Gruyter, 2018; Michael E. Leary-Owhin and John P. McCarthy (eds.), *The Routledge Handbook of Henri Lefebvre, The City and Urban Society*, London and New York: Routledge, 2020.

116 See, most notably, Döring and Thielmann (eds.), *Spatial Turn*; Jörg Dünne and Stephan Günzel (eds.), *Raumtheorie: Grundlagentexte aus Philosophie und Kulturwissenschaften*, Frankfurt/Main: Suhrkamp, 2006; Stephan Günzel, *Raum: Eine kulturwissenschaftliche Einführung*, Bielefeld: transcript, 2017; id. (ed.), *Raum: Ein interdisziplinäres Handbuch*, Stuttgart: J. B. Metzler, 2010; id.

(ed.), *Raumwissenschaften*, Frankfurt/Main: Suhrkamp, 2009; Schroer, *Räume, Orte, Grenzen*; see also Bernd Belina, *Raum: Zu den Grundlagen eines historisch-geographischen Materialismus*, Münster: Westfälisches Dampfboot, 2013; and the excellent volumes by Alexa Geisthövel and Habbo Knoch (eds.), *Orte der Moderne: Erfahrungswelten des 19. und 20. Jahrhunderts*, Frankfurt/Main: Campus, 2005; Alexander C.T. Geppert, Uffa Jensen and Jörn Weinhold (eds.), *Ortsgespräche: Raum und Kommunikation im 19. und 20. Jahrhundert*, Bielefeld: transcript, 2005.
117 Bachmann-Medick, *Cultural Turns*, ch. 6: 'The Spatial Turn'; Löw, *Sociology of Space*.
118 Rau, *History, Space, and Place*.
119 Stock (ed.), *Uses of Space*. Two other important volumes in the field also focus predominantly on the early modern period, namely Beat Kümin (ed.), *Political Space in Pre-Industrial Europe*, Farnham: Ashgate, 2009; and Marko Lamberg et al. (eds.), *Physical and Cultural Space in Pre-Industrial Europe: Methodological Approaches to Spatiality*, Lund: Nordic Academic Press, 2011 – the former pivoting to England, the latter to Scandinavia.
120 Matthew Johnson, 'Living Space: The Interpretation of English Vernacular Houses', in Stock (ed.), *Uses of Space*, pp. 19–42, here p. 23; see also id., 'What Do Medieval Buildings Mean?', *History and Theory* 52, 2013, 380–99; id., *Archaeological Theory: An Introduction*, 3rd ed., Hoboken: John Wiley & Sons, 2020, first published 1999, pp. 132–55 ('The Material Turn'); as well as Thomas F. Gieryn, 'What Buildings Do', *Theory and Society* 31, 2002, 35–74.
121 Robert J. Mayhew, 'A Tale of Three Scales: Ways of Malthusian Worldmaking', in Stock (ed.), *Uses of Space*, pp. 197–226; see also id., *Malthus: The Life and Legacies of an Untimely Prophet*, Cambridge, Mass.: Harvard University Press, 2014; a helpful introduction to the analytical concept of scale is Andrew Herod, *Scale*, London: Routledge, 2011; from a historical perspective, see also the *AHR* conversation 'How Size Matters: The Question of Scale in History', *American Historical Review* 118, 2013, 1431–72.
122 Michael Heffernan, 'The Spaces of Science and Sciences of Space: Geography and Astronomy in the Paris Academy of Sciences', in Stock (ed.), *Uses of Space*, pp. 125–50; see also the instructive volume by Koen Vermeir and Jonathan Regier (eds.), *Boundaries, Extents and Circulations: Space and Spatiality in Early Modern Natural Philosophy*, [Cham]: Springer, 2016.
123 Examples of this more specialised research offer the important volumes and special issues, respectively, by Natalia Aleksiun and Hana Kubátová (eds.), *Places, Spaces, and Voids in the Holocaust*, Frankfurt/Main: Campus, 2021; Kathryne Beebe, Angela Davis and Kathryn Gleadle (eds.), *Space, Place and Gendered Identities*, special issue of *Women's History Review* 21/4, 2012; Sebastian Dorsch and Susanne Rau (eds.), *Space/Time Practices and the Production of Space and Time*, special issue of *Historical Social Research* 38/3, 2013; Beat Kümin and Cornelie Usborne (eds.), *At Home and in the Workplace*, special issue of *History and Theory* 52/3, 2013; Simone Lässig and Miriam Rürup (eds.), *Space and Spatiality in Modern German-Jewish History*, New York: Berghahn, 2017; Holt Meyer, Susanne Rau and Katharina Waldner (eds.), *SpaceTime of the Imperial*, Berlin and Boston: De Gruyter, 2017; Sagi Schaefer, Galili Shahar and Teresa Walch (eds.), *Räume der deutschen Geschichte*, special issue of *Tel Aviver Jahrbuch für deutsche Geschichte* 49, 2021; Karl Schlögel (ed.),

Mastering Russian Spaces: Raum und Raumbewältigung als Probleme der russischen Geschichte, Munich: Oldenbourg, 2011.
124 See chapters 7, 8 and 11 in Warf and Arias (eds.), *Spatial Turn*.
125 Schlögel, *Im Raume lesen wir die Zeit*, pp. 70, 503 (*dass man hinausgeht, sich in Bewegung setzt und vom Hochsitz der Lektüre herabsteigt*).

Part I
Working with sources

1 Maps

Bernhard Struck and Riccardo Bavaj

The title of this chapter is 'maps'. In fact, however, the chapter is about 'mapping' – or, to be more precise, 'counter-mapping'. In his recent book *Cartography: The Ideal and Its History*, Matthew Edney encourages historians of maps to avoid phrases such as 'maps *are*'. He suggests that generic and normative claims about the 'real' nature of maps can be discarded in favour of a much closer focus on the *process* by which maps are created. Historians of maps should seek to understand, and to explain, how and why people have engaged in 'certain kinds of mapping', which agendas they have pursued, and the material and discursive preconditions on which their map-making has been based.[1]

This chapter is concerned with 'counter-mapping' – a term first advanced in 1995 by rural sociologist Nancy Lee Peluso. She aimed to describe the mapping strategies of local activists in contemporary Indonesia. These strategies were designed to counter government-sponsored forest mapping, which was carried out for the purpose of land use planning and natural resource exploitation. The term was also applied to the 'counter-appropriation' of official mapping techniques. The goal here was to 'territorialize' customary claims to natural resources, and to lend further legitimacy to those claims.[2]

More recently, the concept of 'counter-mapping' has been taken up by map historians such as Catherine Dunlop and Julie MacArthur. Dunlop has used the term in the context of 19th- and early 20th-century cartographic battles between France and Germany over Alsace and Lorraine. As she has shown, this involved various kinds of 'counter-mapping' by classroom teachers, geographical societies, and amateur map-makers. MacArthur's adaptation of the concept focused on the formation (ethnogenesis) of the Luyia community in 20th-century Kenya. Her study reveals that mapping was not the sole preserve of colonial powers. In fact, it also served as a 'tool for ethnic patriots' in, for example, classrooms or courtrooms. The counter-mapping activities of these 'patriots' had two purposes: to enact both a 'counterhegemonic' broadside against the state, as well as a 'generative' production of new political imaginations.[3]

In conceptual terms, this chapter draws on these studies, but it directs the discussion on counter-mapping to early 19th-century Poland. More recent scholarship has conceptualized 'Congress Poland', established after the Congress of Vienna (1815), as a 'colonial space'. This space was co-produced by

DOI: 10.4324/9780429291739-3

the imperial powers of Prussia, Austria, and Russia, through and after the partitions of Poland-Lithuania of 1772, 1793, and 1795.[4] The key primary source selected for this chapter is the *Atlas statystyczny Polski i kraiów okolicznych* ('Statistical atlas of Poland and neighbouring countries'), published in 1827.[5] This atlas will be placed in a historical context characterized by increased political suppression, and will be interpreted as an anti-imperial, anti-colonial move.

The chapter will first discuss the authorship of this atlas, which was published anonymously, with no indication of year or place of publication. It will then elucidate some of the cartographic and statistical background behind what was then a new type of atlas, before offering some thoughts on the atlas's materiality. Finally, the chapter will provide a closer reading of the visual rhetoric of some of the maps, as well as their spatial effects. Crucial here is an exploration of the political rationale informing the atlas. This rationale may most plausibly be found in a visualization, however ambiguous, of a future Poland: both of a sovereign state and of a Polish nation. Both of these spatial imaginaries – state and nation – were in flux in 19th-century Poland, and the two were not necessarily seen as congruent, quite contrary to the 'Western' ideal-type of a modern nation-state.

Authorship, map type, and materiality

The origins of the *Atlas statystyczny Polski* are shrouded in mystery. Determining its authorship is anything but straightforward. One finds various names in the literature, as well as in the archives. Sometimes, the name of Stanisław Plater (1784–1851) pops up. In 1827, Plater also released the *Atlas historique de la Pologne* (which was published in Poznań, outside of Congress Poland on Prussian territory).[6] The version of the *Atlas statystyczny Polski* used here (Figure 1.1) is held by the Berlin State Library (*Staatsbibliothek*). It

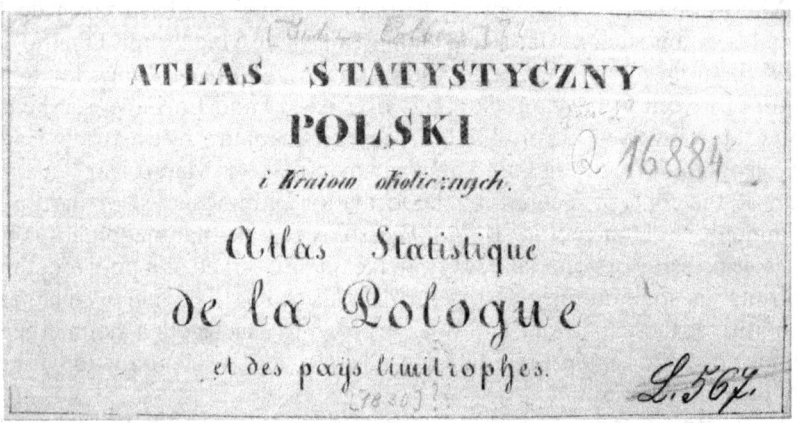

Figure 1.1 Cover sheet, *Mappa Polski* [1827].

Source: *Atlas statystyczny Polski i kraiów okolicznych*, Staatsbibliothek Berlin/Kartenabteilung Q16884.

includes a hand-written note on the cover which reads: '[Julius Colberg]??' This is a reference to Juliusz Kolberg (1776–1831) who, again in 1827, authored the *Atlas Królestwa Polskiego* ('Atlas of the Kingdom of Poland'). This was produced in Warsaw at a short-lived Lithographic Institute (*Instytut Litograficzny Szkolny*). Archival staff at the Berlin State Library were even unsure about the date of publication: '[1830]??' In fact, they misdated it by several years. Still, their hunch with respect to its authorship was closer to the mark.

Indeed, and especially with respect to visual rhetoric, there are some striking similarities between the *Atlas statystyczny Polski* and the *Atlas Królestwa Polskiego*. Both were statistical atlases, and both were published bilingually, in Polish and French (the lingua franca of 19th-century European elites). By contrast, the *Atlas historique de la Pologne* was only published in French. It also looks rather different from the other two publications. The tireless endeavours of Polish map historians such as Janina Piasecka and Waldemar Spallek, as well as the metadata of various Polish archives, have helped to resolve some of the mysteries around the *Atlas statystyczny Polski*. We can now say with some certainty that it was printed in Warsaw by lithographer Józef Slawinski, and that the person behind this endeavour was, in fact, Stanisław's elder brother Ludwik Plater (1775–1846).[7]

Ludwik Plater was born in Krasław (in today's Latvia), just a few years after the First Partition of Poland-Lithuania. He was thus raised in a border region of the Russian Empire. He was descended from a family of Lithuanian nobles with a long pedigree of holding various official positions in the old Commonwealth. He took part in the Polish Uprising of 1794, which sought to re-establish an independent state, and which was crushed by the Russian and Prussian armies, leading to the Third Partition of 1795. In post-1815 Congress Poland, Plater held a position in the state department for revenue and customs. He was also founding director of the 'school of forestry' at the University of Warsaw, which was established two years after the University's foundation in 1816. After participating in the failed uprising of 1830–31, he (and many of his like-minded compatriots) fled to Paris, where he was to spend the next ten years in exile. From the point of view of his involvement in the *Atlas statystyczny Polski*, it is significant that, prior to his exile, Ludwik Plater had been a member of the Warsaw Society of Friends of Science (*Towarzystwo Warszawskie Przyjaciół Nauk*). This was one of the earliest academic societies in Poland, founded in 1800 but dissolved by the Russian authorities in 1832. Plater was part of a committee within this academic society which was working towards a statistical description of Poland.[8]

The *Atlas statystyczny Polski* includes cartographic representations of print culture, education, economic production, religion, and language. It constitutes an early example of statistical mapping. It is emblematic of a cartographic turning point, which corresponded to a larger shift in the evolution of both statistics and geography. The late 18th and early 19th centuries saw a transformation in these areas from predominantly narrative, discursive accounts to a greater emphasis on numbers, charts, and graphs. Cartography became a key instrument in this burgeoning visualization of statistical representations.⁹ At the time of the *Atlas statystyczny Polski's* publication, conventions of statistical mapping were in their incipient stages. The heyday of statistical cartography – with heightened thematic variation, and a visual rhetoric of dots, splotches, colour shading, and cross hatching – still lay in the future.¹⁰

In terms of the materiality of its production process, too, the *Atlas statystyczny Polski* is situated at a particular historical juncture. The atlas was produced lithographically. This means that the underlying base map was printed from a stone or a metal plate, as opposed to being etched onto a copper plate, which would have been more labour-intensive and costly.¹¹ Lithography was invented in the late 18th century; from the 1820s, it became more widely used. This more affordable technique was a crucial precondition for the production of an 'underground' atlas such as the *Atlas statystyczny Polski*.

With its 45 × 35.5 cm dimension, this is a small and unassuming publication. It contains only six maps, and it is bereft of any accompanying narrative text. This is in contrast to many other atlases produced in the early 19th century. For example, the above-mentioned *Atlas historique de la Pologne* by Stanisław Plater comprises a dense 12-page account in addition to ten maps. Plater's volume charts the expansion and contraction of the Polish-Lithuanian Commonwealth from its height in the 17th century to its disappearance.¹²

The paper quality of the *Atlas statystyczny Polski* is coarse and cheap. It was likely printed on hand-made or self-made paper. In fact, this was a makeshift product, hinting at the atlas's clandestine, samizdat-like nature, and the subversive political agenda behind it. This much was also reflected in a rather dismissive review by Prussian geographer Heinrich Berghaus, who found the atlas to be of 'no great geographical value'. Its size and material quality certainly paled in comparison with the more comprehensive, elaborate, and much pricier *Administrativ-Statistischer Atlas vom Preussischen Staat* ('Administrative-statistical atlas of the Prussian state'). This was published around the same time as the *Atlas statystyczny Polski*. Indeed, this second publication was also reviewed by Berghaus in the very same journal issue in which he made his snide remarks about Plater's 'not very nice' atlas.¹³

Incidentally, Berghaus was also unsure about the author of the *Atlas statystyczny Polski*, which he mistakenly believed to have been published in 1830. As he observes in his review, rumour had it that the atlas's creation could be attributed to the same 'Count Plater' who authored the *Atlas historique de la Pologne* (i.e. Stanisław).¹⁴ And yet, despite the shadowy origins of the *Atlas statystyczny Polski*, authorship of an atlas is rarely so difficult to establish. Indeed, particular names can even become synonymous with particular cartographic products. Berghaus himself is a good example of this. The *Physikalischer Atlas* ('Physical atlas'), produced by August Petermann's Gotha-based Perthes publishing house, became known simply as 'the *Berghaus*'.¹⁵

But even in such cases of demonstrable authorship, where a name came to stand for scientific authority, one should perhaps sound a note of caution. This is especially so with respect to more extensive, and highly collaborative, cartographic endeavours. Here, the question of authorship tends to be rather more complex. In his *Cartography in France* (1987), Josef Konvitz explores the making of the 'map of Cassini', the *Carte Générale de la France* (1747–1818). The origins of this map date back to the 1660s, to Louis XIV and his First Minister of State, Jean-Baptiste Colbert. The cartographic survey this map was based on occupied four generations of the Cassini family.[16] But this raises a thorny question: To whom should authorship of the 'map of Cassini' actually be attributed? Does it belong to the Cassinis and the team of surveyors who provided the scientific expertise? To the French state that initiated and sponsored it? Or to other relevant and involved bodies, such as the *Académie des sciences*? After all, this last institution had a vested interest in pursuing related scientific endeavours, such as measuring the Paris meridian (a line of longitude running through France's foremost observatory).

The question of authorship, and of the production process, is thus clearly more complex than it might seem, even in the case of high-profile maps such as those by the Cassinis. But authorship remains particularly opaque with respect to 'underground' maps, such as those in the *Atlas statystyczny Polski*. We are left with very little knowledge of the sources which informed this document. And as we will see, there were very good reasons to remain silent about where, when, and by whom this atlas had been produced.

An act of Polish counter-mapping

Scholars such as Brian Harley have opened numerous avenues to the reading of maps as 'texts'. These avenues include an analysis of the colour-coding of maps, of the use of place names, and of 'silences and secrecy', i.e. a given map's inclusion and exclusion of specific spatial information. Maps do not *mirror* territorial realities; they *create* them in a two-dimensional form.[17]

The *Atlas statystyczny Polski*, for example, effectively makes a political argument about Poland's past, present, and future. It can be read as an act of 'counter-mapping'. Deliberately and subversively, the atlas aimed to contest the post-1815 territorial status quo by calling into question the imperial expansionism of Prussia, Russia, and Austria. It carved out and evoked the space of an independent state and a Polish nation.

All six maps in the *Atlas statystyczny Polski* include the same main title in the top left corner: *Mappa Polski i krajów okolicznych* ('Map of Poland and neighbouring countries'). The wording is pertinent, because it places Poland on par with its 'neighbouring countries'. There is no indication that the 'neighbour' to the south was, in fact, the Habsburg monarchy, let alone that Poland was itself in personal union with an empire, that of tsarist Russia. In terms of territorial status, the term *kraj* ('country', or 'land') had the advantage of being ambiguous and open to interpretation.

The full wording of the map's title is also significant. It points to the map's 'special indication of fortified towns, canals, paved roads', expressly 'based on the most recent facts'. In this way, Poland is presented as a developed country, traversed by abundant infrastructure, territorially bounded but well connected to its 'neighbouring countries'. The map invokes the declared authority of evidence-based scientific validation.

All six statistical maps operate with the same underlying base map. This base map conjures up the spatial imaginary of the Polish-Lithuanian Commonwealth as it existed prior to the First Partition of 1772. This state, certainly an empire in its own right, stretched from the Baltic Sea to Eastern Galicia (in today's Ukraine), and from Poznań to Vitebsk (in today's Belarus).[18]

All of this indicates the importance of reading every aspect of a map's visual rhetoric, including titles, place names, font sizes, and colours. Together, these elements form a visual argument, one concealed behind a veneer of apparent objectivity. The following section zooms in on four maps that carved out Polish spaces along administrative, economic, religious, and linguistic parameters.

Territorial silences and economic prowess

The first map (Figure 1.2) is concerned with 'administrative partition and key places', as we can see from its title in the top right-hand corner. This map shows the territorial status quo of Poland after 1815. In the centre of the map is the '*Królestwo Polskie*', the Kingdom of Poland, also known as Congress Poland, which was established at the Congress of Vienna in 1815. Its external borders are demarcated in red and are clearly distinguishable from those of

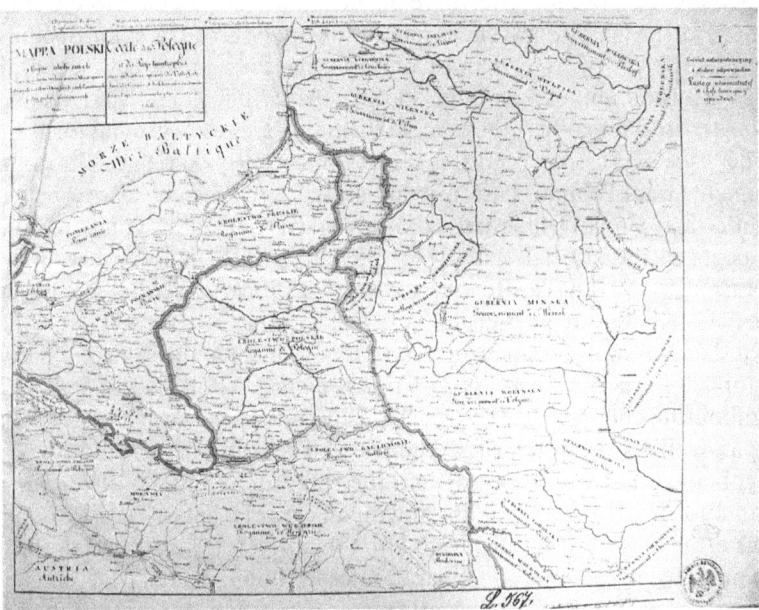

Figure 1.2 Map of 'administrative partition and key places', *Mappa Polski* [1827].

Source: *Atlas statystyczny Polski i kraiów okolicznych*, Staatsbibliothek Berlin/Kartenabteilung Q16884.

its 'neighbouring countries'. The Prussian borders of the Hohenzollern monarchy are blue, those of the Habsburgs are yellow, while those of imperial Russia are green. The red territorial borders of Poland on this map insinuate that we are dealing here with an independent country. In this way, the map conceals the fact that, at the time the atlas was published in 1827, the political and territorial status of the 'Kingdom of Poland' was highly precarious. There existed no international border between the kingdom and the Russian Empire. Indeed, during this period, Poland was subjected to increasing tsarist influence. The historian Brian Porter-Szücs puts it rather bluntly: 'There was no Poland on the map between 1795 and 1918.'[19]

To be sure, the Congress of Vienna had originally granted political autonomy to Poland. The new kingdom was united to, but not an integral, administrative part of, the Russian Empire. The Poles were allowed their own parliament (*Sejm*), and Tsar Alexander I, who also acted as 'King of Poland' between 1815 and 1825, was a fairly liberal ruler by the standards of his times. However, while in principle the *Sejm* had control over the budget, as well as over the Polish army, Russia repeatedly interfered, with the army under the strict command of the tsar's less liberal-minded brother, Grand Duke Konstantin. Moreover, from 1819, restorationist tendencies were beginning to make their presence felt across much of continental Europe. In Poland, freedom of the press was curbed, and censorship was imposed. In 1825, Alexander's death and the failed Decembrist revolt triggered another wave of repression. The *Sejm*, which had not been convened since 1820, was ordered to close its meetings to the public. After the uprising of 1830–31, all legal and constitutional autonomy was suppressed: the *Sejm* was abolished and the constitution suspended. 1831 thus marked the end of Polish statehood for almost a century, until the foundation of the Second Polish Republic in November 1918.[20]

And yet, a key focal point of this 1827 map is Warsaw. The name of this city is underlined in bright red. So too is the old Polish capital of Cracow, then the political centre of the 'Republic of Kraków'. This had been established in 1815, on the border to what had become Austrian Galicia as early as 1772. Conspicuous by their absence, or at least marginalized, are the imperial powers of the region. The Russian Empire, on stark display in so many other maps of the period, is curiously invisible. Half of the map shows areas that were part of Russian territory, and yet the reader will search in vain for the word 'Russia'.[21] Equally noteworthy is the absence of any further colour-coding, apart from the colours applied to borders. Usually, maps of the time would have rendered political territories colourfully visible: green for Russia, blue for Prussia, yellow for Austria, and so on.[22] In this respect, the lack of colour-coding has the striking effect of, to all intents and purposes, wiping (especially) Russia from the map.

To be sure, Austria and Prussia are not entirely absent. They are clearly marginalized, however. The word 'Austria' is relegated to the bottom left corner of the map. There is no mention of an 'Austrian Monarchy', only of the 'Kingdom of Hungary' and 'Kingdom of Galicia'.[23] Prussia, as the third partitioning power, is similarly side-lined. As a result of the partitions, the

Prussian monarchy had vastly expanded its territory and almost doubled its population at Polish expense. And yet Prussia is cartographically represented here as a small 'Kingdom of Prussia' north of Poland on the Baltic coast. All the other Prussian territories we see on the map appear as 'Pomerania', 'Silesia', and the 'Grand Duchy of Posen'. This is certainly in line with the map's remit of displaying administrative units, but the visual effect is striking. Cartographically, Prussia appears compartmentalized – we might even say that it appears 'partitioned'. While the light-blue colour-coding of borders creates a certain degree of cohesion, this is countered by the prominent black font of territorial place names. The map refers to four 'kingdoms': Poland, Prussia, Galicia, and Hungary. Poland appears on this map as the largest of the four, and indeed, as central to the region.

The second map (Figure 1.3) deals with 'manufactories and factories'. Across this map, a number of towns are underlined in different colours. This serves to indicate production sites of linen, cloth, glass, and paper, as well as salt mines, tanneries, steel mills, and foundries. Two aspects are noteworthy about this particular map. First, it appears to highlight the economic prowess of the 'Kingdom of Poland'. This is entirely in keeping with the theme of placing Poland on par with its neighbouring lands, such as economically vibrant Silesia ('Szląsk'). The reality, however, was rather mixed. Poland faced the myriad economic difficulties of a truncated and land-locked economy, and of a customs border with the Russian Empire.[24] And yet the map

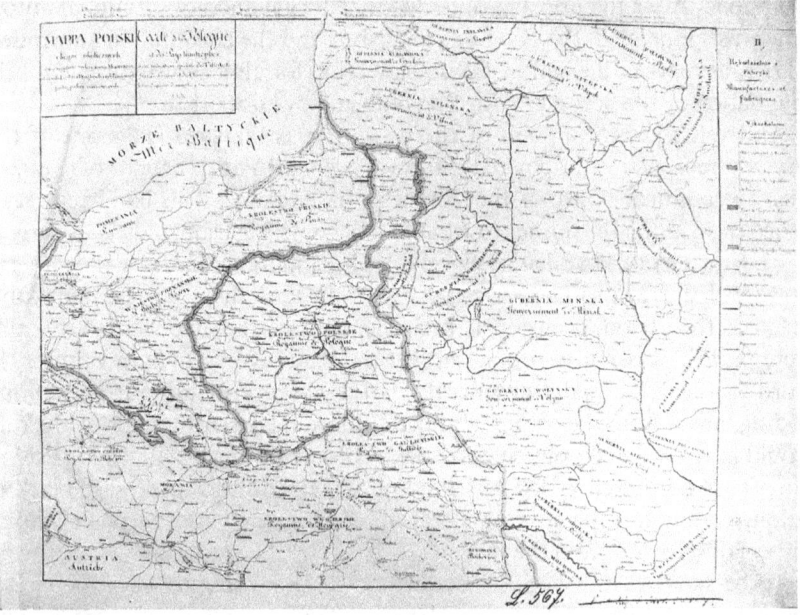

Figure 1.3 Map of 'manufactories and factories', *Mappa Polski* [1827].

Source: *Atlas statystyczny Polski i krajów okolicznych*, Staatsbibliothek Berlin/Kartenabteilung Q16884.

depicts an area apparently teeming with a visually striking density of underlined sites of economic production – 42 in total. Most of these relate to steel mills and foundries, some refer to cloth and linen factories.

Clearly, then, this map aimed to present the Kingdom of Poland as a realm of thriving economic activity. But there is a second point to make here. The map may also have served as a reminder of the economic space of the old Commonwealth. It designates important sites of economic activity beyond the borders of the Kingdom of Poland, especially in Galicia in the south and along the Vistula in the north, with the port city of Gdańsk as a vitally important trade hub. This served as a reminder of the riches of a 'golden age' in Polish history: the famous Wieliczka salt mines near Kraków, one of the largest salt mines in Europe, alongside another eight salt mines scattered across Galicia. The names of Tarnopol and Snialyn in eastern Galicia are marked as sites of tanneries, and the map reader's eye may wander further east and spot five other tanneries in the neighbouring districts of 'Wolynska', 'Podolska', and 'Minska'. The old Commonwealth's 'phantom borders' emerge.[25]

The past and future of a nation

The final two maps surveyed here offer even more explicit challenges to the territorial status quo. They concern religious denomination (Figure 1.4) and language (Figure 1.5). The map on 'religious rites' lists four denominations. First on the list is Roman Catholic (pale pink), followed by Greek Catholic, or 'Uniate' (grey-brown), Russian Orthodox (which features here as 'Greek-Russian', and is coloured green), and finally, Protestant (blue). At the centre of the map lies the territory of the Kingdom of Poland – and it is almost homogenously coloured pale pink, except for a south-eastern stretch of territory which is marked as Uniate (i.e. Orthodox in rite and Catholic in doctrine, placed under the authority of Rome, and named 'Uniate' after the Union of Brest in 1595–96).

Significantly, the Roman Catholic territory that dominates the map extends far beyond the actual boundaries of the Kingdom of Poland. We are presented with a broad Roman Catholic realm stretching south from the Baltic coast down to the Grand Duchy of Posen. To the upper right of the map, another vast, pale pink strip covers large areas of the former Grand Duchy of Lithuania (including Plater's home town). Also coloured pale pink are the southern part of Silesia, and the western part of Galicia. The latter had been annexed by the Habsburgs in 1772 and had once belonged to the Commonwealth. Another former Commonwealth territory is Catholic Warmia, which features here as an island of pale pink surrounded by Protestant blue in the 'Kingdom of Prussia' (i.e. East Prussia).

Overall, then, the map presents a large, homogenous Roman Catholic area. Of course, this is not congruent with the former Polish-Lithuanian Commonwealth. But neither is it congruent with the status quo of the post-1815 Kingdom of Poland. Especially when seeing the maps on religious

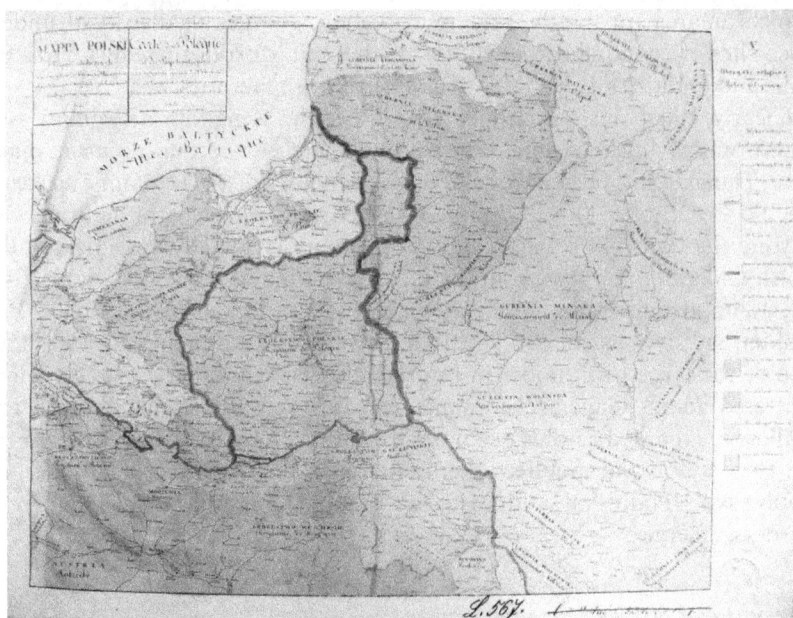

Figure 1.4 Map of 'religious rites', *Mappa Polski* [1827].

Source: *Atlas statystyczny Polski i kraiów okolicznych*, Staatsbibliothek Berlin/Kartenabteilung Q16884.

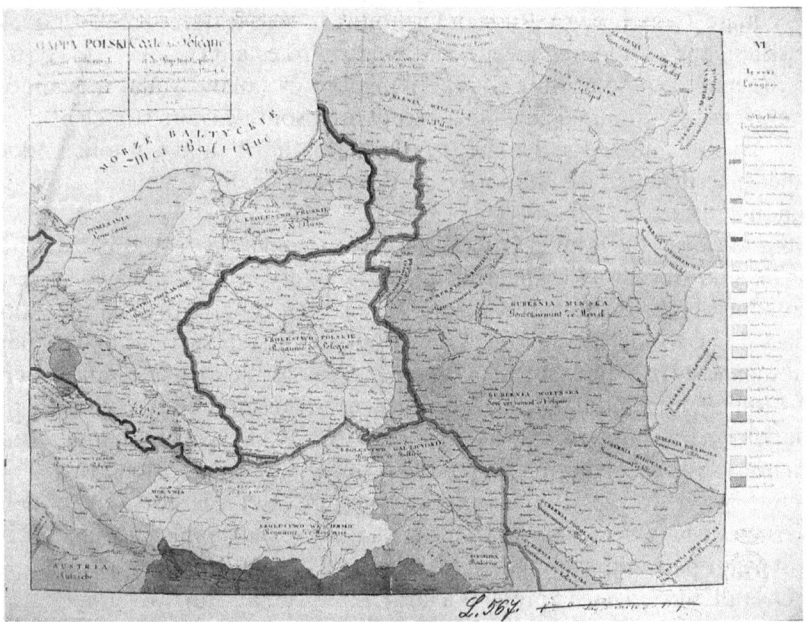

Figure 1.5 Map of 'languages', *Mappa Polski* [1827].

Source: *Atlas statystyczny Polski i kraiów okolicznych*, Staatsbibliothek Berlin/Kartenabteilung Q16884.

denomination and language *in relation to each other*, where *both* Roman Catholic and Polish-speaking areas are coloured pale pink, an alternative vision of a Polish nation emerges, one that gradually associated Polishness with Roman Catholicism.

During the early modern Commonwealth, the notion of Polishness had been strongly linked to the social category of the nobility. This was the *szlachta*, which made up around 7 per cent of the overall pre-1772 population. The *szlachta* represented the Commonwealth politically in the *Sejm*, and it also elected the king.[26] The Third Partition of 1795 ended the existence of the Commonwealth, and while it certainly did not end the social influence of the *szlachta* (especially that of a small yet powerful group of landowners), they lost ground politically. At the same time, the Catholic church was gaining in cultural and educational importance. This was particularly so during the reform era of the 1770s and 1780s under Stanisław August Poniatowski. This period saw the establishment of new educational institutions, and education was firmly in the hands of the Catholic church. Increasingly, the older *szlachta* notion of Polishness was complemented by, and gave way to, alternative, socially less exclusive notions of nationhood.[27]

The map's dedicated allocation of a denominational space for the Greek Catholic, or 'Uniate', Church is also noteworthy in this context. As historian Barbara Skinner has shown in *The Western Front of the Eastern Church*, the distinction between Uniate and Orthodox had become increasingly politicized during the course of the 18th century, with Poland-Lithuania favouring the former and Russia the latter. In the wake of the partitions, which brought hundreds of thousands of Uniates under Russian rule, Uniate parishes were subjected to growing suppression under Catherine II. Attempts to bring them under Orthodox jurisdiction re-intensified after Alexander's death in 1825, eventually leading to the 1839 Uniate Conversion.[28] The cartographic inclusion of a conspicuous Greek Catholic terrain, visually depicted in a way that underlined its association with Roman Catholicism, had the subversive effect of countering Russian attempts to push the confessional fault line further westward. Moreover, the continuous grey-brown swath of Uniate land, spanning various territorial borders, served as a reminder of the pre-1772 status quo ante, when most of the Uniate church had still been united in one polity: Poland-Lithuania.

The cartographic argument made by the mapmaker about the possible future of the Polish nation is rounded off with a final map on the spatial distribution of languages. The legend in the margin of the right-hand side of this map lists 11 languages. Unsurprisingly, the first one is Polish, which is indicated in pale pink. This is followed by Ruthenian, or *Język Ruski* (comprising today's Belarusian, Ukrainian, and Rusyn) in grey-brown, then Lithuanian in beige, Latvian in pale orange, and German in blue. These, in turn, are followed by Russian (*Język Rossyiski*) in green, Wallachian (spoken in today's Romania) in grey beige, Hungarian in grey, Slovak and Czech in light yellow hues, and Wendish (i.e. Sorbian) in orange.

Of course, a variety of linguistic overlaps characterised this region. At the local level, several languages, or dialects, were spoken. Linguistic boundaries were fluid.[29] And yet this map presents homogenous language regions and establishes a visually palpable hierarchy. It is no coincidence that the main languages of the former, multi-linguistic Commonwealth are listed first. They also dominate the map: Polish, Ruthenian, Lithuanian, and Latvian. Most readers of the map would surely be able to distinguish between these languages. Nonetheless, the pale earth tone colours in which they are displayed are sufficiently similar to create an impression of visual cohesion. This is in stark contrast to the distinct blue for German, light yellow tones for Czech and Slovak, orange for the small Sorbian-speaking area in the west, and of course, in clear contrast to the Russian green 'borderland' in the east. What therefore emerges from the map are the contours of the former Polish-Lithuanian Commonwealth, with a Polish-speaking area that is both visually foregrounded and sufficiently similar in its colour-coding to blend in with other areas of the old Commonwealth – reminiscent also of hybrid self-conceptions such as *Natione Polonus, gente Ruthenus* ('Polish by nationality, of the Ruthenian people').[30]

Another key contrast emerges from this map. The post-1815 Kingdom of Poland is juxtaposed against both the vast, visually cohesive terrain of the former Polish-Lithuanian Commonwealth, and the expanse of Polish-speaking areas to the west, north and south of the Kingdom. These include the Grand Duchy of Posen (part of East Prussia), Upper Silesia, and the western part of neighbouring Galicia.[31]

At a mere glance, suggestively and tentatively, this map alludes to similar arguments as those being propounded in historical works and political pamphlets by Romantic nationalists such as Joachim Lelewel, the renowned Polish historian of the 19th century. Lelewel's vision of Poland was firmly rooted in the old Commonwealth, as demonstrated in his works on ancient Lithuania and on the Polish constitutions. Like Ludwik Plater, Lelewel was forced into exile after the uprising of 1830–31. He joined a Polish émigré organization. As Monika Baar writes, this group 'advocated a free, independent and multinational Poland with pre-partition boundaries, a republican government and a democratic society'.[32] Naturally, the *Atlas statystyczny Polski* did not express such desires quite so eloquently, or explicitly. Nonetheless, it did advance a visual argument – distinctive and powerful enough to suggest anonymous publication.

Conclusion

Maps invite us to think about their materiality, genre, and authorship, as well as the context of their creation. Most importantly, they invite us to investigate the visual arguments they advance. They accomplish this by means of spatial homogenization, thematic selection, colour-coding, naming, and 'silencing'. Many of the analytical strategies pursued in this chapter are surely applicable to a wide range of maps. And yet, the *Atlas statystyczny*

Polski remains a particularly instructive case. This is partly due to the historical setting of the atlas's creation, as well as its samizdat-like nature.

This chapter has interpreted the *Atlas statystyczny Polski* as an act of counter-mapping. It was authored by Ludwik Plater, a reform-oriented member of the lesser nobility. Plater had participated in the Polish uprising of 1794. In 1830, he took part in yet another uprising, before going into exile in Paris. His atlas posed a challenge to the territorial and political status quo. The Kingdom of Poland had been founded in 1815 as the truncated part of a rump state, the Napoleonic satellite of the 'Duchy of Warsaw' (1807–13). After the Congress of Vienna, this territorial unit was formally endowed with political autonomy. However, it found itself increasingly subjected to Russian rulership, especially after Alexander's death and the crushed Decembrist revolt of 1825. The atlas was published two years later. Through means of cartographic visualization, it made the case for an independent Poland on par with its 'neighbouring countries'.

The atlas's key method for accomplishing this was through the interlacing of various temporal layers. These were, mainly, the pre-1772 Polish-Lithuanian Commonwealth, visualized primarily in the form of a base map, alongside potential *futures* for a Polish nation and a sovereign state. The spatial imaginaries conjured up here were no mere manifestations of nostalgia. This early example of statistical mapping was demonstrably authored by a modernizing mindset, as can also be gleaned from the overt focus on infrastructure. The atlas's genesis lies in a crucial transition period. At this time, Polishness was becoming gradually associated with Catholicism. Language, too, was slowly gaining traction as a constituting factor in socially less exclusive, less nobility-centred notions of Polish nationhood.

Needless to say, this six-map atlas does not amount to an elaborate, clear-cut statement about Poland's future. It allows for multiple readings, and it is open to interpretation. An oft-cited phrase from Joachim Lelewel is: *Polska tak, ale jaka?*, meaning 'Poland yes, but what sort of Poland?'.[33] The atlas's six *Mappy Polski* pose precisely this question, and they suggest several possible answers. These answers are framed within the geographical contours of Poland's pre-partition past, and they are replete with territorial and national potentialities for a viable Polish future – one no longer dictated by the political ambitions and economic desires of surrounding empires.

Notes

1 Matthew H. Edney, *Cartography: The Ideal and Its History*, Chicago and London: University of Chicago Press, 2019, pp. 236–7. We would like to thank Alex Burkhardt, Catherine Gibson, Tomasz Kamusella, and our co-editor Konrad Lawson for help, comments, and insights.

2 See Nancy Lee Peluso, 'Whose Woods Are These? Counter-Mapping Forest Territories in Kalimantan, Indonesia', *Antipode* 27, 1995, 383–406, here 384; the term 'counter-appropriation' is already prominent in id., *Rich Forests, Poor People: Resource Control and Resistance in Java*, Berkeley: University of California

Press, 1992; see also Dorothy L. Hodgson and Richard A. Schroeder, 'Dilemmas of Counter-Mapping Community Resources in Tanzania', *Development and Change* 33, 2002, 79–100; Joel Wainwright and Joe Bryan, 'Cartography, Territory, Property: Postcolonial Reflections on Indigenous Counter-Mapping in Nicaragua and Belize', *cultural geographies* 16, 2009, 153–78.

3 Catherine Dunlop, *Cartophilia: Maps and the Search for Identity in the French-German Borderland*, Chicago and London: University of Chicago Press, 2015, esp. p. 11, and n. 2; Julie MacArthur, *Cartography and the Political Imagination: Mapping Community in Colonial Kenya*, Athens: Ohio University Press, 2016, pp. 20–1.

4 See, especially, Kristin Kopp, *Germany's Wild East: Constructing Poland as Colonial Space*, Ann Arbor: University of Michigan Press, 2012; see also Philipp Ther, 'Beyond the Nation: The Relational Basis of a Comparative History of Germany and Europe', *Central European History* 36, 2003, 45–73.

5 A digital version of the atlas, with some variations in the right-hand margin of the maps, can be accessed from the website of the Digital Library of the University of Wrocław. Available HTTP: https://www.bibliotekacyfrowa.pl/dlibra/publication/30529/edition/38573/content?format_id=1 (accessed 20 May 2021).

6 See, for instance, Steven Seegel, *Mapping Europe's Borderlands: Russian Cartography in the Age of Empire*, Chicago and London: University of Chicago Press, 2012, pp. 97–101.

7 See Janina Piasecka, 'Polskie atlasy geograficzne do połowy XIX wieku', *Polski Przegląd Kartograficzny* 17/4, 1985, 169–90, here 182–3; id., 'Atlasy statystyczne Królestwa Polskiego od Platera (1827) do Romera (1916)', in Lucyna Szaniawska and Jerzy Ostrowski (eds.), *Z Dziejów Kartografii*, vol. 10: *Kartografia Królestwa Polskiego 1815–1915*, Warsaw: Biblioteka Narodowa, 2000, pp. 127–37, here pp. 127–8; Waldemar A. Spallek, *Polskie szkolne atlasy geograficzne 1771–2012*, Wrocław: Instytut Geografii i Rozwoju Regionalnego Uniwersytetu Wrocławskiego, 2018, pp. 37, 52, 57, 86, 200–3, 381, 416; see also Bożena Łazowska, 'Statystyka na ziemiach polskich pod panowaniem pruskim', *Wiadomości Statystyczne* 63/5, 2018, 78–102, here 91–2. Spallek has also provided the cartography to Marcin Wodziński's pathbreaking *Historical Atlas of Hasidism*, Princeton: Princeton University Press, 2018. For relevant metadata see n. 5, as well as https://www.wbc.poznan.pl/dlibra/publication/558130/edition/475263/content (accessed 20 May 2021). There is, however, still some variation across the metadata of Polish archives, some of which indicate *Stanisław* Plater as author.

8 See the literature cited above in n. 7.

9 See Lars Behrisch, 'Statistics and Politics in the 18th Century', *Historical Social Research* 41, 2016, 238–57; Anne Godlewska, *Geography Unbound: French Geographic Science from Cassini to Humboldt*, Chicago and London: University of Chicago Press, 1999; Arthur H. Robinson, *Early Thematic Mapping in the History of Cartography*, Chicago and London: University of Chicago Press, 1982, esp. pp. 155–88; on the history of statistics more broadly, see Alain Desrosières, *The Politics of Large Numbers: A History of Statistical Reasoning*, Cambridge, Mass.: Harvard University Press, 1998, French 1993; Theodore M. Porter, *Trust in Numbers: The Pursuit of Objectivity in Science and Public Life*, new ed., with a new preface, Princeton: Princeton University Press, 2020, first published 1995.

10 See, for instance, Jason D. Hansen, *Mapping the Germans: Statistical Science, Cartography, and the Visualization of the German Nation, 1848–1914*, Oxford: Oxford University Press, 2015; Morgane Labbé, *La nationalité, une histoire de*

chiffres: Politique et statistiques en Europe Centrale (1848–1919), Paris: Presses de Sciences Po, 2019; Vytautas Petronis, 'Mapping Lithuanians: The Development of Russian Imperial Ethnic Cartography, 1840s–1870s', *Imago Mundi* 63, 2011, 62–75.

11 Colour was still added by hand at the time; 'cromolithography' only came into wider use from the 1830s.

12 Stanisław Plater, *Atlas historique de la Pologne accompagné d'un tableau comparatif des expéditions militaires dans ce pays pendant le XVII-me, XVIII-me et XIX-me siècle*, Poznań: Guillaume Decker et compagnie, 1827. Available HTTP: https://bibliotekacyfrowa.pl/dlibra/publication/117919/edition/108432/content (accessed 20 May 2021). On historical atlases more broadly, see Walter Goffart, *Historical Atlases: The First Three Hundred Years, 1570–1870*, Chicago and London: University of Chicago Press, 2003.

13 Review of *Atlas statystyczny Polski*, in *Kritischer Wegweiser im Gebiete der Landkarten-Kunde nebst andern Nachrichten zur Beförderung der mathematisch-physikalischen Geographie und Hydrographie*, vol. 2, Berlin: Simon Schropp & Co., 1830, pp. 164–6, here pp. 164–5; for Berghaus' review of the Prussian 'counterpart', see ibid., pp. 161–4; for a reprint of the Prussian atlas, see Wolfgang Scharfe (ed.), *Administrativ-Statistischer Atlas vom Preussischen Staat*, reprint, with introduction, Berlin: Kiepert, 1990, first published 1827/28. While the Prussian atlas was more elaborate and comprehensive than the *Atlas statystyczny Polski*, both of them covered the aspects of administrative units, infrastructure, economy, education, religion, and language, and there was no accompanying text in the Prussian atlas either.

14 Review of *Atlas statystyczny Polski*, p. 164.

15 See Jürgen Espenhorst, *Petermann's Planet: A Guide to German Handatlases and Their Siblings Throughout the World, 1800–1950*, vol. 1: *The Great Handatlases*, Schwerte: Pangaea, 2003, pp. 365–83; for a further example from the Perthes publishing house, see Iris Schröder, 'Eine Weltkarte aus der Provinz: Die Gothaer *Chart of the World* und die Karriere eines globalen Bestsellers', *Historische Anthropologie* 25, 2017, 353–76.

16 See Josef W. Konvitz, *Cartography in France, 1660–1848: Science, Engineering, and Statecraft*, Chicago and London: University of Chicago Press, 1987; for more on this, as well as further case studies, see the section on 'Cartographic Representations' in Konrad Lawson, Riccardo Bavaj and Bernhard Struck, *A Guide to Spatial History: Areas, Aspects, and Avenues of Research* (2021), Available HTTPS: spatialhistory.net/guide.

17 J.B. Harley, 'Silences and Secrecy: The Hidden Agenda of Cartography in Early Modern Europe', *Imago Mundi* 40, 1988, 57–76.

18 On Poland-Lithuania, see most recently Richard Butterwick, *The Polish-Lithuanian Commonwealth, 1733–1795: Light and Flame*, New Haven: Yale University Press, 2020; Andrzej Chwalba and Krzysztof Zamorski (eds.), *The Polish-Lithuanian Commonwealth: History, Memory, Legacy*, London and New York: Routledge, 2020.

19 Brian Porter-Szücs, *Poland in the Modern World: Beyond Martyrdom*, Chichester: Wiley-Blackwell, 2014, p. 6; see also Jerzy Lukowski and Hubert Zawadzki, *A Concise History of Poland*, 3rd ed., Cambridge: Cambridge University Press, 2019, first published 2001, pp. 188–97.

20 See Lukowski and Zawadzki, *History of Poland*, pp. 207–8; Piotr S. Wandycz, *The Lands of Partitioned Poland, 1795–1918*, 2nd corr. ed., Seattle and London: University of Washington Press, 1984, first published 1974, pp. 84, 87, 89–90, 97, 122.

21 Admittedly, the 'Russian Empire' does appear, albeit in small font, in the legend that is included in the right-hand margin of each of the following five maps.
22 See, for example, Stanisław Plater, *Atlas historique de la Pologne*.
23 As with the 'Russian Empire', the legend of the other five maps does list both the 'Austrian' and the 'Prussian Monarchy' (again in small font).
24 See Lukowski and Zawadzki, *History of Poland*, p. 194; Wandycz, *Lands of Partitioned Poland*, pp. 79–87.
25 For the concept of 'phantom borders', which has been developed, for varying analytical purposes, in the context of East Central and Southeast Europe, see Béatrice von Hirschhausen et al. (eds.), *Phantomgrenzen: Räume und Akteure in der Zeit neu denken*, Göttingen: Wallstein, 2015; id., 'Phantom Borders in Eastern Europe: A New Concept for Regional Research', *Slavic Review* 78, 2019, 368–89; see also Michael G. Esch and Béatrice von Hirschhausen (eds.), *Wahrnehmen, Erfahren, Gestalten: Phantomgrenzen und soziale Raumproduktion*, Göttingen: Wallstein, 2017; Andrea Komlosy, *Grenzen: Räumliche und soziale Trennlinien im Zeitenlauf*, Vienna: Promedia, 2018, pp. 127–31.
26 See Tomasz Kamusella, *The Un-Polish Poland, 1989 and the Illusion of Regained Historical Continuity*, Basingstoke: Palgrave Macmillan, 2017, pp. 15–36.
27 See Richard Butterwick, 'Catholicism and Enlightenment in Poland-Lithuania', in Ulrich L. Lehner and Michael Printy (eds.), *A Companion to the Catholic Enlightenment in Europe*, Leiden: Brill, 2010, pp. 297–358, here pp. 302, 305–7, 350; id., *The Polish Revolution and the Catholic Church: A Political History, 1788–1792*, Oxford: Oxford University Press, 2012; the link between Catholicism and Polishness is qualified in Porter-Szücs, *Poland in the Modern World*, pp. 18–20; see also Brian Porter, *When Nationalism Began to Hate: Imagining Modern Politics in Nineteenth-Century Poland*, Oxford: Oxford University Press, 2000, ch. 1.
28 Barbara Skinner, *The Western Front of the Eastern Church: Uniate and Orthodox Conflict in Eighteenth-Century Poland, Ukraine, Belarus, and Russia*, DeKalb: Northern Illinois University Press, 2009; see also id., 'Orthodox Missions to the "Ancient Orthodox" Lands in Belarus and the 1839 Uniate Conversion', *Canadian-American Slavic Studies* 53, 2019, 246–62; Larry Wolff, *Disunion within the Union: The Uniate Church and the Partitions of Poland*, Cambridge, Mass.: Harvard Ukrainian Research Institute, 2019 (based on an article from 2002-3).
29 See, for instance, Tomasz Kamusella, *The Politics of Language and Nationalism in Modern Central Europe*, Basingstoke: Palgrave Macmillan, 2009; id., Motoki Nomachi and Catherine Gibson (eds.), *Central Europe through the Lens of Language and Politics: On the Sample Maps from the Atlas of Language Politics in Modern Central Europe*, Sapporo: Slavic-Eurasian Research Center, Hokkaido University, 2017.
30 See Joel Brady and Edin Hajdarpasic, 'Religion and Ethnicity: Conflicting and Converging Identifications', in Irina Livezeanu and Árpád von Klimó (eds.), *The Routledge History of East Central Europe since 1700*, New York and London: Routledge, 2017, pp. 176–214, here p. 185; on the role of languages in Poland-Lithuania, see also Timothy Snyder, *The Reconstruction of Nations: Poland, Ukraine, Lithuania, Belarus, 1569–1999*, New Haven: Yale University Press, 2003, pp. 18–19.
31 On the very gradual, and ambiguous, emergence of regionally inflected Polish identities in Upper Silesia, which had never been part of the Polish-Lithuanian Commonwealth, see Lukowski and Zawadzki, *History of Poland*, p. 199; for later decades, see Brendan Karch, *Nation and Loyalty in a German-Polish Borderland: Upper Silesia, 1848–1960*, Cambridge: Cambridge University Press, 2018.

32 Monika Baar, *Historians and Nationalism: East-Central Europe in the Nineteenth Century*, Oxford: Oxford University Press, 2010, p. 22; see also Jerzy Lukowski, *Disorderly Liberty: The Political Culture of the Polish-Lithuanian Commonwealth in the Eighteenth Century*, London: Continuum, 2010, pp. 259–60; Joan S. Skurnowicz, *Romantic Nationalism and Liberalism: Joachim Lelewel and the Polish National Idea*, New York: Columbia University Press, 1981; Steven J. Seegel, 'Cartography and the Collected Nation in Joachim Lelewel's Geographical Imagination', *Slavica Lundensia* 22, 2005, 23–31.

33 Joachim Lelewel, cited in Norman Davies, *God's Playground: A History of Poland*, vol. 2, rev. ed., Oxford: Oxford University Press, 2005, first published 1981, p. 524.

2 Travel guides

James Koranyi

'What ought to be seen'

Writing in the *Journal of Contemporary History* in 1998, Rudy Koshar launched a famous defence of travel guides as historical source material.[1] Against the prevailing orthodoxy that the advent of travel guides in the 19th century represented a debasement of travelling, Koshar advocated a more complex reading of them. Scholars such as John Urry thought that the shaping of the tourist gaze by travel guides was a mere extension of consumer behaviour.[2] Koshar, by contrast, moved away from a deterministic understanding of travel guides as constrained by market behaviour towards an appreciation of the rich cultural history that lies behind these sources. For Koshar, travel guides lay bare the very fabric of societies. 'What ought to be seen' consisted of places and sites that reflected the societies on display, the travellers and their milieu, and the places and cultures within which these guides were produced. According to Koshar, at the very heart of 19th-century concerns lay nationalism and national identity, and guidebooks facilitated their public promotion.[3]

Other scholars such as William Stowe also drew further attention to the cataloguing effect of travel guides in the 19th century. Murray's *Handbook for Travellers in Central Italy and Rome* explicitly used the word 'catalogue' to list the new and sumptuous museums, which in turn reinforced the burgeoning Italian national ambitions during the *Risorgimento* (1815–60).[4] Museums were quintessentially didactic forums in which national treasures were open to the public and conveyed a national story of historical depth and unity.[5] Stowe also spotted the 'ethnocentricity' of guidebooks. Two of the most well-known guidebooks, *Baedeker* and *Murray*, first published in 1832 and 1836 respectively, have frequently featured in academic studies on travel guides as prime examples of a nationalist framing of treasures.[6] They reveal the ethnocentric and nationalist impulses that emerged in the 19th century and carried over into the 20th century. The gendered nature of guidebooks, too, have been laid bare in studies of those two famous guidebooks.[7] Other aspects have received ample consideration, from imperial claims to leisure and consumption practices.

In all of this, it is striking that most scholarship across disciplines on travel guides takes them at face value. That is, scholars study the text and expose the meaning beneath and behind said text.[8] This is perfectly good historical practice and must be taken seriously as a way of unlocking the cultural and political import of guidebooks. It places guidebooks, as we now know them, in their correct historical context as they gradually took off from the 1830s/1840s. They replaced an older genre of travel writing, which often featured firsthand accounts of journeys to remote and exciting places, though travel accounts continued to co-exist with guidebooks. Guidebooks, however, laid claim to a more authoritative judgement of 'correct' travel whereas individual travel accounts were consumed as literary entertainment. To be sure, such a distinction is problematic in its own right: as Vesna Goldsworthy has demonstrated with respect to the Balkans, literary entertainment was as powerful, if not more so, in fixing the mental maps of south-east Europe in the 19th and 20th centuries.[9] Indeed, the travel networks established in first-hand travel accounts reinforced the ideas of 'what ought to be seen' as much as guidebooks. Yet guidebooks offered a different and perhaps more convincing format suited to an age of mass travel and of more clearly defined national boundaries. In this way, studies of guidebooks have portrayed them as ultimate symbols of a late-modern chronology that allow scholars to explore nationalism, consumption, and modernity.

This chapter will approach guidebooks in a different way, namely by thinking about the *spaces* within guidebooks. It seems an obvious point to make: after all, travel guides dutifully list places and spaces for the traveller and tourist to visit.[10] And yet the spatial qualities of guidebooks have often been overlooked.[11] Focusing on spaces and places – both as they are described and as evidence of mental maps – puts spatial history at the forefront of examining guidebooks. That is not to say that mental maps do not reveal the politics of the day: of course they do. Yet the spatial lens illuminates cultural and political currents in a different way. It uncovers a world not primarily defined by structures, economics, consumption practices, or political networks. Instead, it suggests the ways in which people thought of the world around them in mental maps. Both the individualism of mental maps and the collective imagining of space are defining features in this respect. What emerges then from this chapter are the fascinating cultural and political worlds left behind in guidebooks.

How should travel guides be approached and studied? There are some very fundamental issues that need to be addressed: where were the guides published? How long were travel guides meant to be current? What were their circulation figures? Who was the target audience and who actually read them? How often did they appear? What language(s) did they use? What spaces and places did they concentrate on? And crucially: who wrote them? In all of this, it is important to consider in what sense they are representative (and of what), and how they perhaps differ from other guidebooks.

On a very basic level, travel guides describe physical realities. Explanations of varying length and accompanying maps serve the purpose making the places on display accessible. Accounts of the local landscape – be it urban or rural – explain the places under the spotlight while leaving enough room for the imagination to entice the reader to visit. Illustrations and photographs act as intriguing windows onto the world without revealing too much. Recommendations and catalogues help make sense of the world with bitesize and 'easily graspable facts'.[12] Explanatory glossaries and quick guides to language, etiquette, and everyday infrastructure are meant to ease travel and make it exotic at the same time. In that sense, travel guides are as much about practicalities as they are about the search for adventure and authenticity.

Guidebooks also shape the mental maps of readers in the very choices they make in their focus, lay-out, and style. Focusing on one city only, privileging one place over another in the title, or foregrounding nation-states instead of regions set these mental maps in obvious yet subliminal ways. The balance between text and images as well as the overall length of a book also subtly change the way readers and writers make sense of the world laid out before them. Perceptions of places and spaces are also dependent on the use of pronouns[13]: What does the 'I' in a short vignette on Berlin do for trust, authenticity, and authority? By contrast, an impersonal account may numb the imagination of readers, though of course they might crave the sense of sophistication that such a style might suggest.

Guidebooks suggest, with different levels of forcefulness, 'what ought to be seen'. As William Stowe notes, travel guides comprise all sorts of 'mays', 'wills', 'shoulds', and 'musts'.[14] Such hierarchies give rise to a number of different mental maps. Over a longer period of time, places are reinforced as being 'off the beaten track' or part and parcel of an expected canon.[15] Guidebooks therefore act in a dual function. They prescribe readers where to go while at the same time allowing them to imagine numerous possible worlds. Readers of travel guides browse the books for both 'pleasure and instruction', and it is in this manner that guidebooks must be understood.[16]

There is also the flipside to the mental maps shaped through travel guides, namely, their reception. It is the most challenging aspect of much of historical study, and guidebooks are no exception in that respect. Reviews of all kinds offer some insight into the vexed question of reception, though this may only be a window onto a relatively closed circle of writers. Furthermore, adverts of the travel guides and adverts within the guidebooks themselves are useful hints for establishing who the readership may have been. Publishing runs can be suggestive of a guidebook's success, but this should be treated with some caution. Visitor figures, if available, can at the very least hint at the network of places that travellers sought out, even if the direct connection to guidebooks might be more difficult to prove. Mental mapping, as adopted in historical studies from psychology, is bound up with the experience of travel. Guidebooks both represent experiences of places and pre-empt such experiences in their description of places.[17] Ancillary material such as letters, postcards, photographs, travel accounts, official documents (governmental and other), and increasingly digital/online material can add rich texture to the guidebooks. In combination, these mosaic pieces create a network of writers, readers, travellers, and their places.

Komm Mit: a Romanian German travel guide during the Cold War

If travel guides have been a hallmark of late-modern travel, then the overabundance of material makes the selection of source material particularly interesting. *Baedeker* and *Murray* are conspicuous by their fame and coverage in academic studies.[18] At the other end of the spectrum, contemporary guidebooks such as the *Rough Guide* or *Lonely Planet* series have been strenuously avoided by historians leaving anthropologists, sociologists, geographers, and others to break new ground in research.

This chapter examines a little-known German language guidebook from Romania. *Komm Mit* (*Come Along*) was an annual guidebook for Romania and was published 21 times between 1970 and 1990. Each issue was between 200 and 300 pages in length and consisted of often extensive articles and shorter exploratory essays. Its focus, and more on this later, was entirely on Romania, though the choice of destinations was telling. *Komm Mit* illuminates the large variety among travel guides. Its focus on lengthy essays instead of punchy, immediate information underlines that there is no 'standard' format for guidebooks. *Komm Mit* may not fulfil the expectations of a postmodern travel culture, and yet it is quite emphatically a guidebook of its time and its Cold War context.

To understand the spatial politics of *Komm Mit* some contextual information on the late-Cold War, Romania, and its German minority population is necessary. *Komm Mit* was published against the trend of increasing censorship after the initial thaw under Nicolae Ceaușescu once he became General Secretary of the Romanian Communist Party in 1965. For Romanian Germans, the driving force behind the guidebook, it offered an avenue to experiment with possibilities and with some poetic license while reaffirming their position in Romanian society, which had been increasingly precarious.

Who were these Romanian Germans and why were they in an uncertain situation? Made up mainly of Transylvanian Saxons and Swabians from the Banat, they had become Romanian Germans as a consequence of the enlargement of Romania in 1918. In complex and curious ways, they became staunch allies of Nazi Germany with Saxons in particular leading the way. Perceived as a state within a state, many actively supported the German war effort by joining either the *Wehrmacht* or the *Waffen-SS*. At the end of and after the war, they faced huge repercussions, but unlike their German counterparts in other east-central European areas, they were not expelled. Instead, between 70,000 and 80,000 Romanian Germans were deported to the Soviet Union for up to five years, and only around half of them returned.[19] Gradually, from the 1950s on, Germans were officially rehabilitated and reintegrated into Romanian society. Yet a considerable number of Romanian Germans now also lived in West Germany. The growing émigré community continued to lure the remaining Germans in Romania to Germany with letters, gifts, supplies, and publications.[20] In this paradox, Germans were at once becoming an integral part of the socialist project in Romania while trying to flee from it at the same time.[21] By the late 1970s, the pull factors towards West Germany had become so great and inexorable that large-scale emigration became a fact of life. The communities in Transylvania, the Banat, and elsewhere in Romania began to dissolve in plain sight. The uncertain and divided position of Romanian Germans in Cold War Romania is crucial for understanding and exploring *Komm Mit* as a primary source.

Let us revisit those fundamental questions asked earlier of travel guides. *Komm Mit* was issued once a year between 1970 and 1990. It was published by the publishing house of the official German-language newspaper *Neuer Weg* in Bucharest.[22] The print run was between 20,000 and 30,000 copies each year, and the targeted readership was quite varied. On the one hand, it was directed at the Romanian German community as a whole. Adverts in the *Neuer Weg* are testament to that. However, it was its international audience that made this guidebook so intriguing. While the books were distributed among Romanian Germans in Romania, *Komm Mit* also circulated widely in East Germany and Czechoslovakia.[23] In this sense, *Komm Mit* was part of the Romanian state's attempt at expanding its tourist sector for visitors from abroad. Of all east-central European socialist countries, Romania enticed one of the lowest numbers of tourists. Nevertheless, by the 1980s, a counter-culture of East German backpackers and bikers had emerged that travelled to and through Romania.[24] Romania provided an escape from the everyday for East German visitors. It was both German and quixotic enough for East Germans to feel at home and get lost along the paths in Transylvania, Maramureș, and the Daube Delta. *Komm Mit* acted as a go-between for Germans in the GDR and Romania and thus contributed towards a network in which Romania was discovered spatially by travel, tourism, and in text.

The editors were very much aware of its international profile, particularly because there was a further international audience they were addressing through this informal channel: 'the West'. Romanian Germans had strong ties to their émigré compatriots abroad. The Second World War, emigration, and older familial ties rendered a different view to that of a parochial, east-central European German minority. In practice, Romanian Germans were a global community during the Cold War. The majority of Romanian German émigrés lived in the Federal Republic of Germany, West Germany, but there were sizable communities in Switzerland, Austria, France, North America, and the GDR. On visits from and to relatives from the West, a buoyant exchange of goods and reading matter injected life into the network of Romanian Germans.

Back in Romania, many of the contributors to *Komm Mit* were part of an established network of a German cultural elite in Romania. Names such as Juliana Fabritius-Dancu, Erika Schneider, Helmut Fabritius, Lia Gross, Gerhard Bonfert, Walter Kargel as well as the first editor, Georg Hromadka, appeared time and time again in print. Erika Schneider made her name as an academic in the natural sciences. Juliana Fabritius-Dancu was a famous folklorist and art historian. Georg Hromadka was a big figure in Swabian and Romanian German circles. He had fallen foul of the regime in the 1940s, but rebuilt his career as a writer and a journalist.[25] He emigrated from Romania to West Germany in 1980. Many others, too, disappeared from the table of contents of the guidebook after a while; some had died, but others also emigrated like Hromadka. Nonetheless, *Komm Mit* represented a forum in which the good and great of the Romanian German community were able to exert some degree of control over their dwindling community. They were able to paint a picture of their homeland, which was part of Romania but also discrete enough. Repetition of poems, studies of folklore, and the meticulous mapping of hidden paths around the countryside hardened an image of a German, but locally patriotic Romanian German community and its homeland.

The spatial division of Europe during the Cold War is also significant for reading *Komm Mit* correctly. Romania had enjoyed an initial image boost once Nicolae Ceaușescu took over from Gheorghe Gheorghiu-Dej in 1965. Ceaușescu's refusal to support the Soviet intervention in Czechoslovakia in the summer of 1968 earned him countless plaudits from Western observers. Those positive perceptions of Romania in the non-socialist West soon disappeared. By the late-1970s, Romania's image of being the poster child of socialism had been replaced with it being the ultimate symbol of brutal socialist modernity.[26] Ceaușescu's persona in particular transformed into that of an inhumane dictator. But it was his programme of *sistematizare* (systemization or modernization) that had a deeply detrimental impact on the countryside, and it was precisely this countryside that was exhibited in *Komm Mit*. Its focus on the rural hinterland of Romania, especially Transylvania and Moldova, produced a counterview to that of Ceaușescu's unflinching push for modernization. In this way, the guide acted as a corrective to the view from abroad (and within) that Romania was destroying its own landscapes.[27]

Read accordingly, *Komm Mit* can be understood as a propaganda arm of the communist party and a diplomatic advert for a different kind of Romania: a Romania not just defined by socialist modernity, but also made up of rural spaces that were picturesque, preserved, and symbolic of a far more harmonious landscape throughout the country. Official endorsement of the guidebook was never really concealed. The high quality of paper and print brings home the point that the official landscapes on display had been sanctioned by the regime.[28] Party officials and politicians made an appearance every so often, too: in an interview in *Komm Mit* in 1981, the acting minister for tourism, Costache Zmeu, prefaced the travel guide by praising the opportunities Romania offered for an 'active' holiday.[29] Indeed, the very first article of the first edition in 1970 by Georg Hromadka was entitled 'Rumänische Landschaft': Romanian landscape.[30]

Komm Mit was unmistakably part of contemporaneous, socialist Romania. The relative freedom German contributors had in writing articles was offset against the imperative to frame Romania's spatial politics in accordance with the regime. Picturesque landscapes existed for the benefit of workers, allowing them to become healthy contributors towards the great socialist project. The fact that *Komm Mit* appeared in German made the outward-looking message particularly clear. Undoubtedly shaped by top-down processes, *Komm Mit* was neither overtly countercultural nor oppositional and instead reflected, at least to some extent, the ambitions of the state apparatus to project a certain image.

Celebrating the small scale

Yet this is an incomplete reading of this travel guide's spatial dimensions. The numerous contributions reveal the mental maps of Romanian Germans during the late-socialist period. Occasionally subversive in its selection, *Komm Mit* gave a voice to Germans in Romania to map their own homeland as they imagined it. For a start, the places and landscapes were noticeably German, though they were politically correct enough to please the arms of censorship. As banal as it may sound, even the use of German names for towns, places, and regions was daring. Official newspapers used German names very sparingly. The *Volkszeitung*, published in Braşov/Kronstadt between 1957 and 1968 as an organ of the 'regional party committee and Stalin's regional people's council', used exclusively the name Braşov as well as its alternative name between 1951 and 1961, Oraşul Stalin (Stalintown). *Neuer Weg*, the main Romanian German newspaper from 1949, used German labels minimally and with great caution. *Komm Mit* by contrast provided the liberty for contributors and readers alike to indulge in nostalgic *Heimatkunde*. Romanian landscapes were redesigned as *also German*. Occasionally, both Romanian and German appellations were used: readers would travel around Viscri/Deutschweisskirch or Homorod/Hamruden. Often authors only used the German name: Timişoara was simply Temesvar.[31]

Komm Mit presented Romania throughout its 21 issues as a collection of landscapes with a German hue of socialism. Transylvania was singled out for special treatment. *Siebenbürgen* was reclaimed as a German, and also distinctly Saxon, place. *Komm Mit* featured articles, short notes, and illustrations adorned with the hallmarks of Saxon culture: costumes, poems, and most importantly the church fortresses (*Kirchenburgen*). The *Kirchenburgen* had always stood for the resoluteness of the Saxon community in Transylvania since the 12th century. Romanian Germans celebrated their exceptionalism by focusing on their most prominent architectural legacy.

Taking into account the number of articles that made up the 21 volumes of *Komm Mit*, the German map becomes even more obvious: the southern Carpathians received most attention with 187 articles. This region, between Sighişoara/Schässburg and Sibiu/Hermannstadt, forms the heartland of the Transylvanian Saxons. In second place were the eastern Carpathians, a mishmash of Hungarian, German, and Romanian communities. Lastly, 85 and 83 articles featured the so-called 'Romanian Carpathians' and western Carpathians respectively. The 'Romanian Carpathians' included the Black Sea coast, which had been an integral part of the Romanian tourist industry since the inception of the Romanian state in the 1860s. Romanian Germans, however, relegated the importance of the region below Transylvania, skewing the overall view of Romania towards a decidedly German perspective. Bucharest also received very little attention in *Komm Mit*. Romania's capital seemed remote and unconnected to the Romanian German world of church fortresses and medieval towns.

Komm Mit celebrated the Saxon and Swabian past in an uncritical manner that was uncommon in socialist Romania. Quoting the *Siebenbürgerlied* and other historically significant poems seemed unproblematic. They were not obviously coerced into a socialist worldview. Likewise, folklore and peasantry featured as central components of the landscapes readers encountered in the travel guide. Peasants were peasants and not collectivized socialist workers.[32] Ethnographic images alongside descriptions of deep traditions unsettled the notion of socialist modernity.[33] Such exotic glimpses into a world that existed in parallel with the great process of levelling out difference were certainly daring. Many contributions to *Komm Mit* indulged a form of diversity that challenged the narrative of linear progress. Ferdinand Hirm explored with great excitement the 'many costumes [and] many languages' in the Banat.[34] Peasants were of always utmost importance. Writing in 1988, Hermann Fabritius recounted his travels over the Piatra Târnovului south of Sibiu/Hermannstadt.[35] After six days over the mountainous southern Carpathians, Fabritius concluded that the best guideposts were the local peasants.[36] Their deep knowledge and attachment to the land governed the region, and not the contraptions of modernity. Their celebratory representations also alluded to a problematic continuity to National Socialist and Romanian Fascist imagery which was left unchallenged by the communist regime (Figure 2.1).

It is clear from Romanian German correspondence across the Iron Curtain that this deep knowledge of local customs and traditions was vital for maintaining a sense of belonging in the present, too.[37] Intimate relationships were maintained despite geographical dislocation during the Cold War, and it was the Romanian Germans in Romania who were the guardians of knowledge. Despite the real pressures to leave Romania and the obvious comfort in which their émigré compatriots lived in West Germany, they were the ones who still had a close connection to their Transylvanian and Banat homeland. Letters between family members and friends during the Cold War occasionally uncover the need of Romanian Germans in the West to be reminded again of this or that fact about their former homeland.[38] The power relations between émigrés and non-émigrés appeared to be clear only on a superficial level. Romanian Germans in the West were wealthier than their friends and family in Romania. They supplied them with goods, books, and knowledge. Their phones were not tapped. They lived in relative freedom. Crucially, though, they were cut off from their homeland and from knowledge about their homeland. *Komm Mit*, then, reversed that relationship by providing detailed accounts and splendid descriptions of the nature, folklore, and culture of the region.

Readers were invited to go on a journey of all Romanian German highlights. 'Do you know the *Hohenstein*?', Lia Gross asked her audience in 1973.[39] The same issue introduced its readership to all the 'towers along the *Roter-Turm-Pass* (the Turnu-Roşu Pass)' on the edge of the Făgăraş mountains. Regular depictions of the church fortresses reminded its readership of another world that existed in Romania's socialist paradise. Juliana Fabritius-Dancu, writing in 1978, revelled in the glory of the *Burzenland* near Braşov/Kronstadt. Quoting the famous Transylvanian hymn, the *Siebenbürgerlied*,

Figure 2.1 Photograph of woman from rural Caraş-Severin in the Banat.
Source: Walther Konschitzky, 'Der große Tag der Hirten', *Komm Mit '84*, p. 98.

she described the region as 'a sea of undulating ears of wheat'.[40] In over 30 pages, Fabritius-Dancu embarked on a Burzenland journey from castle to castle while giving a succinct history lesson of the region (Figure 2.2).

The spatial politics of the Cold War played a crucial role, and Romanian German mental maps shifted these politics from the abstract towards manifestations. Quite uniquely, Romanian Germans faced pressure to both comply at home and consider emigration from abroad. And yet they obviously enjoyed considerable liberty to express local patriotism. Viewing *Komm Mit* as a useful counter voice to the lure from the West illuminates why Romanian Germans defended so passionately their local attachment. The *Landsmannschaften* – the homeland societies in West Germany – had masterminded German emigration during the Cold War with much effort. Publications, festivities, and a well-oiled apparatus churned out the unambiguous message that Romanian Germans should abandon Romania for their new true homeland, the Federal Republic of Germany. The *Landsmannschaften* were part of the ideological struggle of the Cold War as they

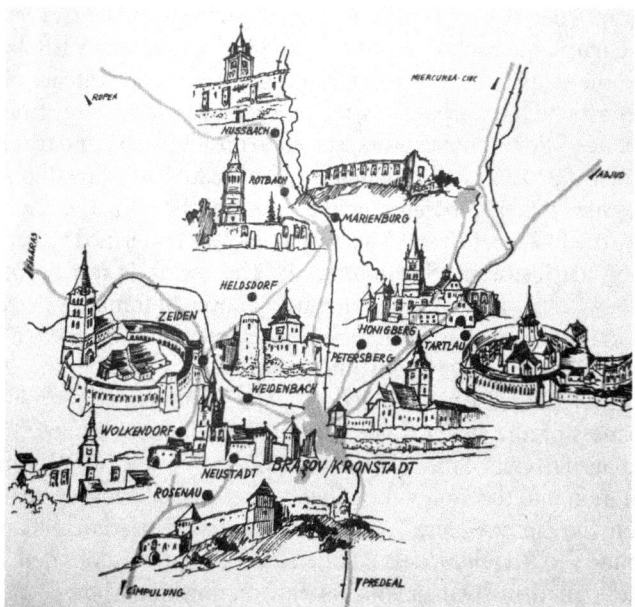

Figure 2.2 Map of the Burzenland.

Source: Juliana Fabritius-Dancu, 'Bauern und Burgen des Burzenlandes', *Komm Mit '78*, p. 67.

insisted on representing the 'better' Germany, couched in a nationalist and conservative rhetoric. The network of writers for *Komm Mit* – along with the local Lutheran Church – sought to neutralize these developments in West Germany by rethinking (parts of) Romania unmistakably as their homeland.

The image created of Romania by the *Komm Mit* authors was also gendered differently. The *Landsmannschaften* were led entirely by men. Their machismo in aggressively lobbying for political and ideological certainty and in intervening in 'big politics' was countered by a very different approach back in Romania. A look at the table of contents of *Komm Mit* over the years reveals a striking gender balance in authorship. Many of the 'star writers' for the guidebook were women, most notably Juliana Fabritius-Dancu, Erika Schneider, and Lia Gross. *Komm Mit* then also unveils an area of spatial politics rarely explored. Away from the Cold War political sphere dominated by men and male language, this travel guide created a gendered version of space through which the Cold War was filtered: inside versus outside, domestic space versus international politics, small-scale maps versus grand-scale political borders, hand-drawn pictures versus manufactured images.

That said, the battle for the soul of the community intensified from 1977 following the now infamous Ceaușescu-Schmidt agreement which officially cost Romanian German potential emigrants.[41] The Cold War mental map was polarized on a big scale: Romania, in *Eastern* Europe, against the *Landsmannschaften*, as symbols of the *West*. The mental map of *Komm Mit*, however, was altogether different. It focused on local landscapes, which were accessible, German, and separate from Cold War politics. The 'big picture' was missing, at least overtly. Instead, the maps, the routes, the articles, and the coverage as a whole tended to zoom in on the small scale. From detailed descriptions and sketches of plants to small routes around the Carpathians, this was a decisive move away from big Cold War geopolitics.[42] Some contributors seemed to positively demand that their readers slow down and focus on the minutiae of local beauty. 'Driving through eastern Maramureș means remembering only fleeting snapshots [of the region]', Claus Stephani warned her audience.[43] This very point was confirmed to her years ago by a visiting art historian. She continued: 'The region is too beautiful to be observed only from a moving car; you must discover Maramureș by hiking…'.[44] Elsewhere, in 1988, Ewalt Zweier admired the remoteness of northern Maramureș to the north of Vișeu de Sus.[45] 'No other mode of transport but the steam train' would do, Zweier noted with satisfaction.[46] The author was especially interested the leisureliness of the steam train's speed: 20 kilometres per hour, no more.[47] Zweier managed to escape the encroaching modernity and Cold War politics on a slow steam train into the mountains. There was even a curious German minority in the region: the Zipser.[48] Here, Romania continued to look different and homely.

Contributors to *Komm Mit* would often get lost in meticulous detail. In 1981, it was Juliana Fabritius-Dancu, as so often, admonishing *Komm Mit* readers to appreciate the famous castles in the *Repser Land* near Rupea in south-eastern Transylvania. Praising the isolated location of Viscri/Deutschweisskirch, Fabritius-Dancu studied local traditional women's costumes before turning her attention towards the details of the church fortress (Figures 2.3 and 2.4).

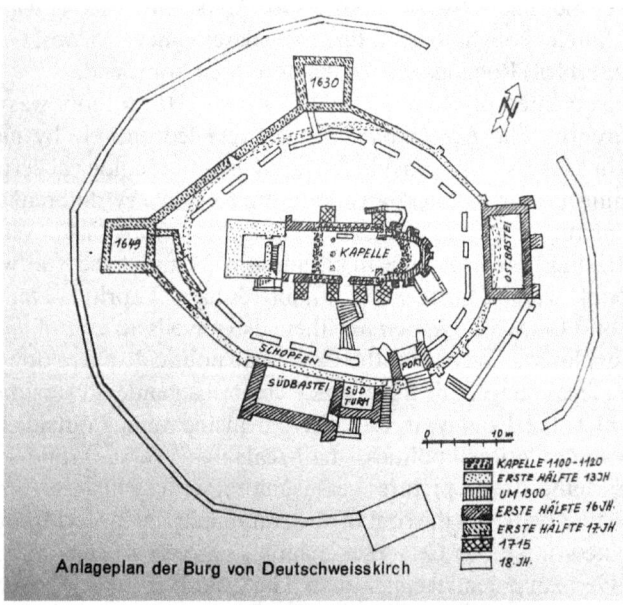

Figure 2.3 Map of the church fortress at Viscri/Deutschweisskirch.

Source: Juliana Fabritius-Dancu, 'Burgen im Repser Land', *Komm Mit '81*, p. 132.

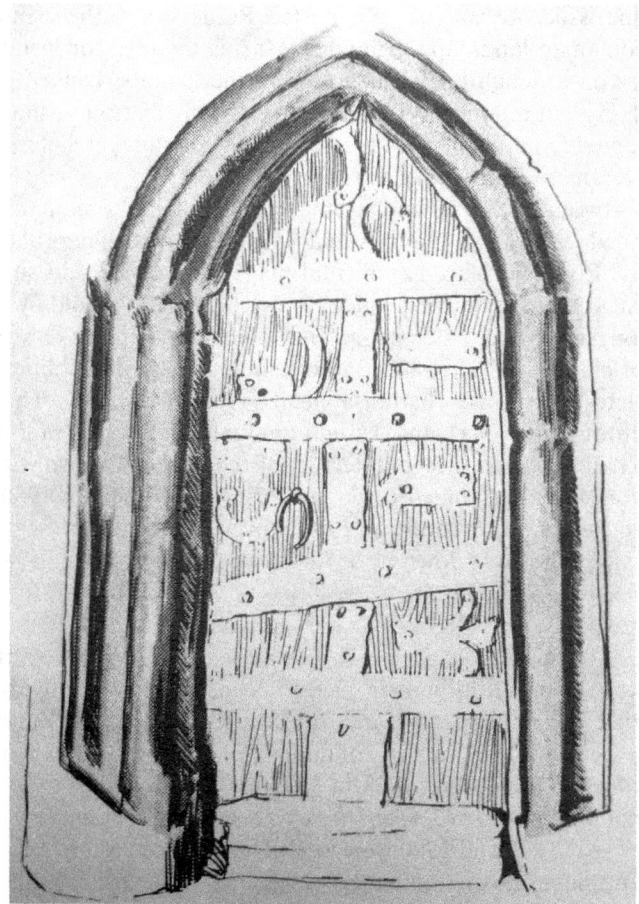

Figure 2.4 Sketch of detail of door.

Source: Juliana Fabritius-Dancu, 'Burgen im Repser Land', *Komm Mit '81*, p. 135.

Komm Mit fled into the details of local beauty away from big politics. Yet even that act of escapism disguised broader spatial claims being made by Romanian Germans. Rohtraut Wittstock-Reich promoted the art museum in Cluj-Napoca/Klausenburg as a 'must see' for art lovers.[49] The museum offered something for everyone: housed in the former 18th century Bánffy Palace, it gave visitors the opportunity to indulge in Habsburg nostalgia. At the same time, Wittstock-Reich took pains to emphasize that the architect, Johann Eberhard Blaumann, who had designed the former palace (1774–86) had been a fine Saxon son of the city of Sibiu/Hermannstadt. Its contemporary holdings included both Transylvanian and Romanian art, a distinction on which Wittstock-Reich insisted. Most importantly, the museum was part of a European palimpsest of European art. But if Romanian Germans were both local patriots and true Europeans, then their commitment to Romania could never be far behind.

The same issue, *Komm Mit '85*, hosted Romanian authors introducing various Romanian landscape highlights. Marina Crainic, for instance, took her readers on a delightful journey around the Danube Delta.[50] Similarly, making an explicit reference to Willi Krämer's contribution to the 1983 edition of the guidebook on hiking trails through Maramureș,[51] Stefan Strătescu corrected Krämer's suggestions by taking the readership on a very Romanian trail.[52] In between the Saxon and Swabian writers, then, Romanian contributors reminded readers that Transylvania was still very much part of Romania. In this way, Romanian German mental maps were varied and superficially paradoxical. Underlining the local as a way of avoiding Cold War politics existed alongside bigger assertions about their rightful place in a European network of culture while being tied into Romanian socialist politics.

By the late-1980s poems had disappeared from *Komm Mit*. The quixotic journeys through the German past had gradually dried up. Traditional stories which had once been a regular feature of the guidebook also vanished. In fact, the travel guide slimmed down considerably by the late 1980s. *Komm Mit* had started out as a book comprising over 300 pages. In the late-1970s, the total page count was down by around 40 pages, though the 1980 edition was another bumper issue of over 300 pages. *Komm Mit '85*, though, was down to 220 pages. The 1988 edition only consisted of 212 pages, and the 21st issue in 1990 a mere 208 pages. The number of individual contributions declined, too. Its first edition from 1970 still consisted of 62 individual items, which varied from poems, to stories, to detailed guides. The first six years until 1976 were similarly rich in volume. After that, the number of articles went down quite noticeably to 28 in 1977 and 26 in 1979, but it stayed at roughly that level until 1990, when there were a mere 20 individual articles.

Both the year 1977 and 1990 were fateful for the Romanian German community. The Schmidt-Ceaușescu agreement in 1977 enshrined Romanian German emigration as an official doctrine in both the Federal Republic of Germany and Romania. Thirteen years later, the end of the Ceaușescu regime made possible the mass exodus of Romanian Germans to reunifying Germany. The reduction of the volume of *Komm Mit* precisely in 1977 and 1990 respectively went hand in hand with crucial political developments. If *Komm Mit* was a product of Ceaușescu's new approach to Romania's image at home and abroad, then its steady demise heralded the end of Germans in Romania. The dwindling of the community was palpable in the changes *Komm Mit* underwent in its two decades. Not only were there simply fewer contributors in Romania, but Germans in Romania had also lost the argument over whether to stay or go. The spatial battle over the homeland seemed a foregone conclusion as the last decade of the Cold War came to a close.

Guidebooks: a palimpsest of mental maps

The history of Romanian Germans in the 20th century is often told backwards. They emigrated *en masse*, and therefore it was a failed community from the outset with a foregone conclusion of dissolution in favour of membership of Germany. Exploring the travel guide *Komm Mit* has complicated that picture. Left behind in these guidebooks are vexed and intricate mental maps of the Cold War, Romanian Germans, and European onlookers. Not all Romanian Germans wanted to leave Romania for West Germany. The travel guide in question speaks volumes as far as that is concerned. The endorsements of the Romanian countryside were not simply deceptive token gestures of individuals trying to negotiate their way through the repressive Romanian state-socialist system. Instead, they were genuine attempts to maintain a sense of belonging and home in Romania against the pressures of Cold War politics. In that sense, *Komm Mit* aided the spatial claims of the Romanian state and challenged it at the same time. Likewise, the potent assumption that the West was best for Romanian Germans came up against opposition in the guidebooks. These guidebooks reasserted the small scale in Romania against the bigger claims of émigrés and their associations, the *Landsmannschaften*.

Guidebooks are more than just the nationalist agendas of a state or a catalogue of 'what ought to be seen'. They reveal complex relationships and conflicting mental maps. Their claims over space transcend straightforward national or social demands. Behind guidebooks are contributors with their own biographies, ideas, and spatial imaginings. They write for an audience, but reach others, too. Their choice of places reflects the concerns and desires of the day, and these do not remain static, either. *Komm Mit*, ostensibly a travel guide for Romania in the late Cold War, is a case in point. It is perhaps appropriate to end with the front cover of the final edition of *Komm Mit '90*. Over 100,000 Romanian Germans emigrated in 1990 alone. Those left behind were isolated more than ever and speaking into an increasingly empty space. Fittingly, the front cover depicted just that: a beautiful alpine landscape of the Carpathians, but with no one left. Even an empty space is loaded with spatial politics.

Notes

1 Rudy Koshar, '"What Ought to Be Seen": Tourists' Guidebooks and National Identities in Modern Germany', *Journal of Contemporary History* 33, 1998, 323–40.
2 John Urry, *The Tourist Gaze: Leisure and Travel in Contemporary Societies*, London: Sage, 1990; John Urry and Jonas Larsen, *The Tourist Gaze 3.0*, London: Sage, 2011.
3 Koshar, '"What Ought to Be Seen"', 325.
4 William W. Stowe, *Going Abroad: European Travel in Nineteenth-Century American Culture*, Princeton: Princeton University Press, 1994, p. 46.
5 Michelle Henning, *Museums, Media and Cultural Theory*, New York: Open University Press, 2006; Tony Bennett, *The Birth of the Museum: History, Theory, Politics*, London: Routledge, 1995.
6 See, for instance, Kathrin Maurer, 'Mit Herrn Baedeker in Grüne: Die Popularisierung der Natur in Baedekers Reisehandbüchern', in Adam Paulsen and Anna Sandberg (eds.), *Natur und Moderne um 1900: Räume, Repräsentationen,*

Medien, Bielefeld: transcript, 2013, pp. 89–102; Stowe, *Going Abroad*, Koshar, '"What Ought to Be Seen"' and Eric G.E. Zuelow, *A History of Modern Tourism*, London: Palgrave, 2016, esp. pp. 76–90.

7 See Stowe, *Going Abroad*, p. 37, Susan Shelangoski, 'Thrills and Quills: Masculinity and Location in Three South African Travel Narratives (1834–1900)', in Kate Hill (ed.), *Britain and the Narration of Travel in the Nineteenth Century: Texts, Images, Objects*, Farnham: Ashgate, 2016, pp. 109–28.

8 There have been some wonderful studies on travel guides that follow the pattern of immersing themselves in the text. See, for instance, Karl D. Qualls, '"Where Each Stone Is History": Travel Guides in Sevastopol after World War II', in Anne E. Gorsuch and Diane Koenker (eds.), *Turizm: The Russian and East European Tourist under Capitalism and Socialism*, Ithaca and London: Cornell University Press, 2006, pp. 163–85.

9 Vesna Goldsworthy, *Inventing Ruritania: The Imperialism of Imagination*, New Haven: Yale University Press, 1998.

10 While I am aware of the conceptual differences and debates between the terms 'traveller' and 'tourist', this chapter uses the two terms interchangeably. Where the source material demands the use of a specific term, it will go with the term used in the source. Scholars have sought to distinguish between 'traveller' and 'tourist' as a way of differentiating out the commodified nature of travel. Accordingly, around the mid-19th century, travellers became tourists. This may make sense as far as the usage of those two terms are concerned, but from a conceptual point of view this chapter takes the view that this distinction rests on a false dichotomy.

11 There are, of course, exceptions to this. Koshar, Urry, and others have in some ways dealt with the spaces within guidebooks, but not in a systematic way. For a spatial approach to Baedeker's guidebooks, see Simon Garfield, *On the Map: Why the World Looks the Way it Does*, London: Profile Books, 2012; see also Herbert Gottfried, *Landscape in American Guides and View Books*, Lanham: Lexington Books, 2013.

12 Stowe, *Going Abroad*, p. 48.

13 See ibid., p. 53.

14 Ibid., p. 45.

15 See, for instance, James Buzard, *The Beaten Track: European Tourism, Literature, and the Ways to 'Culture', 1800–1918*, Oxford: Oxford University Press, 1993.

16 Ibid., p. 44.

17 See Frithjof Benjamin Schenk, 'Mental Maps: The Cognitive Mapping of the Continent as an Object of Research in European History', *European History Online*, 2013. Available HTTP: http://ieg-ego.eu/en/threads/theories-and-methods/mental-maps/frithjof-benjamin-schenk-mental-maps-the-cognitive-mapping-of-the-continent-as-an-object-of-research-of-european-history (accessed 14 April 2018).

18 James Buzard's *The Beaten Track* is a case in point: only two guidebooks are named in the index, namely *Baedeker* and *Murray*. They feature extensively throughout Buzard's book.

19 The others either ended up in West Germany, East Germany, or never returned. For more on both their alignment with Nazi Germany and the repercussions after the war, see Tudor Georgescu, *The Eugenic Fortress: The Transylvanian Saxon Experiment in Interwar Europe*, Budapest: Central European University Press, 2016; Georg Weber et al. (eds.), *Die Deportation der Siebenbürger Sachsen in die Sowjetunion, 1945–1949*, Vols. I–III, Cologne: Böhlau, 1995.

20 From the late 1940s to the mid-1970s, the number of Germans in Romania remained roughly around 350,000 despite a small stream of constant émigrés. This changed dramatically from 1977 when Nicolae Ceaușescu and West German Chancellor Helmut Schmidt reached an agreement to regulate and facilitate German emigration from Romania. By 1989, there were only 200,000 Germans left, half of which left the country in the following two years. Today there are nominally between 30,000 and 40,000 Germans that make up the ageing community.
21 There was even a designated German Anti-Fascist Committee. See Hannelore Baier, 'Das Antifa und der Neue Weg: Einiges aus der Geschichte des Deutschen Antifaschistischen Komitees', *Allgemeine Deutsche Zeitung*, 15 January 2009.
22 For a history of Romanian Germans and *Neuer Weg* until 1971, see Annemarie Weber, *Rumäniendeutsche? Diskurse zur Gruppenidentität einer Minderheit (1944–1971)*, Cologne: Böhlau, 2010.
23 Figures are hard to come by. According to the editor of the *Allgemeine Deutsche Zeitung* – the successor organ of *Neuer Weg* – the majority of guidebooks went to readers in the former GDR and Czechoslovakia respectively.
24 James Koranyi, 'Voyages of Socialist Discovery: German-German Exchanges between the GDR and Romania', *The Slavonic and East European Review* 92, 2014, 479–506.
25 See E.J. Țigla, 'Georg Hromadka zum Gedenken', *Allgemeine Deutsche Zeitung*, 9 April 2015. Available HTTP: http://www.adz.ro/lokales/artikel-lokales/artikel/georg-hromadka-zum-gedenken/ (accessed 14 April 2018).
26 See Cezar Stanciu, 'The End of Liberalization in Communist Romania', *The Historical Journal* 56, 2013, 1063–85.
27 See, for instance, Alina Mungiu-Pippidi, *A Tale of Two Villages: Coerced Modernization in the East-Central European Countryside*, Budapest: Central European University Press, 2013, esp. pp. 155–88.
28 I am grateful to Christian Noack for this observation.
29 'Aktiver Urlaub für Jung und Alt', *Komm Mit '81*, Bucharest: Neuer Weg, 1981, pp. 14–17.
30 Georg Hromadka, 'Rumänische Landschaft', *Komm Mit '70*, Bucharest: Neuer Weg, 1970, pp. 5–8.
31 Franz Liebhard, '"...reich an Pracht und herrlichen Szenerien"', *Komm Mit '73*, Bucharest: Neuer Weg, 1973, pp. 62–7.
32 See, for instance, Walter Konschitzky, 'Der große Tag der Hirten: Zu Gast beim Messen der Schafmilch, dem Fest der kraschowänischen Züchter und Hüter', *Komm Mit '84*, Bucharest: Neuer Weg, 1984, pp. 97–105.
33 Ibid., p. 100.
34 Ferdinand Hirm, 'Viele Trachten, viele Sprachen (...im Banat zu betrachten)', *Komm Mit '74*, Bucharest: Neuer Weg, 1974, pp. 201–4.
35 Helmut Fabritius, 'Hirten waren die Wegweiser', *Komm Mit '88*, Bucharest; Neuer Weg, 1988, pp. 171–7.
36 Ibid., p. 177.
37 See for instance, Koranyi, 'Voyages of Socialist Discovery' and Joachim Wittstock and Stefan Sienerth (eds), *'Bitte um baldige Nachricht': Alltag, Politik und Kultur im Spiegel südostdeutscher Korrespondenz des ausgehenden 19. und des 20. Jahrhunderts*, Munich: IKGS, 2003.
38 The best evidence for this can be found in the *Siebenbürgen-Institut* in Gundelsheim which houses sumptuous collections of personal correspondence of numerous

Transylvanian Saxon individuals. For a comprehensive list, see http://siebenbuergen-institut.de/fileadmin/user_upload/pdf_dateien/Nachlaesse_Homepage_Tabelle.pdf (accessed 14 April 2018).

39 Lia Gross, 'Kennen Sie den Hohenstein?', *Komm Mit '73*, pp. 176–9. The *Hohenstein* is an alpine region to the southeast of Brașov and is known as the Piatra Mare Mountains.

40 Juliana Fabritius-Dancu, 'Bauern und Burgen im Burzenland', *Komm Mit '78*, Bucharest: Neuer Weg, 1978, p. 66. The *Siebenbürgerlied* is the 'national anthem' of Transylvania written in 1848 by Maximilian Leopold Moltke.

41 Hannelore Baier's work has been invaluable in bringing this issue to the fore. Her series of interviews with the former West German middleman are essential reading as far as the West German–Romanian agreement is concerned. See Hannelore Baier, *Kauf von Freiheit: Dr. Heinz-Günther Hüsch im Interview mit Hannelore Baier und Ernst Meinhard*, Sibiu: Honterus, 2013.

42 See, for instance, Walter Kargel, 'Auf Hirtenpfaden zum Godeanu', *Komm Mit '82*, Bucharest: Neuer Weg, 1982, p. 107, Erika Schneider, 'Buche, Beinwell, Bergholunder', *Komm Mit '82*, p. 119.

43 Claus Stephani, 'Fahrt ins Wassertal: "Radlhäuser" und "tie Koffeemiehl"', *Komm Mit '81*, Bucharest: Neuer Weg, 1981, p. 123.

44 Ibid., pp. 123–4.

45 Ewalt Zweier, 'Wo nur die alte Forstbahn fährt', *Komm Mit '88*, Bucharest: Neuer Weg, 1988, pp. 138–46.

46 Ibid., p. 138.

47 Ibid., p. 139.

48 Ibid., p. 142.

49 Rohtraut Wittstock-Reich, 'Europäische Malerei und siebenbürgische Kunst', *Komm Mit '85*, Bucharest: Neuer Weg, 1985, pp. 42–46.

50 Marina Crainic, 'Auf Delta-Entdeckungsfahrt: Fünf Tage in Europas einzigartigem Vogelparadies', *Komm Mit '85*, pp. 153–7.

51 Willi Krämer, 'Wanderwege durch die Maramureș', *Komm Mit '83*, Bucharest: Neuer Weg, 1983, pp. 109–16.

52 St. Strătescu, 'Vom Gutîi zum Tibleș: Eine neue Trasse für Maramureș-Wanderer', *Komm Mit '85*, pp. 158–62.

3 Novels, autobiographies, and memoirs

Sarah Deutsch

Space is made up of places, particular locations, the meanings imputed to them, and the relations in and among them.[1] Space is always already formed, already constituted by relations of power, already filled with meaning. People contest those uses and meanings, even people without resources. In cities without free childcare, for example, low-income mothers leave children at public libraries. To the dismay of public librarians, these women redefine the purpose of the library, of that place, and in so doing, they appropriate public space in their own interests. Though without material resources to build or to direct the allocation of public spending, they redesign the city.[2]

As is particularly clear to women seeking daycare, to homeless people, and to almost anyone with few resources, the relationship between survival and the physical organization of space is intimate. Who owns and controls spaces can determine which spaces are safe for which inhabitants and where people can meet to organize politically in their own interests. But control is never absolute.[3] And space is not static. That set of dynamics makes spatial formations a historical subject.

Space is not a historical actor in its own right. It lacks independent agency. It does not stand outside social forces and power dynamics but is constituted by them, by the way groups of people organize their social, political, economic, and other interactions. This dynamic becomes particularly evident in cities where conscious design so often determined the built environment. Even before the rise of the profession of city planners in the late 19th century, colonial powers and their settlers carefully designed the layout of settlements. Spain placed the church and the military across the plaza from one another; English Puritans in Massachusetts centered the church on the village green, signifying the lack of distinction between sacred and secular rule.[4] To ensure stability and hegemony, the designs made the power structure visible in the built environment. Yet local residents constantly undercut these careful designs, sometimes by explicit revolt but often, like the mothers seeking daycare, simply by the patterns of their daily lives. The constant interplay between different groups of people and city space literally shapes urban history. For all players, the ability to lay claim to certain types of space and the power to shape space – public arenas, housing, and so forth – was crucial to their ability to meet their basic needs and their often less basic desires.[5]

DOI: 10.4324/9780429291739-5

Myriad sources provide a window into the organization of space. If historians only attend to official records and maps, however, they miss the ways in which space gets constructed both materially and conceptually, and, in turn, they miss the power relations encoded in, transgressed, and contested in the built and even the natural environment. They miss the ways in which their subjects must negotiate space in their daily lives – turning libraries into daycare centers – and the ways in which spatial formations impinge on their ability to do so. If historians only use official sources of any kind, they may miss the sense their subjects made of the geography they encountered and helped to construct.

This chapter will first address the spatial fictions in theoretically non-fiction sources and then turn to the benefits of fictional sources and personal narratives for understanding spatial formations. After briefly discussing some examples of such sources for understanding space at different scales, the chapter turns to a more extended case study of one city using multiple fictional and personal narrative sources. In that case, examining spatial formations – both conceptual and material – through those sources makes it possible to change historians' understanding of urban dynamics, in particular the ways in which women and men renegotiated power and status in the late 19th and early 20th century city.

Fiction versus 'non-fiction'

The lines between fiction and non-fiction sources are fuzzy. Most if not all historical sources rely on fictions at least in part. Both photographs and US census records partake of and construct spatial formations – more obviously in photographs, but also in the census' organization of information into households and neighborhoods, domestic and public space. US government instructions to census takers in 1910 told them not to include married women's work unless the woman was the prime earner in the family. When census takers ignored these instructions, they exposed that the stark spatial divide between home and work, that women in the home did not labor, was fiction. In Rio Arriba County, New Mexico, male and Anglo female census enumerators listed ten females with occupations for every one hundred males in the villages they covered. They obeyed the instructions. Sophie Archuleta, on the other hand, herself a public school teacher in the county, whose mother was a seamstress and whose sister and brother performed labor on their farm, listed 79 females with occupations for every 100 males.[6] Even had her mother not worked as a seamstress, she would have contributed to the labor necessary to sustain the farm, as her children did, but the census would have listed the farmer (male head of household) as working, and the 'farmwife' as without an occupation. The records created by compliant census takers helped create, support, and perpetuate a fiction of a stark divide between a non-market domestic space occupied by women and children and a market world in which men maneuvered. This spatial fiction undergirded a fictional normative family with a husband/father capable of supporting the entire family in a domestic space, a refuge from the market, where women and children did not do productive labor. Since the manuscript census records are among the best records we have for people who did not record their own histories, such spatial fictions have become 'fact'.

Such fictions could not only structure households; they could structure regions. Census takers in the US Southwest, land previously claimed by Spain and then Mexico, often ignored the census instructions' racial categories. They failed to match up with local knowledge. Before it became part of the public record, their work was 'corrected'. In 1930, and 1930 only, the federal government made 'Mexican' a census racial category rather than a matter of nativity. Earlier manuscript census records show long columns of 'Mex' or 'M' under 'race'. Through each of these 'M's and 'Mex's, an official had carefully drawn a line, replacing them with 'W' for 'white'. Which was the fiction? And what work did it do? And how does it change our understanding of the spatial formations of the United States, who belonged where, the region and its rightful inhabitants?

It is even easier to see parallels with fiction and memoirs in photographic evidence, even journalistic photographs. In the early 20th century, social reformer Jacob Riis' photographs recorded tenement conditions as well as working-class and impoverished children on the streets, exposing the underbelly of the modern city, the cramped, dark, and dirty spaces that contrasted with the wide boulevards of more prosperous areas. It was an era that embraced the photograph as a mirror of reality, a powerful tool for exposing conditions and generating change. Like all photographers, however, Riis chose his frame, chose what would be in and what would lie outside the frame of the photograph, the relevant and the irrelevant, and he was not above posing his subjects in seemingly spontaneous shots.[7] Deciding the frame helped determine how the viewer saw the space and what it meant. The viewer of the photographs does not see Riis. As with other authoritative work, as historian Hayden White pointed out about conventions of scholarly writing, the absent eye/I allows the illusion of objectivity.[8] The photographer's framing creates a fiction of the meaning and nature of working-class spaces, and not necessarily the same frame and meaning his working-class subjects themselves would have given.

The power of setting the scene

Like the photographer, the memoirist, autobiographer, and story-teller create characters and set scenes. The stories take place, literally. Fiction writers create whole worlds for their characters. Their works have a geography. In setting the scene for their stories and memories, the authors not only describe places, they articulate the sense they make of the places they describe. In their writing, in other words, they display spatial sensibilities. They are attuned to the meaning of the built environment in which they place their characters. They describe safe and unsafe spaces, strange and familiar. They describe places as civilized or savage, and they describe moving between them.

They can also flip our expectations. At the height of the conceptual reorganization of space into a domestic/public binary, Louisa May Alcott crafted her semi-autobiographical novel *Work*. Begun before the US Civil War and completed afterward, Alcott's novel exposes the fictions of the binary. As her heroine moves from one workplace to the next, she reveals bourgeois domestic spaces as not a refuge from the market but equally governed by the market and market relations. When the heroine works as a domestic servant, she finds that cold monetary relations govern the affect between mistress and maid, and that the mistress organizes the household around the marketplace ambitions of the master of the house. In this workplace, women workers, insecure and at the mercy of market and employers, bond with each other, not their mistress. Alcott concludes the novel by revising the geography of separate spheres. Her heroine constructs a greenhouse business run as a collective, a feminized marketplace, creating a place that blurred the boundaries fixed by the cult of domesticity and rejecting the fiction that the home created by capitalism could exclude its dictates.[9]

While Louisa May Alcott's *Work* and its organization of domestic and non-domestic spaces expose the ways the market inescapably infiltrates the home, critiquing separate spheres as a representation of human relations in space/place, Melody Graulich's article on domestic violence uses three semi-autobiographical novels of the US late-19th- and early-20th-century West to expose the way spaces of intimacy and domestic violence were conditioned in part by larger spatial expectations. Men in the novels expect the US West, which they see as a vast open space devoid of other claimants, to provide them with untrammeled personal sovereignty, a fantasy widely promulgated in the promotional and popular literature of the time. Going West to escape their failures and constraints back East, they are impatient of any restraint, whether of economy, land, or family. Their impatience and frustration with the unrealized fantasy manifests in domestic violence.[10] In other words, their expectations of the region – the meaning popularly given to the region – determined dynamics in domestic spaces. In actuality, just as Alcott's home could not escape market relations, the West's domestic spaces could not escape the pervasive violence that shaped the region. There are traces of domestic violence in non-fiction sources of the time and region, but while often horrifying, they lack the fully realized nature of the novels' characters and setting along with the holistic analytical acumen. It is precisely the subjective experience of place and space and the power of its rendition that makes novels and memoirs of the time so valuable to the historian.

Gender and the city

As is evident from the examples above, space is deeply gendered. Those with the power to do so delineate certain spaces as appropriate for women and others as appropriate for men.[11] In the early-19th-century US, before women had the vote, voting sites were often in saloons. Saloons were off limits to respectable women, making it harder to imagine respectable women as voters.

The 19th-century city had no public spaces for respectable women. 'Women of the streets' was a euphemism for prostitutes, as was 'public women'. Women on the streets were presumed to be unchaste, despite the presence of servants and other working women traversing the city. Presumption of chastity laws meant that only chaste women were capable of being raped. Serving girls were doubly vulnerable – vulnerable to the man of the house where they served and vulnerable on the streets. They rarely won their rape cases.

In the late-19th century, middle class and elite women launched efforts to revise this gendered geography and demand access to the public. They began their revision by claiming buildings for their institutions – club houses and women's associations of various kinds – and attacking the presence of prostitutes on the street. By the turn of the century, they increased their ambitions, bolstered by their own changes to the built environment, but they still carried with them their assumptions about working girls' sexualized geography.

Getting at what women thought about urban space requires looking at different sources than for male theorists. Men in the late-19th and early-20th century had recognized theories about the city. They wrote of zones and rings that mapped their understandings of ethnic and racial categories and hierarchies onto streetscapes. Women like prominent US reformer Jane Addams who wrote about the city were not seen as having theories. They were seen as applying the theories of men. In a reverse of the old chestnut on sexist history ('men do, women are'), men thought, and women did. In the US setting, men in universities created theories, and women created settlement houses.[12]

To find women's theories of the city, historians need to look differently. Along the lines sociologist Patricia Hill Collins pointed to in her groundbreaking text on black women's epistemology, historians need to look at what women wrote about how they made their way in the city.[13] Women enacted and performed their theories of the city, and they, along with men outside the university, wrote memoirs, diaries, and fiction that encoded their views of urban spatial dynamics. Using such sources brings challenges not entirely distinct from any historical sources. The views are more open about lacking the pretense of objectivity. Their imagined landscape, like the often staged photos of reformers, may bear a complicated, partial, and sometimes even distant relationship to physical realities of the day. Yet read in tandem with other sorts of sources – maps, census, tax, police, and other government records, photographs, and other texts – they provide a unique window into the lived experience and interior world of a city dweller, and the sense that resident or visitor made of the environment through which they moved.

Remapping Boston's moral geography

Novels of Boston not only give the historian access to diverse mental maps of Boston – the spaces that loomed as significant or disappeared from the consciousness of inhabitants, the imagined distances between places – but also the sense those living there made of Boston's geography, the textured experience of moving through it.[14] When fiction writers set a scene, they set people in place; and the action of the novel requires them to move people through space. With more than one memoir or fictional account, reading them beside the seemingly dispassionate evidence of photos, reports, census records, and other more standard sources and juxtaposing them exposes clashing maps and understandings of urban space.

Like the creators of other sources, the authors are selective. These authors tell a subjective truth. The subject of the autobiography may be virtually unrecognizable to close friends who would remember the subject and circumstances differently. They may also remember the setting differently, just as Jacob Riis' subjects may have remembered the chaotic, crowded lively tenement space not only as sordid, desperate, and suffocating, but as vibrant and full of affective relations.[15]

As with any source, it is best to know as much about the author as possible. Sometimes it is not possible to know much. But knowing the author's social standing and politics can provide clues as to authorial choices, and in turn, the fiction can provide insight as to the author's perspective. In the same way, reading the novel or memoir alongside other contemporary sources – maps, quantitative sources, other texts – can clarify or sharpen what is distinctive about the text's engagement with space and place.

It is possible to glean enough about Boston's neighborhoods in the late-19th and early-20th centuries from census records, fire insurance maps, and reports generated by social reformers to paint a picture of them for the reader. Here's what that picture looks like, without the fictions and memoirs:

> Each district had its own character. Though the West End, like the South End, had once been fashionable, by the 1880s its refurbished old buildings with bay windows, ornaments, bells, and speaking tubes housed the densest population in the city. Blacks, Jews, Irish, Portuguese, and Italians all had their sections. In this labyrinthine district, alley led off alley; narrow passages emptied unexpectedly between high buildings or under them, and the only entrance to a tenement might be underground. Wooden walkways at different heights ran between the dark, crowded buildings.
>
> The dilapidated elegance of the West End and of the bowfront brownstones and avenues in the South End contrasted with the small, dark, cramped buildings that lined the North End's narrow, winding streets. With streets sometimes only six feet wide, sunlight rarely entered the buildings. In the back, buildings were even closer. Enterprising investors had filled the narrow tenement yards with more houses. Even the damp, noisome, and seeping basements had tenants. They suffered the highest death rate in the city. Scarlet fever, diphtheria, pneumonia, and whooping cough plagued the district. Stillbirths were common.[16]

Missing in this rendition is the sense inhabitants made of this landscape. The passage provides a rich description of working-class residential neighborhoods in Boston. It gives little sense of the competing meanings imputed to those spaces by residents and visitors. Social reformers (and seemingly many working-class parents) looked with disdain and even horror at working-class girls enjoying the streets, dating, twirling in dance halls, and dressing with panache. They saw the crowded tenements without separate sleeping spaces, jumbling men and women, unrelated, or even related, sleeping in the same room as each other and as the kitchen, much as (though they seemed not to know it) their English forebears had in colonial times. They saw these spaces as endangering girls' virtue and as signs of women's failure to uphold virtue, particularly when these places served as sites of income production whether through taking in boarders or taking in sewing. Having in the previous century successfully marketed a geographic ideology of separate spheres where the sacred space of the home eschewed the self-interest and competition of the market-place and where relations were governed by affection rather than cash, they saw in the working-class matrons both failed women and a danger to the future of the republic.

These views of urban geography and its moral and immoral spaces were inscribed in social reformers' investigative reports and novels. Working-class women rarely left their own texts. To get at working-class views, historians have become adept at reading working-class experiences through middle-class and elite records. The picture of working-class women's lives rendered through that oppositional reading pits a vibrant sexualized working-class singles culture practiced in new commercialized leisure sites of 'cheap amusements' against a bourgeois set of observers determined to 'uplift' working-class women and endow them with middle-class self-restraint. But while working-girls were highly sexualized read through these sources, it is possible that they are oversexualized by historians for whom those sources are often their only window.[17]

There certainly were hints of an alternative view in non-fiction sources. Young Wellesley College intern Louise Bosworth diligently investigated working girls and their use of space in Boston. When what she found through her investigations and her own experience living among working girls in the South End failed to match the speculations of Harvard University sociologist Albert Wolfe, who saw brothels everywhere; she expressed her puzzlement but bowed to his superior status:

> Albert Wolf, in his study of lodging houses of the South End, gives a comprehensive and searching view of the lodging house problem. The dangers to unprotected girls, as well as the temptations to seasoned lodgers are seen to be very real and far reaching in their effects. That the present investigator has not met with this problem may be explained in several ways still consistent with the existence of the evil.'[18]

Lodging and boarding houses of the time did indeed mix men and women, skilled and unskilled workers without parental supervision. The number of such houses grew dramatically in this period to house the ever larger number of young single men and women migrating to the city for work. Despite the mixing (which rarely if ever crossed racial lines), novels of the time provide a less sinister view of lodging and boarding houses than Albert Wolf's. In her novel *Contending Forces* (1900), novelist and clerical worker Pauline Hopkins described the African American lodgers of Ma Smith's house. They included a student preacher, a female stenographer, two dressmakers, and a pair of 'respectable though unlettered people', an illiterate former cook who ran a laundry business with a similarly illiterate former housemaid. And union activist and journalist Frank Foster in his novel *Evolution of a Trade Unionist* (1901) described the cheap, cheerless white South End boarding house where his hero lived as housing manual laborers, a machinist, an advertising canvasser, a spinster shopkeeper, two tailoresses, a typesetter, and a lady boarder married to a railroad clerk, 'one of that not inconsiderable number of American women who "detested keeping house"'.[19]

Despite the blurring of categories, and the unwomanly refusal to keep house (and thus the Victorian moral geography of the bourgeoisie), neither Foster nor Hopkins portray the sort of moral free-for-all of the social reformers. In Foster's depiction, the mixing in the boarding house is precisely what made the site urban, modern, and even enlightening, rather than dangerous. While social reformers often imbued the site and so its inhabitants with the taint of suspicion, the novelists depicted the inhabitants as respectable, and endowing the site with their respectability. Both types of sources described the same sites, but the geography and the logic that gave the spaces meaning differed.

The moral geography of the working-class youth differed from that of their middle-class observers. The spaces of the city held different meanings and different possibilities for them. In a world peopled by such autonomous individuals, the novels' working girls needed the tutelage of neither parents nor social reformers. They learned best how to navigate the urban landscape from each other. Frank Foster's novel depicts a working-class youth geography more heterosocial than the working-girls clubs reformers favored and less sexualized than the picture painted by the investigative literature. His novel is full of women independent of mind and money who share lodging houses, bicycle excursions, union picnics, ideas, teasing, and friendship with men. They play safely in places often designated as dangerous by reformers. They meet in cafés, at parks, or on the Common, where 'sprucely dressed' retail clerks rub elbows with laborers in overalls and jumpers, cash girls and boys, typists and stenographers. While this spatial logic makes sense as the literally thousands of working girls who hit the streets at lunch time and closing time from their workplaces in offices, department stores, and factories, threatened to swamp the men accustomed to dominating downtown space, it is not a view found in the voluminous social reports collected during the period.

The working class take on safe and unsafe spaces flipped those of the middle-class and elites. The only working woman in Foster's novel to be 'ruined' is the one living with her parents. Her home life (which space the middle class insisted held the best possible preparation for a life of virtue) left her unprepared for negotiating the public world of work and leisure. Seduced by a labor spy whose rooms she visits and who promises marriage, her deathbed testimony exposes the spy and saves the hero. In Foster's depiction, autonomous housing strengthened women, and the city streets posed few dangers to them, whereas dependent women, those living at home, were endangered everywhere.

Foster's novel may, of course, provide as questionable a guide to the moral geography – the terrain of safe and unsafe spaces – of working-class Boston as court records and investigative reports, but there are a few other affirming sources. Mary Jane McGrath Sheldon worked two jobs as a waitress, a worksite moral reformers saw as among the most dangerous since waitresses relied heavily on their ability to please their male customers to earn the necessary tips. She waitressed at Durgin Park during the day, and then at a lodging house among strange men until midnight. In her autobiography, she describes herself as 'without family and home', though she lived with two of her sisters, having left an unhappy situation with her brother's family. And sure enough, lonely and bereft of her parents, McGrath struck up a friendship with a male customer. Together they rented a room for three years, while retaining separate lodgings. But what scandalized McGrath's family was not her keeping company with an unmarried man in a loft every evening while he worked his second-hand printing press and she sewed or entertained friends, but her elopement with him and their marriage outside the Catholic Church, after his business became established.[20]

McGrath's narrative gives no indication that she saw herself morally endangered or even on morally questionable turf at any point except the denouement – the precise point at which middle-class and elite narratives would have seen her as finally reaching solid ground, safely married, in a Protestant church no less. For McGrath and her peers, there was nothing inherently morally dangerous in the city's working-class restaurants, streets, and tenements and nothing inherently safe in living with family. In marrying out of the church and by elopement, McGrath had transgressed moral codes of loyalty to family and church. But at no point does she seem to have supposed that she had transgressed sexual codes in simply keeping company with a man, unchaperoned, in a domestic space devoid of the markers of moral safety parents would have provided, which the social reformers would have seen as an endangering spatial transgression.

Digging into these works for their depiction of particular urban spaces – the lodging house, for example – enables historians to decode both particular spaces and their relation to the city as a whole (the city's moral geography in this case). For historians, both these novels and the autobiography cover the same literal ground as other sources. Even the actions in these sources often mirror those of non-fiction sources. Yet these sources open a different window onto the meaning of those actions and spaces. By attending to what spaces appear in works of fiction and memoirs more generally, for any time and place, attending to how the authors organize space and the sense they make of it, in their explicit and implicit judgments about safety and danger, they alert the historian to a different urban lived experience than is evident in the standard sources usually if not always authored by middle-class and elite men and women.

Reimagining the public city

In smaller towns, as William Dean Howells depicts them in his 1880s divorce novel, *A Modern Instance*, middle-class women, too, enjoyed this level of approved social autonomy – keeping company alone with a man in a confined space – but not, as Howells indicates, in Boston. When Wellesley professor and settlement house co-founder Vida Scudder penned her novel, *A Listener in Babel* (1903), she reflected Howells' view of Boston and did little to challenge the standard view of sexual danger for the city's working girls. Her heroine mapped Boston as she strode from the settlement house in the South End to the Back Bay and back. At first glance, this was a standard depiction of a journey across class and ethnic/racial boundaries as well as across space. The crowded South End she described contrasted with the Back Bay's broad, straight avenues and lower density. Between them lay 'dreary warehouses' which 'reared their immense sinister surfaces against the day'. Once she crossed the railroad tracks into the Back Bay she found silence, 'the dignified city of her youth rose about her – a city of prosperous and pleasant homes, of attractive churches, of noble public buildings, of tranquil sunny streets'. Her heroine concludes, 'poverty and wealth, labor and luxury, connected, or divided, by commerce […]. I have walked straight through our civilization'.[21]

But this journey did challenge the standard urban geography in another way. Scudder did not depict the safely heterosocial geography of the working-class novels, but she did depict a different Boston from the standard fare, and a different geography for its middle-class protagonist. The city streets were not typically seen as safe spaces of lone bourgeois women who wanted to protect their reputations. After all, as mentioned above, 'women of the streets' was a euphemism for prostitutes. Her heroine's uncle objects to her walking to and from the Back Bay and settlement house, and her cousin was scandalized to learn that she spent an afternoon at the Central Labor Union, 'not a spot frequented by young ladies of his acquaintance'.[22] Here Scudder redraws the gendered geography of the city. Where domestic space had been the only legitimated safe space for respectable women in the Victorian city, Scudder's heroine knows no such boundaries.

Scudder uses the novel to redefine, literally, the city's gendered geography. Scudder's protagonist departs from her cousin's home in the Back Bay, a feminine, domestic, private space, and arrives at the female-founded and female-dominated settlement house in the South End. There were homelike aspects to the settlement house, but it was not a home, and Scudder makes the distinction clear. Scudder contrasts her heroine's choice to settle among the poor with her heroine's mother's determination to shut out the world and provide a safe haven and refuge for her heroine's father. The settlement house, by contrast, functions as the center of a reconceived city. The forces of the city – businessmen, workers, and the unemployed of both sexes, welfare workers and philanthropists, educators and doctors, journalists and ladies – all meet in the settlement house. They all meet on female turf. They oppose one another over dinner. The political and labor disputes that paralyze cities of the era here are domesticated, their impact controlled. They do not impede the settlement house from its work finding employment, educating, investigating, and organizing neighborhoods and unions.

The settlement house eradicated bounds between public and private – eradicated the notion of home as refuge from the world outside it and of proper women as limited in their proper sphere to the space within the four walls. Living in the settlement house was a public action. The novel makes clear the settlement house was both part of a city reconceived and a vehicle for building a reconceived city.[23] Scudder used the novel to challenge the gendered map of the city. Scudder also used her novel to construct the city as female space – a female dominated city that appears nowhere on official maps and reports – an alternative reading of the urban terrain. And, indeed, the novel hints at the possibility that women reformers of Scudder's era experienced that feminized landscape, that alternative city, on a daily basis.[24]

In this reading of Scudder's novel, the novel's spatial formation becomes part of the plot-line; it undergirds the possibilities for the heroine. It also undergirds an alternative understanding of the city at the time. Where Scudder has her heroine walk, what she thinks about as she walks, where the action of the novel occurs, makes visible for the urban historian how the author and her heroine negotiate urban geography and the power dynamics it encodes.

Conclusion

The novels and memoir referenced here do not provide a consistent view of place and space. They each disrupt received views in different ways. They disrupt the organization of space into domestic and public, marketplace and non-market. On a grander scale, they disrupt the organization of space by region, lands of opportunity, or fruits of conquest. They expose the ways in which the division of space into safe and dangerous was highly contingent on the identity of the subject. They do so by their subjective depiction of places and spaces in their works, and how their characters move through them and behave in them.

In doing so, they provide a guide to the ways historians can use fictional sources, autobiographies, and memoirs in general to get at issues of place and space and to understand the implications of spatial formations for other historical questions. Both the subjective nature of the novel and its intention to set the scene for readers means the author must be explicit not only in her descriptions but in their meaning. By attending to the ways in which novelists and memoirists select places and spaces in which to set their work and the way they organize the action spatially, and by starting with the map the authors create rather than imposing the action on a map scholars bring to the novel, fictional sources and memoirs can do more than simply disrupt standard geographies. Scholars can find a quite different image of the city or other setting spring to life, a different city-imagined, housing not just a critique of the status quo, but a different lived experience. Armed with that image, scholars can return to other sources with fresh eyes. Reading them anew, different bits of evidence leap out as significant. Scholars can achieve a new understanding of spatial dynamics and what they reveal about lived experience, rewriting key moments in our past.

Notes

1. My ideas about place and space have been influenced by Henri Lefebvre, *The Production of Space*, trans. Donald Nicholson-Smith, Oxford: Blackwell, 1991, French 1974, and have evolved through my own research as well as from other theorists including James Scott, Dolores Hayden, and Michel de Certeau, cited below. For contemporary disputes about how to define these terms, see Maarja Saar and Hannes Palang, 'The Dimensions of Place Meanings', *Living Rev. Landscape Res.* 3/3, 2009. Available HTTP: http://lrlr.landscapeonline.de/Articles/lrlr-2009-3/articlese2.html (accessed July 18, 2018).
2. Melissa R. Gilbert, '"Race", Space, and Power: The Survival Strategies of Working Poor Women', *Annals of the Association of American Geographers* 88, 1998, 595–621, is particularly insightful on the importance of distance from work and childcare for impoverished women and their strategies for staying off welfare.
3. See James Scott, *Seeing Like a State: How Certain Schemes to Improve the Human Condition Have Failed*, New Haven: Yale University Press, 1998.
4. For Spanish, English, and other colonial examples see the various works of Donald Meinig, including the multi-volume *The Shaping of America: A Geographical Perspective on 500 Years of History*, New Haven: Yale University Press, 1986–2004.
5. Useful theoretical approaches to space include Lefebvre, *Production of Space*; any of several works by Michel Foucault, including the two useful anthologies: Donald F. Bouchard (ed.), *Language, Counter-Memory, Practice: Selected Essays and Interviews*, trans. Donald F. Bouchard and Sherry Simon, Ithaca: Cornell University Press, 1977; and Paul Rabinow (ed.), *The Foucault Reader*, New York: Pantheon Books, 1984; Michel de Certeau, *The Practice of Everyday Life*, trans. Steven Rendall, Berkeley: University of California Press, 1984, in which appears the concept of 'lines of desire' – the paths pedestrians create outside the intended pavement lines; Mary P. Ryan, *Civic Wars: Democracy and Public Life in the American City during the Nineteenth Century*, Berkeley: University of California Press, 1997, pp. 4–16 provides a useful overview of theorists of the public; and see Elsa Barkley Brown, 'Negotiating and Transforming the Public Sphere: African American Political Life in the Transition from Slavery to Freedom', *Public Culture* 7, 1995, 107–46 on the ways in which by their practices of democracy in autonomous spaces they transformed the dynamic intended by those who thought they wrote the rules. Gunther Paul Barth, *City People: The Rise of Modern City Culture in Nineteenth Century America*, New York: Oxford University Press, 1980, was an early practitioner of focusing on the connection between different kinds of spaces and the power and culture they instigated and revealed; work on the citing of sports arenas could be seen as a more recent example, e.g., Jerald Podair, *City of Dreams: Dodger Stadium and the Birth of Modern Los Angeles*, Princeton: Princeton University Press, 2017.

6 Sarah Deutsch, *No Separate Refuge: Culture, Class, and Gender on an Anglo-Hispanic Frontier in the American Southwest, 1880–1940*, New York: Oxford University Press, 1987, p. 54.
7 James West Davidson and Mark Hamilton Lytle, *After the Fact: The Art of Historical Detection*, New York: Alfred A. Knopf, 1986, pp. 213–39.
8 See Hayden White, *The Content of the Form: Narrative Discourse and Historical Representation*, Baltimore: Johns Hopkins University Press, 1987; and Hayden White, *The Fiction of Narrative: Essays on History, Literature, and Theory, 1957–2007*, ed. Robert Doran, Baltimore: Johns Hopkins University Press, 2010.
9 Louisa May Alcott, *Work: A Story of Experience*, Boston: Roberts Brothers, 1873; reprint ed. New York: Schocken Books, 1977, with a wonderful introduction by Sarah Elbert.
10 Melody Graulich, 'Violence against Women: Power Dynamics in Literature of the Western Family', in Susan Armitage and Elizabeth Jameson (eds.), *Women's West*, Norman: University of Oklahoma Press, 1987, reprinted from *Frontiers* 7/3, 1984.
11 Feminist geographers have pioneered the study of the relationship between gender, power, and geography. Some of the foundational works include Doreen Massey, *Space, Place, and Gender*, Minneapolis: University of Minnesota Press, 1994; the wide-ranging works by Joni Seager including her *Women in the World Atlas* of various dates; Dolores Hayden's various books including *The Grand Domestic Revolution: A History of Feminist Design for American Homes, Neighborhoods, and Cities*, Cambridge, Mass.: MIT Press, 1982; *The Power of Place: Urban Landscapes as Public History*, Cambridge, Mass.: MIT Press, 1995; Linda McDowell, *Gender, Identity and Place: Understanding Feminist Geographies*, Cambridge: Polity, 1999; Gwendolyn Wright, *Moralism and the Model Home: Domestic Architecture and Cultural Conflict in Chicago, 1873–1913*, Chicago and London: University of Chicago Press, 1980; Helen Lefkowitz Horowitz, *Alma Mater: Design and Experience in the Women's Colleges from Their Nineteenth-Century Beginnings to the 1930s*, New York: Knopf, 1984; Susan Saegert, 'Masculine Cities and Feminine Suburbs: Polarized Ideas, Contradictory Realities', *Signs: Journal of Women in Culture and Society* 5/S3, 1980, S96–S111; Catharine R. Stimpson et al. (eds.), *Women and the American City*, Chicago and London: University of Chicago Press, 1981.
12 See Ellen F. Fitzpatrick, *Endless Crusade: Women Social Scientists and Progressive Reform*, New York: Oxford University Press, 1990, on this dynamic.
13 Patricia Hill Collins, 'The Social Construction of Black Feminist Thought', in Micheline R. Malson (ed.), *Black Women in America: Social Science Perspectives*, Chicago and London: University of Chicago Press, 1990, pp. 297–326; see also her *Black Feminist Thought: Knowledge, Consciousness and the Politics of Empowerment*, Boston: Unwin Hyman, 1990.
14 Perhaps the most well-known mental map is the *New Yorker* magazine cover that portrays the view of the United States from the perspective of New Yorkers – centering a disproportionately large rendition of Manhattan, skipping the Midwest altogether, and quickly arriving at California. The foundational text on mental maps is Kevin Lynch, *The Image of the City*, Cambridge, Mass.: MIT Press, 1960; for a useful encapsulation of interdisciplinary work on conceptual geography see Katherine G. Morrissey, *Mental Territories: Mapping the Inland Empire*, Ithaca: Cornell University Press, 1997, p. 172, 44n.
15 James West Davidson and Mark Hamilton Lytle, *After the Fact: The Art of Historical Detection*, 2nd ed., New York: Alfred A. Knopf, 1986, pp. 213–39.

16 Sarah Deutsch, *Women and the City: Gender, Space, and Power in Boston, 1870–1940*, New York: Oxford University Press, 2000, p. 7.
17 There are myriad examples of scholars who relay the sexualized nature of working-class women's lives including Elizabeth Lunbeck, *The Psychiatric Persuasion: Knowledge, Gender, and Power in Modern America*, Princeton: Princeton University Press, 1994; Kathy Peiss, *Cheap Amusements: Working Women and Leisure in Turn-of-the-Century New York*, Philadelphia: Temple University Press, 1986; Mary E. Odem, *Delinquent Daughters: Protecting and Policing Adolescent Female Sexuality in the United States, 1885–1920*, Chapel Hill: University of North Carolina Press, 1995; Christine Stansell, *American Moderns: Bohemian New York and the Creation of a New Century*, New York: Metropolitan Books, 2000.
18 Bosworth was not alone in noting the absence of evidence for promiscuity and sexual danger despite the prevalence of searching for such evidence on the part of men and women moral reformers; for the quotation see Louise Marion Bosworth, *The Living Wage of Women Workers: A Study of Incomes and Expenditures of 450 Women in the City of Boston*, New York: Longmans, Green, 1911, p. 23, quoted in Deutsch, *Women and the City*, p. 70, and see fn 57 p. 313 for other such sources.
19 Deutsch, *Women and the City*, pp. 92–3 quoting Frank K. Foster, *The Evolution of a Trade Unionist* (n.p. 1901), pp. 1–2, 37, 97–9; see also Pauline E. Hopkins, *Contending Forces: A Romance Illustrative of Negro Life North and South*, Boston: Colored Co-operative Publishing Company, 1900, reprint ed., New York: Oxford University Press, 1988, pp. 85, 102, 104, 106.
20 Mary Jane McGrath Sheldon, 'The Little Print Shop that Grew and Grew and Grew', at Schlesinger Library, Radcliffe College, Cambridge, Mass., quoted in Deutsch, *Women and the City*, p. 101.
21 Vida Scudder, *A Listener in Babel: Being a Series of Imaginary Conversations Held at the Close of the Last Century*, Boston: Houghton, Mifflin, 1903, p. 88, as quoted in Deutsch, *Women and the City*, p. 10.
22 Scudder, *A Listener in Babel*, p. 31 (on her mother) 68, 105, 108, 143, 187, for quotation see 151, quoted in Deutsch, *Women and the City*, p. 12.
23 Deutsch, *Women and the City*, pp. 13–14.
24 One particularly fine example of using such sources to revise our sense of the daily rounds of middle-class women and their significance for organizing is Jessica Ellen Sewell, *Women and the Everyday City: Public Space in San Francisco, 1890–1915*, Minneapolis: University of Minnesota Press, 2011.

4 Newspaper archives

Sherry Olson and Peter Holland

For poets and novelists the landscapes of their childhood have been wellsprings of lifelong creativity.[1] On the farming frontiers of New Zealand in the early-20th century, Jane Mander encountered 'a vibrating silence' and the smell of the bush. 'I discovered the wind'.[2] Such compelling interpretations rely on adult recall and exceptional skills of self-expression. Because they are rare and rooted in time and place, historians are eager to find more immediate as well as more commonplace reports of children's encounters with landscape.

From 1886 into the 1930s, we find in the children's page of the *Otago Witness* large panels of witnesses to how childhood experience was modulated in a particular region, and how a cohort of children interpreted their own spatial history. This newspaper was posted weekly to subscribers living beyond the reach of daily mail deliveries primarily in the South Island provinces of Otago and Southland. Readers included farm families and country storekeepers, road crews, gold miners, and shepherds. Most of the children who wrote to the page were born in New Zealand to emigrants from Scotland and England. The young people ranged in age from 6 to 20, and many, after leaving school at 12 or 13, continued to write, addressing an elusive 'Dot' and their pen pals in print. In this chapter we ask how 'Dot's Little Folk' made use of their surroundings, how they described their landscapes, and how they became attached to the land beyond the homestead.

While these letters may bring to the surface readers' memories of other times and places, our purpose is to demonstrate the rich possibilities of newspaper sources for regional history at a time of intense environmental disturbance. Since the year 2000, the national libraries of Australia and New Zealand have devised workable models for digital reconstitution of newspaper archives. Access and searchability are re-making 'the information landscape within which historians labor'.[3]

The chapter is organized in three parts. The first, grounded in our experiments with the on-line archive PapersPast, identifies challenges we face in working with newspaper sources recycled in a digital medium. The second, covering a 25-year span, 1886–1910, reports what children were saying about the plants and animals they encountered. Their correspondence paints a childscape – a panorama of impressions from which we interpret, in the third, places loaded with meaning and the children's struggle to convey those meanings.[4]

DOI: 10.4324/9780429291739-6

Question, source, and method

To answer a research question, the historian looks for a promising source, but methodical evaluation of that source will alert us to ambiguities, challenge our assumptions, and demand a battery of tests or a complement of other sources. It may even re-direct the entire line of questioning. Making our way into a newspaper archive inspires all the caution and exhilaration of the Victorian child exploring her patch of woods in Southland: 'On the other side of the Rock is a lovely bush with beautiful ferns in it. The bush is so very thick that few people have been in it, and so the ferns are left unmolested'.[5]

The historical appeal of newspaper archives lies in the precision with which time and place are stamped on the documents, making it possible to track events from day to day and over decades, or to discern differences of opinion from place to place. In the colonial ventures of the 19th century, the printing press arrived with the plough, and even in small towns the news – local, regional, and international – was published with a sense of urgency. The rate and frequency with which a local newspaper reported events from other parts of the world allow us to appraise the pulse of commerce and fashion, and situate the community in its national and international networks.[6] For the period we target, scholars have recognized an emergence of ecological concerns on several continents: forest management, for example, soil erosion, and destruction of valued scenery. Scholars are asking to what extent those concerns were influenced by *local* responses to *local* changes: fire and flood, perceived shortage of timber, or anxiety about disease in their own community.[7] Thanks to the rigour of the time-and-place stamp, newspapers are an obvious historical source for the local.

The value of such raw material depends, however, on the continuity and breadth of an archive. The edition 'hot off the press' cooled quickly. Printed on cheap paper, most copies found their way to the outhouse or the stove in a matter of hours or days. In the 1930s, public libraries and historical societies began to rescue newspaper archives by transferring whole pages from crumbling sheets of paper to microfilm, and this effort fostered more systematic strategies of content analysis for political opinion.[8] Since the 1990s, with high-speed computers and internet collaboration, ambitious surges of investment have generated an immense volume of high-resolution images, but the discoverability that the historian craves remains rare because efficient search for words or phrases depends on the further step of costly re-processing of images into text.[9]

Initiatives of public libraries and historical societies are confined by tight budgets and local priorities. Some prominent newspapers maintain their own archives but charge fees for access; display of 20th-century editions is confined by copyright, and provincial papers have been neglected relative to those in cities, despite their distinctive pools of readers. In Australia, for example, Katherine Bode has shown that small-town weeklies, as compared with the metropolitan dailies, were publishing more original fiction by local authors, and were aligned with a different set of commercial networks in

Britain and the USA.[10] From digital text it is easy to collect wordcounts, word sequences, or phrases, but how well does the material available represent 'the press' of a particular decade and region? Joan Judge and Barbara Mittler, by situating a Shanghai newspaper in a larger 'ecosystem' of print media of the 1910s, have shown how its editors employed details of typography, artwork, and advertising to introduce new concepts of modernity and femininity.[11] Whatever the topic of research, the user of a newspaper archive must ask what filters were applied in collection, salvage, and selection for digital display and public access.[12]

Our project has had the advantage of near-complete holdings and full digital coverage of the *Otago Witness* over the quarter-century span of our research interest. Thanks to the intelligent search interface of PapersPast, with instant switch between image and text, we could capture the material we needed by copy-and-paste from the internet. The letters were segregated under the title 'Dot's Little Folk', and have been tapped by other scholars for other themes.[13] The decision of a newspaper to publish letters from children in 1886 was a significant historical event, like the hiring of a woman as a foreign correspondent and the appearance in that decade of regular features titled 'Cycling Notes' or 'Trades and Labour'. Nineteenth-century newspapers offer systematic coverage of many topics, some already classified and others discoverable by keyword ingenuity.[14]

As part of a broader research program directed to farming practices in the region, we assembled a file of 12,000 letters with the intent to discover the tasks children were performing on the farms.[15] By mapping subscribers from lists published 1904–8, we confirmed their concentration on the rural frontier that we had previously targeted from account-books of farms and marketing agents. As one must with any news medium, we probed for the identities of the editors, writers, and readers of the children's page to uncover bias. Five out of every six correspondents were girls; the frequent writers were 11 to 15 years old, and they wrote more letters in winter and on rainy days. To confirm that portrait we had to work our way through a jumble of pen names, variously spelled and often changed or imitated.

In that first search we were looking for 'facts', not 'feelings', but our young informants had more to say. 'I love Green Island'. Or, 'Dear Dot, don't you think this world of ours is very beautiful? This wee corner of it is at any rate'.[16] Taken by surprise, we changed course to follow our young informants into the sensory and relational realm of 'place'.[17] A burgeoning literature asserts the significance of 'thick places' for human creativity, and critics have asked for a better definition of what constitutes a 'thick place'.[18] We propose that the sheer effort needed to produce a 'thick description' – vivid, personal, and rich in contextualization – reflects the writer's own appraisal that a particular landscape is indeed freighted with memories, anticipations, emotions, and social meanings.[19] Our objective comes close to that of Raymond Williams, who sought to uncover 'the structure of feeling' conveyed in literary descriptions of English landscapes over the 19th century as each successive generation reimagined a 'golden age'.[20]

Access to massive chunks of digital text invites experimentation with digital techniques for analysis. To classify tasks on a farm and appraise their seasonality, we used coding software; to categorize emotional expressions, we employed a psychologist's wordlist; and to uncover networking among the young people, we applied an open-source graphics package. Experiments in digital humanities are rapidly enlarging the analytic tool kit.[21] In what follows, however, to explore 'thickness of place' and 'structure of feeling', we rely on the traditional practice of *explication de texte*, or close reading. With rare exceptions, the Little Folk were confined to 450 words. Our strategy was to distill from the letters their interactions with plant and animal species (in part 2) and (part 3) the accounts of particular places on which the young people lavished their quota of words.

Near-home encounters

Within the warm, busy kitchen of the farmhouse, the toddler began to recognize who belongs where. A pet lamb 'on baking days lies in front of the fire and gets its wool singed', and a swan 'hatched by the fireplace keeps by the spade and shovel'.[22] Sometimes there is trespass: Nannie's lambs one hot day got into the spare room, 'jumped up on the beds and were all five chewing their cud there... and I got blamed...'[23] Intruders were bustled out to the yard, a place of greater risk: 'The pet ram will give you a good tumble,' and the old horse, cunning enough to unlatch the gate, makes a nuisance of himself on washing day by drinking the soapy water.[24]

Short-lived pets modeled the harsh realities of population control under strict economic imperative: 'Piggie got so big and fat that father had to kill it'.[25] The children recognized themselves as part of the ecological system: 'I am wearying for a taste of his trotters".[26] '...I must close, I have to go and drown five kittens'. Most of the children's pets were babies of the world's small repertoire of domestic animals, bred from beasts imported from England, Scotland, or Australia, and they supplemented a diet of chaff and table scraps by hunting and foraging: rabbits ate the bark of the apple trees; rats killed the bees; cats kept down the rats but killed the pet goldfinch; ferrets killed chickens and pet rabbits; and the dogs terrorized the cats and sucked hens' eggs. The children grew to recognize predation as well as 'natural' enmities between species, and became fascinated with apparently contrary behaviours. Nannie's little dog slept in the cats' box outside at nights. 'If they were in first she would dig them up with her nose so that they would be on top to keep her warm'.[27] A hen made her nest with the kittens; 'you will hear her calling them sometimes...'[28] (Figure 4.1).

The Little Folk puzzled over the responses of wild as compared with domesticated species. Frequent mentions show their familiarity with more than 50 edible and commercial crop plants, numerous varieties of turnips and grasses, various breeds of poultry, and close to 100 of the species of flowering plants imported from Great Britain 'for mother's garden'. Of endemic native plant species, however, just two dozen received more than a single mention,

Figure 4.1 'Dorothy Fitzgerald reading to her siblings', c. 1905: Older children supervised much of the play and learning of younger children, and organized their adventures on the hills, at the shore or in the bush.

Source: J.E. Fitzgerald Collection, P1998-085/3-283, courtesy of the Hocken Collections – Uare Taoka o Hākena, University of Otago.

among them the showy flowering rata and mistletoe, three tree species prized for firewood, and the cabbage tree, its dead leaves heard clapping in the dusk: 'Every time I think of it I get quite shaky, it was so uncanny'.[29]

Without tutelage in the names and growth habits of the indigenous plants, the children were nevertheless intrigued, continually bringing home bits of the wild and carrying out experiments in rearing them in pots, shells, and tin cans. Evelyn described the young duck hatched by placing eggs of a wild duck under a hen. 'All the other hens flocked around, and showed a great deal of curiosity when they saw this duck. The yellow hen seemed never to grow tired of gazing at this rare-looking chicken.'[30] 'In the caves we found lovely maidenhair ferns; but they died before we reached home'.[31] Their instinct was nevertheless to capture what they could. 'When once I commenced pulling them I couldn't stop, for as soon as I left one clump I found another prettier one'.[32]

In their nests, hidey-holes and play-places the children created eventful worlds.[33]

> One day I went up to a sawmill here and pegged off dredging claims in the sawdust which is carried away from the mill by water. A large piece of bark off a birch served as a water race. Will you take some shares in it, Dot?[34]

While most of those projects were short-lived, about one in 100 of the young correspondents watched with patience and reported why and where. One of the girls fretted over the kidney ferns she had found 'away back in the bush. I am afraid they won't grow. I think they want to be in some place where the water is just dripping on them all the time, and not keeping them too wet and damp'.[35] When Larrikin Tom asked how to rear goldfinches, Monte reported having discovered foods they would accept: lint, grass seed, thistle seed, rose, and cabbage blight.[36] Those were writers who accounted for the rarely mentioned plants, and who sent the editor bugs and ferns for identification. Another was Bramble, who in 1905 was trying to raise a kiwi:

> For a cage I put it in a piano case laid down flat, the slanting part I covered with some strong wire netting. The food I gave it was worms, slugs, and white pine grubs.... All day long it would sleep, and when it used to get wakened up it would tumble about the box as if it had a fit: but at night it changed its antics, for all the time it was a continual thump, thump.[37]

A decade later Bramble would have had no chance to capture a kiwi, or even catch a glimpse of one. The native ground-living birds were already becoming rare. Bramble also described a kakapo (a large parrot) that loved potatoes and made a noise like a pig grunting. 'In shape and build it is like the kiwi, but in all other things it is very different'. The distinction lay in how and where the animals lived:

> ...the kiwis always live near swampy land, while the kakapo loves the rough, rocky land. In summer its home is on the mountains, but in winter, as soon as it begins to get cold, they all make their way down to the lower country. Its plumage is lovely, being a beautiful greenish yellow colour, and when the sun is shining on it it makes it go all colours.[38]

Who were the exceptional observers? Several of them lived in a village of fishermen, others in a remote and newly settled block near Greytown, and some on an island or isolated promontory with a lighthouse. Many were bird-nesters who roamed widely on their own.

The majority of Little Folk had less sustained encounters with wild species, but nevertheless reported feelings of empathy that led them to ethical puzzles. 'I keep no pet animals or birds. I think the creatures are better running or flying wild than shut up in nasty little boxes'.[39] This line of thought was emerging in the English-language press worldwide,[40] and the editors of the *Otago Witness* promoted stories of kindness to animals. In the children's own letters, however, the sense of responsibility generated by care and affection was matched by episodes of carelessness, acquisitiveness, and violence. A youngster relished a gory scene between ferret and rabbit, then took home a baby rabbit to make a pet of it, and a boy of 11 described his initiation into hunting by one of the station hands.

As we were going down a hill the dogs hunted two wild pigs out of the fern, and they ran into a little clump of manuka. The man I was with fired several shots… As we had only a few charges left we went away, leaving them half-dead… We got three full-grown rabbits and four young ones. We put the young rabbits in with a lot of ferrets, by which they were soon killed.[41]

Responding to government bounties for control of pest species of birds, the Little Folk made excursions into the hills and bushland: 'I got 8s 6d for eggs last year, at 2s a hundred'.[42] At the same time they sympathized with the birds:

We had a goldfinch sitting in one of our trees. She had four eggs under her. One day a big bird came down from the bush and sucked or broke the eggs. The two birds cried all day when they found out their loss. It has done the same thing for three years. Other birds' nests have been destroyed by it. We have always seen how sorry the birds were.[43]

The logic of robbing nests and trapping rabbits arose from the ecological predicament of the human being, and those encounters raised a tangle of questions about relatedness, right and wrong, the lovely and the dreadful. In a world of change and competition, where do I belong? Dot's correspondents elaborated their feelings about the natural world in fantasies that typically began 'Dear Dot, I am a Shag', and then re-enacted the chain of species relationships. 'The weather is very stormy, and I can hardly see the fish swimming in the water, so I eat poisoned rabbits'.[44] A Little Grey Rabbit speaks up: 'Dot, do you think it was right of the Acclimatisation Society to import my ancestors, and now hunt us down?'[45] Children took active roles in the poisoning campaigns: 'When the snow comes down I always lay poisoned wheat on a small piece of land for the birds to eat'.[46] They worried, however, as the effects of poison spread through the food web: 'I am going about picking up all the dead ones I find, and burning them. I am in terror in case some of my cats get poisoned'.[47] At Taioma, 'I had a little lamb, but it got poisoned. I was very sorry about it'.[48] And inevitably, 'there was a little boy who died not far from where we live. He got poisoned with eating poison that is used for killing rabbits. He was only 4 years old'.[49]

Venturing beyond the farm

For rural children, enrolment in school, at age 5 or 6, launched a career of exploring, supervised and spurred on by older brothers and sisters. The journey to school was a daily routine they endeavoured to fill with novelty and sociability, all the while maintaining tension between security and daring. 'We get stinging nettles on the way to school and sting each other with them'.[50] High in the walls of rock where Irema found lovely white flowers, 'we got into trouble for climbing down the side of a cliff. Dad said we would break our necks, and mother made us promise not to go down there again'.[51] When released from school and chores, children in sixes and sevens set off with bread, cake, and kettle. Their accounts of 'boiling the billy' reveal three preferred places: the beach, a high point, and the bush. These three habitats elicited the thick descriptions that signal their importance in memory and feeling.

The environment of the beach engaged all the senses.[52] 'We ran about on the sand with the waves of the sea rolling and tumbling about our feet'.[53] They gathered cockles and mussels and sea urchins 'of a sort of a greenish colour with little bristles all over them'.[54] Among the letters describing their 'grand fun playing houses in and out among the rocks',[55] we find reports of injuries, rescues, and drownings. When no danger threatened, the children courted it: 'I thought I should like to see how my horse took to the breakers'.[56] Another group, with the breakers 'dashing in almost to our feet' climbed a zigzag wooden staircase to the sandy cliffs above: '...just as we got to the top there was a great crack, and the whole embankment and the race were carried out to sea'.[57]

The prime attraction of high points lay in the effort to get there. 'Up and up we climbed, until reaching the top we turned westward and scrambled through native boxwood nearly as high as ourselves'.[58] Rimbecco described her first view from a hill that overlooked the little village and schoolhouse of Houipapa: 'The many different paths and roads wound in and out as far as eye could see, like long, thin, silver ribbons'.[59] Miniaturized, composed, and connected, the scene shows an investment of human labour.[60] Arthur, looking down the steep, rocky sides of an abandoned gold battery occupied by birds and rabbits, describes having felt 'a giddy sensation come over us.... these buildings have stood for years, rotting in rain and sun, with their ground floors covered with dirty water pools and floating lumps of timber'.[61] In these expeditions the children were appropriating a history: 'Now turn your eyes towards the village', announced Vice-Admiral Togo from a perch in a tall tree on a hilltop above the Taieri Plain, '...all over you see smoking hollows.... people are afraid to walk over the land, and some are even in fear of the land falling away from under their houses'.[62]

The third kind of place, native woodland, offered the most varied opportunities and evoked the most acute recall:

> I must now tell you about the bush. It is very lonely and quiet. The wind never comes there, and it is always warm. The bush is so thick that you cannot go far, and if you did succeed in getting in far, you would be lost.[63]

Its depth and intricacy compelled constant alert: 'There are lots of wild cattle in the bush, but they keep out of sight.' 'I put my hand into a bees nest.'

Some youthful activity was purposeful and rewarding – eeling, fishing, felling a tree for honey, or pulling out dead wood for the stove – but play was often the sole objective: 'We play nothing at school, so we got a number of boys together and had a hunt. It was a great hunt, too, jumping races, falling into water, sliding down cliffs, pouncing through broom, jumping pieces of gorse, and falling over stumps!'[64] Sometimes a holiday picnic was made memorable by an unexpected scramble:

> The higher up Mount B we went, the worse it got, and we scrambled through the dripping scrub like bush rangers. We came upon a quarry... as dark and damp as could be, with great stones all over... On the way back it began to rain, so we started to run... We went rushing along, jumping over logs, slipping down banks, getting caught in brambles.[65]

In these narratives the Little Folk employ a very small number of descriptors — beautiful, lovely, delightful, glorious, fine fun — then compensate with an explosion of verbs indicative of the high energy levels of the young.[66]

Activities in the bush were often spontaneous. 'We made fern hats and laughed at each other.'

> ...we looked about for some vines for skipping ropes... we found some beauties. After skipping a while, we made a fire to boil the billy with. It happened that we made the fire under a tree, and the consequence was that the tree caught alight. We had the tea made, but we got such a fright that we couldn't drink any. So we threw it all over the tree to quench the fire.[67]

Another group, having found some large stones, 'set these rolling down into the gullies, then crashing into trees and logs that stood or lay in their path'.[68] As youth reached 20, their farewell letters to the page vividly recalled the fires, scrambles, and rescues as 'days of pure enjoyment... But now those days are faded and past'.[69] They reported feelings they could not explain, of belonging to some marvellous whole. These feelings they associated with those thick places that constituted the childscape — a landscape they now perceived as never to be recovered. Dolly, recalling mountains mirrored in Lake Wakatipu, insisted on the inadequacy of words: '...it overwhelms you with a feeling that though you were to live for thousands of years you would never forget it, and neither would you ever be able to express your feelings.'[70]

The nostalgia the adolescents expressed after leaving home was aggravated by recognition that the landscape itself was undergoing change. Tramways brought timber from tracts of indigenous forest to sawmills; patches of forest were being cleared for farming, and cattle, even horses, were wintered there and left to fend for themselves. Again the young people raised ethical questions: at Houipapa:'it seems a terrible shame the way the bush is being burnt... trees felled and left to rot...'[71] Among other threats to their native places, Punch reported that the Awamoko Stream was very yellow because the mining claims at Livingstone River were working regularly.[72]

The habitat these children occupied was intensely practical, consistent with the British vision of empire as appropriation. To see was to possess.[73] The landscape was not yet encrusted with myth or story or verse. The most frequently reported religious event was the harvest festival when parishioners rejoiced in their 'Home' plants and animals: wheat, flocks of sheep, the fatted calf, and the green hill far away. Missing in the new land were the sacred wells of Ireland and the disquieting 'little people' of Britain, the giants, the druids, and the standing stones.[74] The Protestant mission, with the priority of the Word, had early and firmly rejected consecration of earthly places. New Zealanders retained a large share of native place names and many Little Folk chose Māori words as pen names, but myth was reduced to the fanciful. In stories published in the *Otago Witness*, Māori fairies were disempowered and their moral interventions warped. The fairy stories proffered by the Little Folk themselves were curiously alien. The only exceptions we discovered were those of a Southland girl who signed herself Pixie:

> After dark the bluebells ring and we assemble together for some fun. We first gather by the hollyhocks to give an account of our day's work, then off we go. We used to catch the rabbits, and have races round the flower-beds, and jumps over the shamrock borders; but the people who live in the big house in the garden did not like their flowers being eaten off by our steeds, so they set traps and caught and killed most of them... There has been a fall of snow on the mountains lately, and at night we get the paradise ducks to carry us there on their backs, so that we can have some fun snowballing each other, and when we are tired of that we have dances in the shelter of the rocks.[75]

Pixie subsequently described a moonlight trip to the other side of the Waiau River: '...it was splendid sailing over the hills, looking down on the silvery waters of the creeks or at the mountains ahead of us, with the bush growing half way up their sides, and their dazzling white tops.'[76] Aided by a soaring imagination, Pixie visualized a rich tapestry for the South Island, a landscape fractured and parcelled into Then and Now, There and Here. Her imagined ride on the back of a paradise duck points to the importance of bird life to her peers. New Zealand had no large native land mammals, and flightless as well as winged birds filled many niches, making the landscape alive with birds soaring, running, feeding, nesting, and wading.[77] Concern for native birds is arguably the starting point of the preservation movement in New Zealand,[78] and adoption in 1900 of the kiwi as the badge of identity of Dot's Little Folk coincides with other indications of the iconic status attributed to this bird as a symbol of national identity. Another bird rarely seen but often mentioned by the children was the night-calling morepork: 'It is wonderful what a deal of feeling these screeches of theirs may express, Dot'.[79]

Although the South Island Māori population was small (perhaps 1,500) with children attending largely separate schools, some of the birdnesters and fishermen among the Little Folk were probably Māori. 'You were asking where seagulls build their nests... They build on the coasts, sometimes away down the cliffs, in small caves, and I like going down many feet, swinging in midair, with a rope round my waist'.[80] Tom's account of muttonbirding records Māori practice:

On summer evenings we go round the cliffs, and when we come to a hole we put a stick in, and they bite it, then we pull them out and kill them. We string them on our backs with a flax, and work our way carefully along, for if we lose our footing we will fall down 500 ft or 600 ft and get killed on the rocks below.[81]

Tom described a cave at the foot of the cliffs: 'They make a noise like someone in trouble… When they come flying over your head you would think it was some witch flying away with someone. Sometimes they hit you in the face with their wings.' The young Pākehā – of the first generation born in New Zealand – were short on technical and cultural baggage, names and legends, but were as susceptible as the young Māori to the invitation of the birds.

Conclusion

Dot's Little Folk loved riddles and shared an inexhaustible stock, but we have seen that they confronted other riddles which did not have solutions printed at the bottom of the page – riddles of life-and-death, the cruelty of nature, and the cruelty in human nature. Their discoveries belong to the universal experience of middle childhood (ages 6 to 12), with a progressive awareness of order and interplay of all that was living and growing around them. We interpret this awareness as the grounding of ecological thinking. Children as young as 8 or 9 years were attentive to relationships between species, including their own. Their questions about predatory or mothering behaviours, about display and concealment, or about fright and grieving, display intuitions of an ecological system, with its dynamics of mutual adjustment. These intuitions arose in instants of wonder, in a sudden fleeting intimacy with the beautiful or the awful.

On leaving home, still struggling to find the words, young people acknowledged the landscape of their childhood as a paradise lost: 'I tried to drown the voice of memory'.[82] Despite experience so widely shared, their attempts to communicate a structure of feeling took forms specific to their generation, to a local fauna and flora, and to an intensely local landscape – the smoking hollow and the flood at the doorstep. For our attempts at 'doing' spatial history, the 6-year-old – relishing the tickle of the blade of grass or the scramble through dripping fern – offers us a model for an elastic and inclusive sense of the space we share.

The *Otago Witness* provided a window into childhood experiences on the farming frontier. What we discovered of the local landscape and the local colouring of child life is a mere taste of the possibilities of historical research using a newspaper archive. The regional weekly we tapped offers raw material for other topics, such as the progress of epidemics in school districts, trials of the latest farm machinery, emergence of cooperative dairy factories, or farmers' responses to unions of shearers and miners. The on-line resource PapersPast compiles other newspapers, including partners and rivals of the *Otago Witness*, all of them expressive of local conditions, local opinion, and a local structure of feeling. A digital newspaper archive of this kind, like a spill of WikiLeaks, or a well-catalogued library of state papers, invites its users into collaboration for wider sampling, comparative analysis, and imaginative scholarship.

Notes

1 Edith Cobb, *The Ecology of Imagination in Childhood*, Dallas: Spring, 1993; Louise Chawla, *In the First Country of Places: Nature, Poetry, and Childhood Memory*, Albany: State University of New York Press, 1994. Often cited of a large array are William Wordsworth's *Preludes* and Walt Whitman's *Song of Myself*. The authors are grateful to the personnel of the Hocken Collections, University of Otago, and of the National Library of New Zealand site PapersPast; to Professor Tony Binns for helpful comments on a draft, and to their university departments.
2 Jane Mander, *The Story of a New Zealand River*, London: John Lane, 1920, p. 9, and *The Passionate Puritan*, London: John Lane, 1922, pp. 84 and 22; also Phillip Steer, 'Jane Mander, 1877–1949', *Kōtare* 7, 2007, 37–54.
3 Lara Putnam, 'The Transnational and the Text-Searchable: Digitized Sources and the Shadows They Cast', *American Historical Review* 121, 2016, 377–402, here 379.
4 For a cross-cultural perspective with an alternative definition of the term, see Jay Griffiths, *Kith: The Riddle of the Childscape*, Milton Keynes: Penguin, 2013.
5 Loaf, 28 November 1900. Citations from 'Dot's Little Folk'. Available HTTP: https://paperspast.natlib.govt.nz by date and search strings in the *Otago Witness* for a name, pen name, or expression. Age is given where known.
6 Eric Pawson, 'Time-Space Convergence in New Zealand, 1850s to 1990s', *New Zealand Journal of Geography* 94, 1992, 14–19.
7 Paul Star, 'Regarding New Zealand's Environment: The Anxieties of Thomas Potts, c. 1868–88', *International Review of Environmental History* 3, 2017, 101–38; James Beattie and John Stenhouse, 'Empire, Environment and Religion: God and the Natural World in Nineteenth-Century New Zealand', *Environment and History* 13, 2017, 413–46; Thomas Dunlap, *Nature and the English Diaspora: Environment and History in the United States, Canada, Australia, and New Zealand*, Cambridge: Cambridge University Press, 1999.
8 Mark West and Linda Fuller, 'Toward a Typology and Theoretical Grounding for Computer Content Analysis', in Mark West (ed.), *Theory, Method and Practice in Computer Content Analysis*, London: Ablex, 2001, pp.77–94.
9 PapersPast and its partners perform 'blocking' to structure newspaper layout (columns and headings) prior to automated optical character recognition, but they do not follow up with high-cost post-processing corrections. Jenny Presnell and Sara Morris, 'The Historical Newspaper Crisis: Discoverability, Access, Preservation, and the Future of the News Record', 14pp. International News Media Conference April 27–28, 2017, Reykjavík. Available HTTP https://kuscholarworks.ku.edu/handle/1808/24599 (Accessed 23 June 2018); Kenning Arlitsch and John Herbert, 'Microfilm, Paper, and OCR: Issues in Newspaper Digitization', *Microform & Imaging Review* 33, 2004, 59–67.
10 Katherine Bode, 'Fictional Systems: Mass-Digitisation, Network Analysis, and Nineteenth-Century Australian Newspapers', *Victorian Periodicals Review* 50, 2017, 100–38.

11 Joan Judge, *Republican Lens: Gender, Visuality, and Experience in the Early Chinese Periodical Press*, Berkeley: University of California Press, 2015; Barbara Mittler, 'Imagined Communities Divided: Reading Visual Regimes in Shanghai's Newspaper Advertising (1860s–1920s)', in Christian Henriot and Yeh Wen-hsin (eds.), *Visualising China: Moving and Still Images in Historical Narratives*, Leiden: Brill, 2013, pp. 267–377. The 2013 volume offers further examples of the importance of the material object, which we might miss if we rely entirely on digital copy.
12 Bob Nicholson, 'The Digital Turn, Exploring the Methodological Possibilities of Digital Newspaper Archives', *Media History* 19, 2013, 59–73.
13 Keith Scott, *Dear Dot – I Must Tell You – A Personal History of Young New Zealanders*, Auckland: Activity Press, 2011, attentive to adolescent trajectories and subsequent experiences in World War I. For alternative sources on the emotional world of New Zealand young people, Chris Brickell, 'Histories of Adolescence and Affect: Setting an Agenda', *History Compass* 13, 2015, 385–95.
14 The authors have probed the same newspaper archive for mechanical innovations and energy flows on the farming frontier: Peter Holland and Sherry Olson, 'The Farmer's Cutting Edge in Southern New Zealand 1864–1914', *International Review of Environmental History* 6, 2020, 27–56; and 'Appetite for Grass: Re-engineering Landscapes of Otago and Southland 1864–1914', *New Zealand Geographer* 76, 2020, 237–46.
15 Sherry Olson and Peter Holland, 'Conversation in Print among Children and Adolescents in the South Island of New Zealand, 1886–1909', *The Journal of the History of Childhood and Youth* 12, 2019, 219–40. The four batches 1886–96, 1899–1900, 1904–6, and 1908–9 include 4,000 different writers.
16 Emily, age 10, and Ellie, 13, Tomahawk, 10 August 1888.
17 Tim Cresswell, *Place: A Short Introduction*, Oxford: Blackwell, 2004; David Crouch, *Flirting with Space: Journeys and Creativity*, Farnham: Ashgate, 2010; Edward Relph, *Place and Placelessness*, London: Pion, 1976.
18 For example, Edward Casey, 'Between Geography and Philosophy: What Does it Mean to Be in the Place-World?', *Annals of the Association of American Geographers* 91, 2001, 683–93; challenged by Cameron Duff, 'On the Role of Affect and Practice in the Production of Place', *Environment and Planning D: Society and Space* 28, 2010, 881–95.
19 Clifford Geertz, *Thick Description: Toward an Interpretive Theory of Culture*, New York: Basic Books, 1973, pp. 3–30.
20 Raymond Williams, *The Country and the City*, Oxford: Oxford University Press, 1973.
21 See Tim Sherratt, 'A Map and Some Pins'. Available HTTP: http://invisibleaustralians.org/blog/author/tim-sherratt/.
22 Essie, 11 October 1905; Katie Weir, Gimmerburn, 26 July 1894.
23 Nannie, 26 July 1900.
24 Tin Tacks, 24 August 1904; Petunias, 6 July 1904.
25 Willie, age 12, 1 October 1886.
26 Pussy, 15 February 1900.
27 Nannie, 26 July 1900.
28 Edith Lavinia Weir, Naseby, 16 November 1888.
29 Moto Rimu, 14 December 1904; see also Philip Simpson, *Dancing Leaves, The Story of New Zealand's Cabbage Tree, Tī Kōuka*, Christchurch: Canterbury University Press, 2000. For the plants and animals named, see Gerard Hutching,

The Natural World of New Zealand, London: Penguin, 1998. Available HTTP: https://teara.govt.nz/en; https://en.wikipedia.org. The letters do not always give enough detail for accurate species identification.

30 Evelyn Kinghorn, Purekireki, 28 June 1894.
31 Nahua, Oreti caves, Centre Bush, 1 February 1900.
32 *Vanity Fair*, 3 October 1900.
33 Jerome Bruner, 'Life as Narrative', *Social Research*, 2004, 71, 691–710 (first published 1987); Gaston Bachelard, *L'air et les songes: Essai sur l'imagination du mouvement*, Paris: J. Corti, 1990; David Sobel, 'A Place in the World: Adults' Memories of Childhood's Special Places', *Children's Environments Quarterly* 7, 1990, 5–12.
34 Mousie, 16 April 1900.
35 Connecticut II, 8 September 1909.
36 Monte, 27 January 1909.
37 Bramble, 28 June 1905. The kiwi eventually escaped and was killed by the dogs.
38 Bramble, 28 June 1905.
39 George M'Rae, 10, Hokonui, 30 December 1887.
40 Darcy Ingram, 'Wild Things: Taming Canada's Animal Welfare Movement', in Joanna Dean, Darcy Ingram and Christabelle Sethna (eds.), *Animal Metropolis, Histories of Human-Animal Relations in Urban Canada*, Calgary: University of Calgary Press, 2017, pp. 87–115.
41 James Coleman, age 12, Avondale Station, Marlborough, 15 October 1886.
42 8 shillings 6 pence. George Buchanan, Newstead, 15 July 1887.
43 M.N., 12 Port Molyneux, 28 December 1888.
44 The Shag, Waitahuna, 18 July 1895.
45 A Little Grey Rabbit, 16 June 1892.
46 Smut, 20 July 1904.
47 Tin Tacks, 24 August 1904.
48 Ellen Millar, Taioma, 12 November 1891.
49 Ham and Eggs, 26 May 1909.
50 Mabel Black, 18 November 1887.
51 Irema, age nine, 7 July 1909.
52 Erwin Straus, *The Primary World of Senses: A Vindication of Sensory Experience*, London: Free Press of Glencoe, 1963.
53 Visit to Ashgrove, from Milton.
54 W.Y.M., Oraki, 12 December 1900.
55 Ethel, age eleven, Riverton, 24 July 1890.
56 Maid of Orleans, 11 January 1900.
57 Ricordi, Orepuki, 13 April 1904; see also Wilful Lassie, 10 August 1904.
58 Viola Cornuta, 31 March 1909.
59 Rimbecco, 26 September 1900.
60 On symbolic representations of time-space, see Susanne Langer, *Feeling and Form*, New York: Charles Scribner's Sons, 1953.
61 Arthur, 6 July 1904.
62 Vice-Admiral Togo, 15 June 1904.
63 George M'Rae, ten, Hokonui, 30 December 1887.
64 Soprano, 11 November 1908.
65 Mag, 26 September 1900.
66 For an analysis of ecology in terms of behavioral energetics at differing levels of organization, see Cobb, *Ecology of Imagination*, p. 18.
67 Herpasilate, 13 July 1904.

68 *Vanity Fair*, 3 October 1900.
69 *Cornish*, 26 April 1905.
70 Dolly Poppelwell, Dunedin, 3 May 1894.
71 Mistletoe, Houipapa, 22 February 1900.
72 Punch, Ngapara, 9 October 1890.
73 Giselle Byrnes, *Boundary Markers: Lands Surveying and Colonisation of New Zealand*, Wellington: Bridget Williams Books, 2001, p. 41.
74 On sacred space see Alexandra Walsham, *The Reformation of the Landscape: Religion, Identity and Memory in Early Modern Britain and Ireland*, Oxford: Oxford University Press, 2011.
75 Pixie, 27 May 1908.
76 Pixie, 12 August 1908; also 15 September and 10 November 1909.
77 Michael J. Stevens, 'Kāi Tahu Writing and Cross-Cultural Communication', *New Zealand Journal of Literature* 28/2, 2010, 130–57; Margaret Rose Orbell and Carolyn Lagahetau, *Birds of Aotearoa: A Natural and Cultural History*, Auckland: Reed, 2003; and Bachelard, *L'air et les songes*.
78 Paul Star, 'Native Bird Protection, National Identity and the Rise of Preservation in New Zealand to 1914', *New Zealand Journal of History* 36, 2002, 123–36.
79 West Coaster, 2 December 1908.
80 Maori Masher, 10 March 1909.
81 Tom, 17 August 1893. Cross-referencing names of Little Folk 1905–9 uncovered networks of Māori youth on Ruapuke Island and at Karitane.
82 Cooee, 21, 21 December 1904.

5 Architectural drawings

Despina Stratigakos

In simple technical terms, a floor plan can be described as 'a scale drawing showing the arrangement of one level of a building projected onto a horizontal plane'.[1] Typically, the plan indicates the position of walls, windows, doors, and stairs, as well as the dimensions of the rooms. Fixed features, such as sinks and toilets, are often included, and sometimes the intended placement of moveable furniture is also illustrated. These interior spaces are viewed from above, as if the upper sections of a building were removed and a horizontal cut was made through the building's solids and voids. In this sense, a floor plan may seem straightforward in the information it presents.

Scholars used to texts and images as their primary sources, however, may find floor plans difficult to read. The conventions used to indicate physical elements in an abstract two-dimensional drawing – such as the thickness of lines, gaps, dashes, numbers, letters, and symbols – require some practice to grasp. Moreover, floor plans are created for a broad range of structures, from parliament buildings and temples to shops and houses. Any such plan inevitably changes over time from what the creator intended, layering uses and meanings. Finally, floor plans do not tell us about the lived experience of architectural spaces in the way that a diary or family photo album can, even if words and images must also be interpreted. Nonetheless, plans are deeply meaningful, encoding in their spatial configurations social values and meanings. As Bill Hillier and Julienne Hanson have argued, 'the ordering of space in buildings is really about the ordering of relations between people'.[2] As a map of those spatial and social dynamics, a floor plan can provide insights unavailable from other sources.

For example, the layout of a floor plan establishes sightlines, determining who or what is visible. A floor plan's circulation paths indicate how a body moves through the space, and where a person might be encouraged to linger or rush, feel comfortable or uneasy. The compression or expansion of space, floor levels and transitions, axial alignment, and the quality of light can also have a powerful psychological effect on the user, and undergird relations of power or intimacy. Finally, a floor plan tells us about the shape and dimensions of spaces, thickness of walls and partitions, cardinal orientation, adjacencies, sequences, and symmetries or asymmetries in the layout. In these spatial elements and patterns, we may find previously unknown pieces of a socio-spatial history. For example, when archaeologist Mary Shepperson

DOI: 10.4324/9780429291739-7

cross-referenced architectural plans of the Mesopotamian city of Ur with natural light conditions, she discovered that the layout of residential areas created 'shadow networks', which protected inhabitants from the fierce summer sun. Residents could carry out domestic work in their courtyards or traverse the city without ever being directly exposed to the sun. Moreover, all inhabitants shared access to this remarkable urban microclimate, suggesting an egalitarian attitude toward a critical resource: shade.[3]

Although Shepperson based her analysis on contemporary drawings by archaeologists, evidence of architectural plans as a representational form exists in Mesopotamia from the late third millennium BCE.[4] In ancient Greece and Rome, architectural sketches have been found carved into floors and walls, such as the incised 'blueprints' discovered by Lothar Haselberger on the unfinished inner walls of the Hellenistic temple of Apollo at Didyma (in modern Turkey).[5] The earliest known architectural plan on parchment dates to the ninth century, the so-called Plan of Saint Gall, which depicts the layout of a large and well-organized Carolingian monastery. The term 'blueprint' derives from Sir John Herschel's introduction in 1842 of the cyanotype printing process, which allowed for the reproduction of architectural drawings. The term is still used, despite the adoption of whiteprints after the Second World War. Today, architectural plans are created with the use of computer software and viewed digitally.

Architectural historians have long turned to architectural plans as historical sources. Because the plans' meanings are not stable over time, historians must take into account the context in which such representations are created. An architectural plan might not be intended as a guide for construction. The Plan of Saint Gall, for example, has been described as 'a memory map for scriptural truths', a visualization of Benedictine Rule: 'thus navigating one's way around the Plan is akin to reading the chapters of Benedict's *Regula*'.[6] An architectural plan may be a political or ideological instrument. Of the design of Jeremy Bentham's Panopticon, Michel Foucault wrote that it 'must not be understood as a dream building; it is the diagram of a mechanism of power reduced to its ideal form ... a figure of political technology'.[7] Architectural plans may be interpreted on symbolical, conceptual, and literal terms. A comprehensive reading of the architectural plan of Daniel Libeskind's Jewish Museum in Berlin, for example, demands an awareness of all three levels at which the design communicates to the viewer.[8] Thus, a historian decoding the layered spatial meanings of a plan will read it in relation to a wide range of other sources – spanning religious texts, sunlight charts, penal codes, excavated objects, oral histories, and a great deal more – to understand the assemblage of lines, dashes, and symbols.

In this chapter, I explore how floor plans, as part of a broader toolkit of sources, including photographs and memoirs, help us to understand the visitor experience of one of the most elite and powerful spaces of the Third Reich: Adolf Hitler's Berghof, his mountain retreat in the Bavarian Alps (Figure 5.1). Although the residence was ostensibly a place of retreat, where the dictator went to recover from his political struggles, the Berghof actually functioned as an unofficial chancellery and military headquarters. Hitler spent almost a

104 *Despina Stratigakos*

Figure 5.1 Heinrich Hoffmann, postcard of the Berghof, c. 1936.
Source: 260-NSA-37, National Archives, College Park, Maryland.

third of his time in power at the Berghof, conducting party and government business as well as a global war from its domestic spaces. Politicians, diplomats, religious leaders, generals, royalty, and press barons, among other influential figures, visited Hitler at the Berghof. A hybrid of private and public spaces, the home enabled mundane and ceremonial functions, as well as secrecy and public exposure. For the German people, Hitler's life in the mountains was both inaccessible – protected by guards and electrical fences – and immediate, rendered highly visible by Nazi propaganda and an eager celebrity media. Until 1939, international press reports about the off-duty Führer were often positive in tone, presenting him as a genial host and good neighbor.

Given the veritable parade of dignitaries to Hitler's mountain home in the 1930s, it is not surprising that there are a number of sources available to the historian seeking to understand their spatial experiences. A small number of foreign visitors published their impressions of the house, both positive and negative, after the war in the form of memoirs.[9] Contemporary propagandistic materials – official photographs, books, and merchandise – are abundant, but present life at the Berghof in highly idealized terms. Eva Braun's home movies and photo albums, which focus on the residence's outdoor spaces, are more candid, yet still opaque. Journalistic accounts have to be taken with a grain of salt: because of the fascination with Hitler and his private life during the Third Reich, even reputable newspapers and magazines published reports from dubious sources. One British journalist repeatedly sold stories recounting fictional visits to the mountain chalet of his 'good friend' Adolf Hitler.[10]

Thus, despite a relative abundance of materials, the Berghof's floor plans, preserved in Munich archives, have an important role to play in forming a more holistic and reliable picture of visitors' experiences. Importantly, two sets of plans exist for the Berghof: one created for the residence's renovation in 1935–36, and the other for an unrealized expansion planned around 1939. I begin by describing how visitors encountered and moved around the Berghof, drawing on floor plans and other sources. I then conclude with a close reading of the unbuilt plans for the residence that Hitler envisioned for himself as the head of a global empire. Taken together, the built and unbuilt plans reveal new aspects of Hitler's self-image and how he wished to be seen by visitors to his home.

The Berghof as performance space

Hitler cared deeply about his domestic spaces.[11] He himself created sketches for the 1935–36 expansion of Haus Wachenfeld, the modest cabin he had purchased in 1933, into the monumental residence he renamed the Berghof. These sketches were adapted by Alois Degano, a Bavarian architect who signed the finished drawings. Hitler also worked closely with Gerdy Troost, a Munich-based designer and powerful tastemaker during the Third Reich, who created interiors for Hitler's public and private buildings. At the Berghof, they collaborated on the home's furnishings as well as on the selection of art for the Great Hall, where Hitler was most often photographed. To his guests, Hitler presented himself as a cultured statesman, with the Berghof acting as the stage set for his performance.

The arrival of distinguished guests to the Berghof was a carefully choreographed ritual that led from the driveway to the Great Hall, the main public room of the house. Visitors arrived by car on the northern side of the house, where they were greeted on the stairs by its master. Together, to the accompaniment of a rolling drum, they climbed a broad flight of steps to a terrace, where they walked past Hitler's black-uniformed SS honor guard, who presented arms, turned right under a covered walkway, and then entered a vestibule through a heavy oak door. The low-ceilinged and dimly lit lobby area featured red marble columns and vaulted arches that reminded one visitor of 'a cathedral crypt'.[12] Toilets and coat racks were located here as well. As the ground-floor plan shows, from the garderobe, the Great Hall could be approached in two ways (Figure 5.2). Continuing to the end of the vestibule,

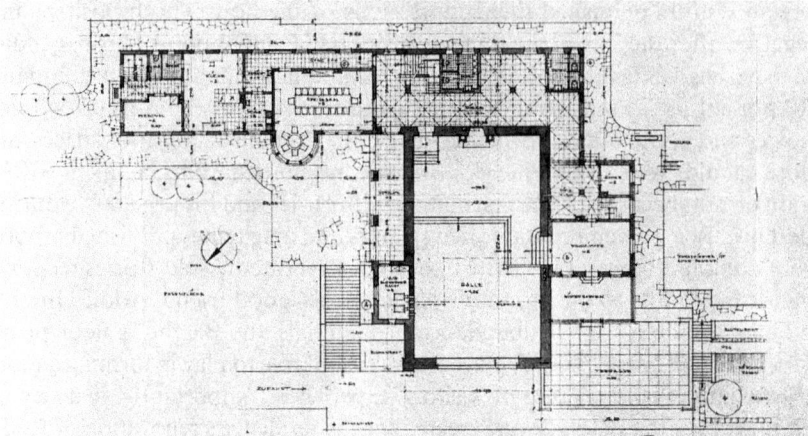

Figure 5.2 Alois Degano, ground-floor plan for the expansion of Haus Wachenfeld, dated 16 November 1935 and approved by local building authorities on 22 January 1936.

Source: BPL Berchtesgaden 1936–14, Staatsarchiv, Munich.

visitors could have turned right, proceeded down a corridor, and come to another small lobby, from which they would have entered a small living room. A large rectangular opening covered with a heavy curtain reminiscent of a theater, and foreshadowing the performance of the statesman within, admitted the visitor to the Great Hall. Alternately, by retracing their steps, they could have entered the Great Hall more directly, through a door immediately to the right as one entered the house. In each case, the progression from low-ceiled, more compact, and darker spaces to the vast, window-lit expanse of the Great Hall made the psychological impact of the room all that much greater. Thus, as the floor plan demonstrates, the architecture of the house mounted a powerful 'reveal' as visitors encountered its most prestigious space, the Führer's audience room. Moreover, the plan provides the historian with evidence that this experience – meant to induce anticipation and awe in the visitor – was encoded into the building's structure by the patron and the architect. With this piece of information, we begin to understand how Hitler intentionally manipulated the psychological potential of his domestic spaces to his advantage.

The Great Hall was an oversized rectangle of a room with a floor on two levels. It measured over 42 feet wide by 74 feet long and 18 feet high, the size of a small gymnasium, and was intended as a multi-functional representational space where Hitler would receive foreign and domestic guests, hold meetings with his ministers and generals, host official functions, entertain, and socialize. On the southern end of the room was a grand marble fireplace, and opposite it a panoramic window measuring over 28 feet long by 12 feet high (Figures 5.3 and 5.4). Hitler, it has been said, pioneered the work-from-home movement, and the Great Hall was at the center of his intention to rule an empire from the comfort of his living room sofa.[13]

Architectural drawings 107

Figure 5.3 Heinrich Hoffmann, postcard of the Great Hall, c. 1936. The marble fireplace is visible at the far end of the room. To the right, through the opening, is the small living room. At the far right, a tapestry hides a movie screen.

Source: Heinrich Hoffmann Photographic Archive, Bavarian State Library, Munich.

Figure 5.4 Heinrich Hoffmann, view of the window in the Great Hall, c. 1936. Note the large map table in front of the window, and to the left, the globe satirized by Charlie Chaplin in his 1940 film, *The Great Dictator*.

Source: Heinrich Hoffmann Photographic Archive, Bavarian State Library, Munich.

Precedents for such a room can be found in medieval great halls. These multi-purpose communal spaces displayed, through design elements such as the length of the room, height of the ceiling, or size of the central hearth, their lord's status. Hitler may have drawn on this domestic model, associated with feudal power, to reinforce his connection to a local legend: folktales about Mount Untersberg, framed by the Great Hall's majestic window, claimed that the court of Emperor Charlemagne (or, sometimes, Barbarossa) slept within the mountain, awaiting the sign that would awaken the king and his knights to fight a cataclysmic battle that would usher in a glorious new Reich. As Hitler told Albert Speer, 'you see the Untersberg over there. It is no accident that I have my residence opposite it'.[14] The view of the mountain brought the medieval warrior-king's symbolic presence into the room, but perhaps the Great Hall itself was meant to mirror his imagined court buried deep within the mountain.

At the same time, Hitler may have had another, more modern, architectural precedent in mind for his Great Hall, in keeping with his vision of himself as an artist-politician. Namely, he may have looked to the great artists' ateliers of the later 19th and early 20th centuries: large spaces that encompassed professional, domestic, and social activities. Hitler admired Hans Makart's famous Viennese studio and its accumulated layers of 'memories', although he himself sought to instantly fabricate self-mythologizing spaces.[15] A celebrated 1885 painting of Makart's studio before its sale, created by Rudolf von Alt, depicts one of the huge historical canvases at which the artist excelled. Believing that politics was art, Hitler may have conceived the large window in the Great Hall, which pictured 'German' nature at its most eternal and invincible, as his own canvas, and beneath it, the fourteen-foot long Untersberg marble and oak table, where he studied maps with his generals, as his easel.[16] This piece of furniture also evokes the high table of medieval great halls, at which the lord took his meals, conducted business, and dispensed justice.

The great window was the showcase of the room, and not only because of its size and magnificent views. The window, made of a single pane of glass, could be rolled down like a car window, disappearing into the floor and transforming the window frame into an enormous void. Unity Mitford, a fanatical English follower who visited Hitler at the Berghof in July 1938, told her sister Diana, wife of Oswald Mosley, the leader of the British Union of Fascists, that the effect was surreal, 'more like an enormous cinema screen than like reality'.[17] Thomas Jones, a former senior civil servant and deputy secretary to the cabinet, accompanied former British Prime Minister David Lloyd George, an admirer of Hitler, on an ostensibly private visit to the Berghof in September 1936, not long after the renovation had been completed. As Jones reports, the Great Hall made a strong impression,

> but what fascinated us all was the vast window at the north end, or rather the absence of it, for it had been wound up or let down out of sight into a groove and there was nothing between us and the open air and sky and mountains and a view of Salzburg in the distance.[18]

Indeed, it inspired Lloyd George to tear out a wall in the library of his country house at Churt in Surrey and replace it with a similar panoramic window.[19]

Hitler expected his visitors to be dazzled by this architectural feat of commanding nature, no matter the circumstances. In November 1937, Lord Halifax, then lord president of the council in Neville Chamberlain's administration, visited Hitler at the Berghof to address Anglo-German relations, which had deteriorated as Hitler's hopes for a British alliance faded and Germany's expansionist rumblings pushed the limits of British tolerance for the Nazi regime. Amid the tense discussions, Hitler 'insisted' on a performance of his window's vanishing act for the British delegation. Ivone Kirkpatrick, the head of chancery at the British embassy in Berlin, who was present at the talks, later recalled that 'a couple of stalwart S.S. men doubled into the room, fixed things like motor-car starting-handles into the sockets and wound violently. The whole structure sank noiselessly into the floor, giving the room the appearance of a covered terrace.' Kirkpatrick was unmoved and left the Berghof with the memory of Hitler having behaved like a sullen, spoiled child.[20]

Hitler's power to order limitless space was also on view inside the Great Hall. Gerdy Troost embraced the idea of the panorama as the defining experience of the capacious hall. Writer Matthew Stadler has argued that 'Hitler loved panoramas' and

> his favorite rooms were organized around their panoramic views .... Where there was no view out, Hitler organized interior views – the rooms became stage sets for the dramas of the Führer. Their vastness compelled guests to watch rather than touch or feel the closeness of sociality. This spectative logic was the only one Hitler understood.[21]

As photographs (Figure 5.3) and floor plans indicating furniture placement reveal, the room's long walls became their own kind of panoramas. Furniture was arranged along or near the walls so as to create elongated views with nothing to block (or draw) the eye at the center of the room. The intended effect was not a concentration of attention, but its dispersal as the eye roamed over the broad scenes, whether of the mountain peaks seen through the window or the interior vistas of furniture and art.

The room's emphasis on visuality, what Stadler describes as a 'spectative logic', was also present in its multiplicity of screens and canvases, real or allegorical. Near the great window, a 17th-century Gobelin, itself a woven canvas, hid the openings of an external projection booth, and on the opposite side of the room, another tapestry concealed the screen. Before the war started, Hitler would watch films here nightly with his guests and house staff. The panoramic window could itself be experienced as a movie screen, as captured in Mitford's description of the cinematic unreality of the open-air view. The paintings and sculpture that filled the southern end of the room similarly encouraged a museal gaze in the Führer's 'best' room, characterized by an attentive and reverential form of looking.

As is often the case in traditional art museums, the Great Hall's privileging of vision resulted in a lack of attention to the corporeal. Indeed, a number of memoirs comment on the body's unease in its spaces. Jones described how when Hitler received Lloyd George at the Berghof, 'Hitler sat on an easy chair and L.G. uncomfortably on a couch which had no back'.[22] In fact, the couches in the Great Hall did have backs, but the seats were so deep that guests were forced to perch on their edges. Speer, too, criticized the uncomfortable sofas and the 'inept arrangement of the furniture', which discouraged a common conversation and reduced each person to talking 'in low voices with his neighbor'. He also noted how gasoline fumes blew into the Great Hall when the window, which was situated directly above the garage, was open. While Speer blamed Hitler's poor design skills for assaulting his visitors' noses, it also testifies to the dictator's focus on visual effects.[23] Other public parts of the house displayed a similar emphasis, including the external and internal vistas incorporated in the long and narrow dining room, and the outdoor terrace, on which Hitler installed a telescope.

Comfort in the Great Hall was, in any case, secondary to its main purpose: to impress. In this room, Hitler greeted kings and princes, prime ministers and marshals, religious leaders, secretaries of state, and ambassadors.[24] It was where he negotiated with the powers of Europe that stood between him and his vision of a Greater German Reich. Like the renovation itself, the Great Hall was meant to convey the 'new' Hitler – not the ex-corporal who roused rebels in beer halls or the dictator who cut down his opponents in cold blood, but rather a powerful, cultivated, and, above all, trustworthy statesman. It was the stage on which he performed this new role and invited others to respond to him accordingly. Jones, describing the meeting between Lloyd George and Hitler, conveyed the intended effect of the room when he noted in his diary how it 'made a great impression at once upon the visitors and gave a sense of dignity to the proceedings'.[25]

In inclement weather, tea, if offered, was served in the Great Hall, as it was to Chamberlain on 15 September 1938. The British prime minister had travelled to the Obersalzberg to discuss terms that would diffuse Hitler's threat to invade Czechoslovakia, leading two weeks later to the concessions of the Munich Accord. The *Anglo-German Review*, a British journal devoted to fostering 'good understanding and co-operation between Great Britain and Germany' and to delivering the message that the Nazi government desired peace, featured the historic refreshments 'in the Führer's famous chalet' on the cover of its September 1938 issue, making a choice thereby to emphasize Hitler's hospitality over his bullying tactics, which had forced the sixty-nine-year-old Chamberlain to the Alps in a desperate bid to avoid war (Figure 5.5).[26]

If invited to join the Führer for lunch, the visitor reentered the vestibule and turned east into the dining room (Figure 5.2). As the furniture illustrated on the floor plan shows, the long and narrow room, perpendicularly positioned to the Great Hall, could seat 18 at the main table and six in the semicircular alcove, the latter a cozy spot where early risers could breakfast (Hitler took his breakfast separately, in his bedroom). The shape of the room and of the main table tell the historian that this was conceived of as a formal space

Architectural drawings 111

Figure 5.5 Photograph of Neville Chamberlain (left of Hitler) and other guests having tea with Hitler in the Great Hall of the Berghof, as seen on the cover of the 24 September 1938 issue of *Anglo-German Review*. The visit occurred on 15 September 1938, when the British prime minister travelled to the Obersalzberg to discuss the international crisis brewing over Hitler's insistence on invading Czechoslovakia.

that communicated restraint and power. Considered in conjunction with the Great Hall, it suggests that the power dynamics experienced in the audience room would have been maintained here, if on a different scale. In other words, whatever psychological effect was constructed for the distinguished visitor in the Great Hall was not meant to be relieved over lunch: the physical staging of the visit remained formal and controlled. Hitler sat in the middle of the long table with a view of the Untersberg, and he and his guests were served by SS men in white uniforms, which added to the formal effect. According to Speer, the decoration of 'the dining room was a mixture of artistic rusticity and urban elegance of a sort which was often characteristic of country houses of the wealthy'.[27] Meals were prepared in the kitchen adjacent to the dining room.

The floor plan of the dining room and of its furniture placement can be thought of as the description of a stage set meant to enable a particular kind of performance. Analyzing the plan, the historian derives clues as to what was intended and encouraged. At the same time, the historian cannot deduce the tone of a specific luncheon from the floor plan alone. The rigid formality of the dining room could have been undermined, for example, by raucous jokes or behavior. Indeed, the expectations a visitor may have had based on the visual and architectural appearance of a room could have been used against him or her to create a psychological shock when expectations and experiences did not align.

Now and then, official visitors would climb the stairs in the lobby to the second floor, walk down an impressive, art-lined corridor that measured 17 feet in width, and enter Hitler's private study for more confidential discussions. The spacious room, with three French windows leading onto a balcony and built-in book shelves on either end, was located directly above the Great Hall. At the center of the room was Hitler's desk, behind which hung oil portraits of his parents (based on photographs). Across from it was a cream-colored tile stove, painted with monochromatic green figures by Munich artist Sophie Stork. On the western side of the room, a sitting area was arranged in front of a fireplace, over which hung a painting of a young Frederick the Great by Antoine Pesne.[28] A contemporary German design journal described 'the impression of warm domesticity' conveyed by this room, although it must not have felt that way to Kurt Schuschnigg, the Austrian chancellor, who was browbeaten here by his host for hours on 12 February 1938 in an effort to make him agree to violate his own country's sovereignty.[29]

Significantly, all three of the public or semipublic spaces in the Berghof used for important visitors – the Great Hall, dining room, and study – were those most closely associated in traditional domestic architecture with masculinity. These were typically the representational spaces of the house, where business was conducted or the owner's wealth and power were displayed. When the Berghof was completed in early July 1936, it had only been two years since the 'Röhm Putsch', or Night of the Long Knives (30 June to 2 July 1934), when Hitler had political adversaries and old enemies murdered, including many SA leaders and its chief, Ernst Röhm, whom Hitler accused of planning a coup. Röhm's homosexuality was also repeatedly alluded to as a justification, and on 28 June 1935, the first anniversary of the purge, the regime broadened the legal definition of punishable homosexual offenses and intensified its persecutions of so-called sexual degenerates. These actions helped to alleviate suspicions that Hitler shared Röhm's sexual orientation, although whispered rumors persisted.[30] In the rooms that Hitler chose to display himself to those he most wished to impress, he thus communicated not only his power and cultivation, but also his 'correct' masculinity, an important part of the appearance of 'normality' that he performed in the mid-1930s. In the apparent absence of a flesh-and-blood female companion (Germans would only learn about the existence of Eva Braun after the war), sensual representations of women in the Great Hall, placed prominently near the fireplace and often visible in published photographs – such as the reclining nude goddess in *Venus and Cupid* by the Italian Renaissance painter and Titian pupil Paris Bordone, and Anselm Feuerbach's 1862 portrait of his model and muse *Nanna*, said to have been Hitler's favorite painting – further reinforced Hitler's heterosexual bachelor domesticity as well as his cultural sophistication.[31]

Revelations of the unbuilt Berghof

In 1939, Hitler commissioned a further expansion of the Berghof, which extended the main wing of the house eastward and added a new bay window as well as a separate driveway and entrance for deliveries.[32] But it appears that originally he had had a far grander scheme in mind, which would have added a second, massive wing to the eastern side of the house. A little-known and largely unpublished portfolio of drawings in the Bavarian Central State Archive in Munich contains unrealized designs for the proposed addition.[33] While these remained paper schemes, they provide important insights into how Hitler imagined his domestic self and how he wished it to be perceived by others.

The collection of unbuilt designs contains numerous elevations, sections, and floor plans representing different versions of the proposed new wing. Although unsigned and undated, they were created around 1939, when Hitler extended the eastern wing of the house. The quantity and quality of the work reveal that the idea of an expansion on a grand scale was not a mere passing whimsy, but a carefully considered project, with considerable resources expended on its conceptualization and planning. If completed, it would have significantly transformed the architectural form and experience of the house.

Envisioned was a second wing that would have projected eastward from the Berghof's main building, but would have been flush with its façade. It would have stood parallel to and directly in front of the eastern wing that was completed in 1936, blocking its views of the mountain and creating a courtyard. This layout also would have restructured the visitor's experience of arrival by creating a grand new entrance, with a circular roundabout or plaza for cars, at the far eastern end of the wing.[34] In one version of the unbuilt floor plans, visitors climbed a flight of broad steps to enter a large entrance hall (Figure 5.6). Turning right, they passed through a room with an intricately arched ceiling into a cloakroom with adjacent bathrooms. From here, visitors walked the length of the wing before arriving at the Great Hall, which they would have entered through a new door located directly beside the large window. This extended door-to-Führer path, intended perhaps to build suspense, recalls the processional route that Speer created in the New Chancellery in Berlin, although without the same type of architectural splendor along the way. Instead, visitors would have walked past a long panorama of mountain views, glimpsed through a series of windows, before entering the Great Hall and experiencing the largest panorama of all. Adjacent to the hallway would have been five enfilade rooms (creating another interior panorama), intended

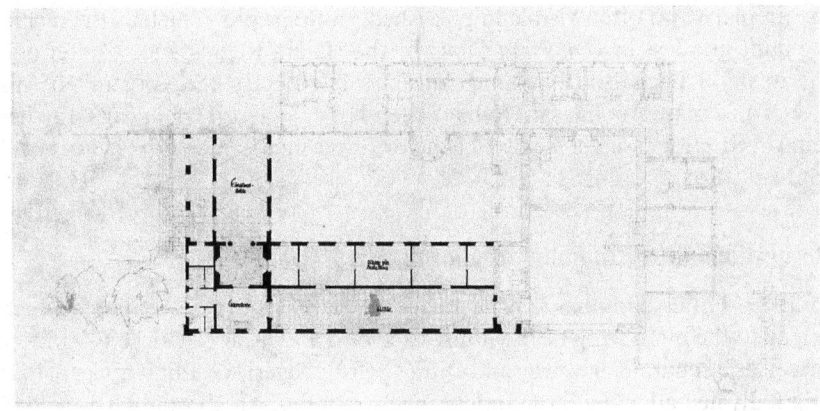

Figure 5.6 An unsigned ground-floor plan for the expanded Berghof, not built, n.d.
Source: OBB KuPL 5454, Bayerisches Hauptstaatsarchiv, Munich.

for the use of the visitors' entourage. Thus the ground floor of the second wing would have been devoted to visitors, and especially to shaping a more dramatic journey of arrival.[35]

But the portfolio of unbuilt drawings contains a bigger surprise. Given the attention lavished on the Great Hall in the prewar press about the Berghof, it is astonishing to discover that Hitler had planned a showpiece that might have overshadowed it: namely, a monumental library on the second floor (Figure 5.7). As marked on the unbuilt floor plans, the library would have been a grand chamber – far longer than the Great Hall – on two levels, with a capacity to hold a remarkable 61,000 books, comparable to some public and college libraries. The entrance to the library would have been from the eastern end of Hitler's study. A door would have opened onto the upper level of the library, from which a monumental staircase would have led to the floor below (a smaller staircase was planned for the opposite side of the room as well).[36] The centrality of the library in this conception of domestic identity suggests Richard Wagner's home, Wahnfried, and his prized book collections as a possible model.[37] Typically associated with masculine power and creativity, the library would have bolstered Hitler's masculine image. But above all, it would have presented him not only as a powerful leader, but also a cultivated individual. While this was similarly achieved through the attention to artwork and music in the Great Hall, the inclusion of a major library in the house would have strongly reinforced the image of the Führer as a learned man. Indeed, if the project was developed shortly before the launch of an expansionist war, it suggests another way in which Hitler intended to present himself as the new Charlemagne of Europe. Toward the end of the eighth century, the Carolingian emperor compiled a great and influential court library, which historians credit with preserving and diffusing much of the heritage of the classical world.[38]

Unfortunately, the portfolio does not contain any conclusive evidence as to why Hitler abandoned the idea of a second eastern wing. The multiple

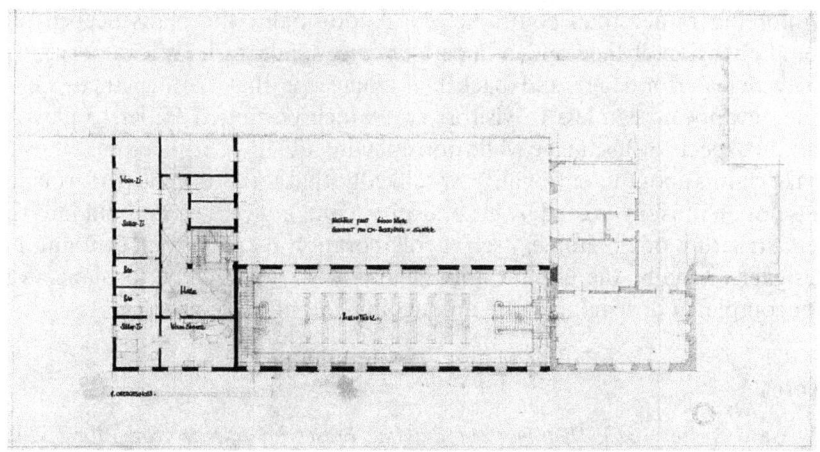

Figure 5.7 An unsigned second-floor plan for the expanded Berghof showing the library (*Bibliothek*), not built, n.d. Immediately to the right of the library staircase was the entrance to Hitler's study.

Source: OBB KuPL 5458, Bayerisches Hauptstaatsarchiv, Munich.

versions drawn of the northern façade reveal the designer struggling to balance the forms and proportions of the extension. One proposed solution would have built a house-like facade and terrace on the eastern end of the wing to mirror those same structures on its western side.[39] Other versions tried to distract from the massiveness of the wing by adding balconies, bay windows, or Bavarian-style painted decoration to the facade.[40] But no matter how many variations were created, it was clearly difficult to avoid a bastion-like effect, and Hitler surely would not have wanted to evoke the nearby Salzburg castle, which represented all too clearly autocratic rule. Looking back in 1942 on the construction of the Berghof, Hitler confessed that when he visited the building site in 1935, 'the dimensions of the house made me somewhat afraid it would clash with the landscape. I was very glad to notice that, on the contrary, it fitted it very well. I had already restricted myself for that reason – for, to my taste, it should have been still bigger'.[41] Perhaps Hitler later regretted his initial caution, only to again decide, when expanding the house in 1939, in favor of a less grandiose option. The slope of the mountainside also may have been a factor in deciding to scrap the plans.[42] In any case, the existence of these unbuilt drawings reveals that Hitler struggled with how to best present his domestic self to the world. On the eve of war, the Führer, it seems, was still casting about for a fitting setting and legend for his anticipated role as emperor.

As a historian of Hitler's domestic spaces, I seek in my work to complicate our views of National Socialist architecture by expanding our understanding of the regime's ideological spaces. In the analysis of the role of the Berghof within this revised history, its floor plans represent a critically important and unique source. They allow us to grasp the residence as an integrated whole serving both political and domestic functions, and to see these as mutually reinforcing rather than contradictory. Additionally, the plans deepen our comprehension of the house as a performance space, with Hitler at its center. They reveal frontstage and backstage zones for that performance, which made it appear seamless to visitors and which permitted Hitler to play up certain aspects of his image while downplaying others. The plans enable us to make claims about intentionality: specifically, that this domestic performance was not an illusion created by his publicity team, but rather was built into the very structure of the home, itself a collaboration between the client and his designers. Finally, the unbuilt floor plans reveal that this performance was not completed or fixed, but was an ongoing and mutable process.

Notes

1 Ward Bucher (ed.), *Dictionary of Building Preservation*, New York: John Wiley and Sons, 1996, p. 188.
2 Bill Hillier and Julienne Hanson, *The Social Logic of Space*, Cambridge: Cambridge University Press, 1984, p. 2.
3 Mary Shepperson, 'Planning for the Sun: Urban Forms as a Mesopotamian Response to the Sun', *World Archaeology* 41, 2009, 363–78.
4 Béatrice André-Salvini, 'Seated Statue of Gudea: Architect with Plan', in Joan Aruz (ed.), *Art of the First Cities: The Third Millennium B.C. from the Mediterranean to the Indus*, New Haven: Yale University Press, 2003, p. 427.
5 Lothar Haselberger, 'The Construction Plans for the Temple of Apollo at Didyma', *Scientific American* 253, 1995, 126–33.
6 Lynda L. Coon, *Dark Age Bodies: Gender and Monastic Practice in the Early Medieval West*, Philadelphia: University of Pennsylvania Press, 2011, p. 165.
7 Michel Foucault, *Discipline and Punish: The Birth of the Prison*, trans. Alan Sheridan, London: Penguin, 1977, p. 205.
8 Paul B. Jaskot, *The Nazi Perpetrator: Postwar German Art and the Politics of the Right*, Minneapolis: University of Minnesota Press, 2012, pp. 127–65; Gavriel D. Rosenfeld, *Building after Auschwitz: Jewish Architecture and the Memory of the Holocaust*, New Haven: Yale University Press, 2011, pp. 179–98; Bernhard Schneider, *Daniel Libeskind: Jewish Museum Berlin: Between the Lines*, Munich: Prestel, 1999.
9 See, for example, Thomas Jones, *A Diary with Letters, 1931–1950*, London: Oxford University Press, 1954; and Ivone Kirkpatrick, *The Inner Circle: Memoirs of Ivone Kirkpatrick*, London: Macmillan, 1959.
10 Despina Stratigakos, *Hitler at Home*, New Haven: Yale University Press, 2015, pp. 194–220.
11 Ibid. My analysis of the visitor experience of the Berghof explored in this chapter draws from chapter 4: 'From Haus Wachenfeld to the Berghof: The Domestic Face of Empire', pp. 68–106.
12 Jones, *Diary with Letters*, p. 248.
13 Arthur H. Mitchell, *Hitler's Mountain: The Führer, Obersalzberg and the American Occupation of Berchtesgaden*, Jefferson: McFarland, 2007, p. 44.
14 Albert Speer, *Inside the Third Reich*, trans. Richard and Clara Winston, New York: Touchstone, 1997, p. 86.
15 H.R. Trevor-Roper (ed.), *Hitler's Table Talk, 1941–1944: His Private Conversations*, trans. Norman Cameron and R.H. Stevens, New York: Enigma Books, 2008, p. 184.

16 Ralph Gleis (ed.), *Makart: Ein Künstler regiert die Stadt*, Munich: Prestel, 2011, p. 27; Otto K. Werckmeister, 'Hitler the Artist', *Critical Inquiry* 23, 1997, 270–97; Frederic Spotts, *Hitler and the Power of Aesthetics*, New York: Overlook, 2002, pp. 43–94; Eric Michaud, *The Cult of Art in Nazi Germany*, trans. Janet Lloyd, Stanford: Stanford University Press, 2004, pp. 1–73; Christa Schroeder, *He Was My Chief: The Memoirs of Adolf Hitler's Secretary*, trans. Geoffrey Brooks, London: Frontline, 2009, p. 164; Drawing of the table, Berghof portfolio of drawings, Gerdy Troost Papers, Ana 325A.V.4, #2, Bavarian State Library, Munich.
17 Unity Mitford to Diana Mitford, 18 July 1938, *The Mitfords: Letters between Six Sisters*, ed. Charlotte Mosley, New York: Harper, 2007 p. 127.
18 Jones, *Diary with Letters*, p. 249.
19 Albert James Sylvester, *The Real Lloyd George*, London: Cassell, 1947, p. 235.
20 Kirkpatrick, *Inner Circle*, 96–7; on Halifax's visit to Germany, see David Faber, *Munich, 1938: Appeasement and World War II*, New York: Simon and Schuster, 2008, pp. 9–45.
21 Matthew Stadler, 'Hitler's Rooms', *Nest: A Magazine of Interiors* 22, 2003, 64, 66.
22 Jones, *Diary with Letters*, p. 244; on the couches, see also Schroeder, *He Was My Chief*, p. 164.
23 Speer, *Inside the Third Reich*, pp. 90, 86; on the visuality of the Great Hall, see also Lutz Koepnick, *Framing Attention: Windows on Modern German Culture*, Baltimore: Johns Hopkins University Press, 2007, pp. 163–99.
24 For a list of prominent foreign visitors to the Berghof, see Volker Dahm et al. (eds.), *Die tödliche Utopie: Bilder, Texte, Dokumente, Daten zum Dritten Reich*, Munich: Institut für Zeitgeschichte, 2008, pp. 133–4.
25 Jones, *Diary with Letters*, p. 249.
26 C.E. Carroll, 'Editorial Notes', *Anglo-German Review*, 1, 1936, 5; *Anglo-German Review*, 2, 1938, cover; on Chamberlain's visit to Berchtesgaden, see Faber, *Munich 1938*, pp. 272–96.
27 Schroeder, *He Was My Chief*, p. 155; Gerdy Troost, response, Karen Kuykendall questionnaire, no date, Karen Kuykendall Papers, MS 243, Special Collections, University of Arizona, Tucson, Arizona; Speer, *Inside the Third Reich*, p. 88.
28 Birgit Schwarz, *Geniewahn: Hitler und die Kunst*, Vienna: Böhlau, 2009, pp. 124–7.
29 'Die Innenräume des Berghofes', *Innen-Dekoration* 49, 1938, 60–1, 63.
30 Lothar Machtan, 'Hitler, Röhm and the Night of the Long Knives', *History Today* 51, 2001, 17; Bella Fromm, *Blood and Banquets: A Berlin Social Diary*, New York: Kensington, 2002, p. 92.
31 Schwarz, *Geniewahn*, pp. 159–64.
32 Josef Neul, *Adolf Hitler und der Obersalzberg*, Rosenheim: Deutsche, 1997, p. 85.
33 OBB KuPL 5441, 5450-5476, Bayerisches Hauptstaatsarchiv, Munich.
34 OBB KuPL 5441, 5471, Bayerisches Hauptstaatsarchiv, Munich.
35 OBB KuPL 5454, Bayerisches Hauptstaatsarchiv, Munich.
36 OBB KuPL 5456-5459, Bayerisches Hauptstaatsarchiv, Munich.
37 Heinrich Habel, *Festspielhaus und Wahnfried*, Munich: Prestel, 1985, pp. 536–7.
38 Bernhard Bischoff, *Manuscripts and Libraries in the Age of Charlemagne*, trans. Michael Gorman, Cambridge: Cambridge University Press, 2007, p. 76.
39 OBB KuPL 5468, Bayerisches Hauptstaatsarchiv, Munich.
40 OBB KuPL 5462-5463, 5465-5467, Bayerisches Hauptstaatsarchiv, Munich.
41 Trevor-Roper, *Hitler's Table Talk*, p. 161.
42 Ulrich Chaussy and Christoph Püschner, *Nachbar Hitler: Führerkult und Heimatzerstörung am Obersalzberg*, Berlin: Links, 2007, p. 111.

Part II
Exploring spaces

6 Ships

Matthew Ylitalo and Sarah Easterby-Smith

Introduction

Spatial history offers tremendous opportunities for maritime social history. As social historians, we seek to identify and recover lives, experiences, and cultural practices of people who lived and worked in maritime environments. In this chapter we argue that a spatial approach can afford alternative ways of understanding the maritime activities and experiences of those at sea. Furthermore, we make a case for interpretively redeploying sources that are generally underused in maritime social history: ships' official documents.[1] Reading ship records alongside more commonly consulted materials like diaries and visual sources can clarify how shipboard space was used, who used it, and how and in what circumstances the use of that space might be altered.

Underpinning our chapter is an understanding that space has had a historically important effect on work and life at sea.[2] We focus on the seafarers who worked or travelled aboard a particular type of 19th-century British merchant vessel, an Arctic whaler. Our analysis considers the fluctuating physical and cultural spaces that existed within the S.S. *Esquimaux*, a Scottish whaling barque which operated during the second half of the 19th century. Drawing inspiration from Michel Foucault's notion of the 'heterotopia', we examine how during a typical voyage the crew would construct and experience multiple kinds of space within the same vessel. Some of the real or perceived spaces that emerged aboard the *Esquimaux* offered exciting possibilities for new social relationships. Other kinds of space, however, significantly constrained the behaviour of the crew, restricting their activities to those prescribed by a distant bureaucratic elite.

What did Foucault mean by 'heterotopia'? Like his contemporary Henri Lefebvre, Foucault considered how physical spaces have defined historical social relations, and how those social relations, equally, have defined our perception of space. He did so by examining the interplay between representations of space, meanings invested in space, and the actual experiences of the people who lived within or passed through certain spaces. Observing that these sets of relations have changed over time, Foucault noted that historical epochs have thus been defined by different attitudes towards space. Space thus has its own history. Structuralist thinkers identified, furthermore, that

DOI: 10.4324/9780429291739-9

cultures tend to conceive of space in terms of binary oppositions. In most modern European cultures, for example, private space has often been opposed to public space; spaces for leisure are countered by spaces for work.[3] Foucault went further and identified other spaces that operated as counter-sites, which he labelled 'heterotopias'. They were spaces where a culture's normal binary oppositions 'are simultaneously represented, contested and inverted'.[4]

Foucault's work acted as the inspiration for this chapter because he suggested that maritime or riverine vessels were paramount examples of spaces in which normal relations are reversed or disrupted:

> The boat is a floating piece of space, a place without a place, that exists by itself, that is closed in on itself and at the same time is given over to the infinity of the sea .... The ship is the heterotopia *par excellence*.[5]

Although Foucault did not set out a method for analyzing these multi-layered spaces, the notion of heterotopia offers several possibilities for thinking critically about maritime social history. The portrayal of the heterotopic ship reminds us, for example, that while these vessels were self-reliant contained spaces, they were also wholly dependent on and subject to the environments that surrounded them. Ships, then, are spaces that are simultaneously secluded *and* open, vulnerable to incursion from the natural and human worlds.

How might attention to the shifting uses of space and the multi-layered perceptions of space enhance our understanding of life at sea? A brief consideration of the working life of the *Esquimaux* offers an initial insight.

Built in 1865 by Alexander Stephen and Sons of Dundee, Scotland, the *Esquimaux* (Figure 6.1) embodied the type of hybridity that it also 'lived'. Constructed of wood and heavily reinforced, the *Esquimaux* used both sailing masts and a 70-horsepower auxiliary steam engine. The flexibility of this design allowed the vessel to sail or steam through open water as well as navigate through pack ice. Furthermore, during each distinct phase of a voyage, the crew manipulated the spaces of the ship to suit their particular needs. Every year between 1866 and 1900, the *Esquimaux* set out on a transatlantic voyage from Dundee, usually in late January or early February, as a sea-going vessel, just like other regular merchant marine vessels (Figure 6.2). Upon reaching Newfoundland, however, the crew would physically transform the barque. Whaleboats were removed, punts were added, extra planking was laid down, and rough wooden bunks suitable for sleeping up to 300 additional Newfoundland 'swilers' were constructed between decks. Between March and April, the *Esquimaux* worked as a 'sealer', steaming through fields of ice in its search for harp and hood seals on the floes off the Newfoundland and Labrador coasts.[6] By May, the ship would have been converted once more into a 'whaler' for spring and summer whale hunting in the Davis Strait and Baffin Bay along the west coast of Greenland. In autumn, the *Esquimaux* changed form and techniques so that its crew could 'rocknose' off the ice floes mainly along the east coast of Baffin Island. After finishing the hunt for bowhead and right whales and anything else of value that

Figure 6.1 Whaling barque *SS Esquimaux* at ice edge, 1890s, with her reflection visible in calm water in the foreground. Note the steam funnel as well as the 'crow's nest' (lookout platform) fastened to the main mast, with man inside.

Source: The McManus Art Gallery and Museum, Dundee, #1968-78-2. Reproduced courtesy of Dundee Art Galleries and Museums.

they could catch or obtain, the *Esquimaux* would stow its crow's nest, whaleboats, and hunting paraphernalia, rig the vessel for going to sea, and return to Dundee, most often terminating the voyage in October or November. The malleable design of vessels like the *Esquimaux* meant that its crew could convert their barque quickly and with relative ease. Each mutation suited it precisely to a new economic activity and different environmental conditions.

Researching how the *Esquimaux*'s crew responded to these shifting spaces poses a significant challenge. The first section of our chapter considers the kinds of sources that can help to re-create the experiences of sailors and others living or working within the space of the ship. We then examine the analytical possibilities offered by Foucault's notion of heterotopia, discussing the significance of shipboard mobility and addressing two key features that characterize British maritime heterotopia: the physical malleability of shipboard space and 19th-century bureaucratic efforts to regulate and control that space.

Sources

In theory, each voyage for every British merchant vessel that sailed after 1850 was officially documented for the Board of Trade. At every major port

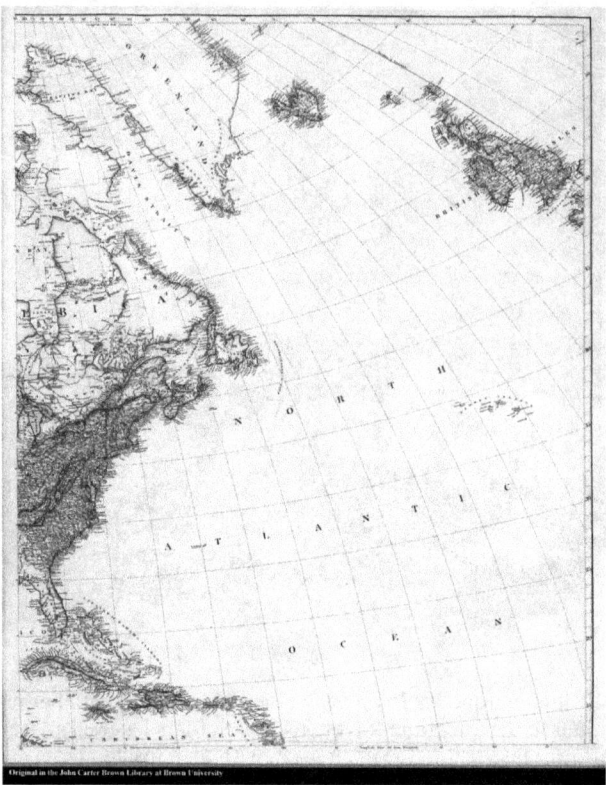

Figure 6.2 Map of the North Atlantic, showing Newfoundland, Baffin, and Greenland. Dundee, which is on the east coast of central Scotland, is indicated on the map but is not named. The city is due west from Bell Rock, on the north bank of the inlet (Firth of Tay).

Source: Aaron Arrowsmith, *Map of America*, London, 1822. The John Carter Brown Library, JCB Map Collection, C-6921. Reproduced courtesy of the John Carter Brown Library at Brown University.

throughout the British Empire, the Mercantile Marine Office (MMO) served as the administrative clearinghouse for all seafarers and apprentices going to sea. The superintendent of an MMO verified competency certifications for masters and mates, witnessed the engagement or discharge of foreign-going crews, oversaw the payment of seafarers' wages, inquired into deaths and injuries at sea, and upon completion of the voyage, received and reviewed all official ship documents before forwarding them to the Registrar General of Shipping and Seamen (RGSS) in London.

At sea, shipmasters were required to maintain 'Crew Agreements' and official ship logs, among other documents.[7] Crew Agreements were government-arranged contracts between the master, who represented the interests of the shipowners, and the hired crew.[8] Official ship logs recorded any incidents that occurred during a voyage which affected the well-being of the crew or craft.

Official Notices featured select instructions, policies, or announcements often extracted from parliamentary acts and published with Crew Agreements and official ship logs by the Board of Trade. These documents can appear rather tedious: they contain pages of information but lack overtly narrative elements. As a result, merchant marine Crew Agreements have been underused by social historians.[9] However, ships' administrative documents constitute some of the most prolific and complete records available in maritime research, and they offer valuable insights into crews' lives. They are key sources for understanding how shipboard space was perceived and inhabited.

Maritime social historians have tended to seek out narrative sources, especially diaries, journals, and correspondence, to access the social histories of seafarers and their passengers.[10] A notable corpus of such materials exists for the latter part of the *Esquimaux*'s life. After more than 30 years' service as a commercial whaler, in 1899, the *Esquimaux* was purchased by Andrew Barclay Walker (1865–1930) and re-designated as a 'pleasure yacht'. That year, Walker hired a truncated whaling crew to take him and several associates on an Arctic sport-hunting venture. He published his diary ten years later as *The Cruise of the Esquimaux*.[11] Written narratives such as this are wonderfully evocative of life at sea, but also carry their own interpretive challenges. Walker's publication, for example, claims an authoritative position but presents a distinctly non-mariner's point of view.

In what follows, we read Walker's published diary against the official maritime records relating to the *Esquimaux* and its crew. The empirical novelty of our approach lies in our insistence that self-evident narrative sources such as those produced by Walker should be combined with a ship's administrative documents. The narrative elements of the latter are less obvious but provide insight into the worlds of the non-hegemonic members of a ship's crew.[12] By contrast, relying on a single kind of source can obscure or flatten the dynamism that characterized shipboard life, creating a static perception of the ship. Moreover, our analysis shows how the concept of heterotopia can further an understanding of the relationship between shipboard space and mariners' lives. The documents relating to the whaler *Esquimaux* reveal a world characterised by its mobility, by its malleability and by evolving forms of discipline.

Ships as mobile spaces

Foucault emphasized that mobility was one characteristic particularly conducive to the development of heterotopic spaces. After evocative descriptions of how all sorts of vehicles could act as heterotopias, he concluded that '[i]n civilizations without boats, dreams dry up, espionage takes the place of adventure, and the police take the place of pirates.'[13] Hyperbole aside, Foucault's contention was that ships are heterotopia 'par excellence' partly because of the cultural significance attributed to their mobility. Mobility affects how such spaces have been understood; that cultural imagining, in turn, could affect how a space was inhabited. Culturally loaded responses to ships range from highly romanticized fictional re-constructions of life at sea to the much more mundane realm of bureaucratic regulation.

The idea of what comprises a 'ship' can often become blurred or inverted. In our sources, the 'space' on the ship was often represented as not only the physical vessel but also the people who worked within it. After all, an ocean-bound vessel is nothing without its crew. The notion that ship and sailors were a single entity was so commonplace that it was rarely commented on explicitly, even among the numerous sources pertaining to the *Esquimaux*.[14] The men (and occasional women) who lived and worked onboard formed part of that heterotopic space, being both part of the very fabric of the ship and distinct from it.

Narrative sources such as the journals kept by passengers tend to present maritime voyages as largely mundane affairs, characterized by repetition. Andrew Barclay Walker, for example, simply noted 'same again' for days where activities and climatic conditions appeared to him to be simply repeated. In 1899, Walker and his companions travelled from Florida to Newfoundland to join the *Esquimaux* for its transit north to the Davis Strait. Though shorter than a full transatlantic crossing, the experience still gave Walker and his friends a sense of what sea life could be like. After a week, he wrote,

> Wednesday, May 3rd. – Hard gale. Most uncomfortable. Hove to. Very heavy sea. Four oil bags out. Snow.
>
> Thursday, May 4th. – Terrific gale. Hove to. Four oil bags out. "Esquimaux" behaving splendidly. Snow.
>
> Friday, May 5th. – A repetition of yesterday – if possible, worse ... Mr. McKenzie [surgeon] prostrate in his bunk; Cecil [Hammond, ship's purser] recovered.
>
> This is getting monotonous. I had a nasty fall to-day in the saloon; one of the fixed (!!) chairs was carried away and sent me head foremost into the rifle rack ... Persuaded McKenzie to get a breath of fresh air for five minutes.[15]

As Stephen Berry has noted, the constant shift work undertaken by crew stood in stark contrast to the passivity experienced by ships' passengers.[16] While Walker rode out storms below decks, describing his days in terse details, the crew would brave the hazardous conditions and continued to work topside.

The disparity between the worlds of seafarers and those of their passengers clearly influenced the kinds of accounts produced about a journey. Banal phrases such as 'same again', or 'this is getting monotonous' belie a life of activity, tension, pleasure, encounter, hard work, and occasional recreation that has been much less represented because it seemed too 'mundane' to comment on. By taking into account the significance of a ship's specific environment, we can challenge assumptions about the apparently humdrum lives of the crew, apprehending fluctuations in life (and death), and thus filling in some archival blanks. Foucault's notion of heterotopia encourages us to consider ships in ways that stretch beyond basic observations about their mobility. Indeed, if we take mobility as a given, we can then pay attention to other kinds of physical or perceived spaces aboard. As we will see in the next two sections, sailors existed within a physical world in which dynamism and change were so normal they were barely worthy of remark. They also lived within a space bounded by unseen rules and regulations where behaviour was monitored and recorded.

Ships as malleable spaces

The *Esquimaux* was a mobile space sailing within a dynamic environment. As an ocean-going vessel, the barque was distinct from Foucault's other kinds of itinerant space, such as trains and aircraft, because it had to be self-sufficient over extended periods of time. During a lengthy voyage, all manner of human activity was accommodated within a finite space. A vessel's real value, then, rested in its malleability, its capacity to adapt to any situation. As a whaling vessel, the S.S. *Esquimaux* was primarily a place of work deeply invested in notions of productivity through resource extraction. On multiple occasions on each voyage, the crew would reconfigure the spaces aboard the ship to meet the changing needs of the vessel and the people onboard.

The *Esquimaux*, then, was not a stable entity: its internal structure and overall appearance were altered frequently in response to shifting environmental conditions and human needs. An obvious example would be the rapid adaptations made during storms. In rough weather, the ship's space took on a vertical nature: crewmen scrambled up masts to reef or unfurl sails; others manned the deck pumps to remove water from the bilge. Below decks, yet more laboured to secure ballast in the hold. Living within a space that might be changed often and at great speed defined the social conditions experienced by the crew.

The official documents produced when the *Esquimaux* was operating as a commercial whaler allow us to address some of the social consequences brought about by this state of flux. These must be read with care, however, for if taken at face value, official documentation could suggest that crewmembers' responsibilities (and therefore, lives) were defined entirely by their contracted function. The Crew Agreement signed for the 1872 voyage shows that less than half the crew were hired as Able-Bodied or Ordinary Seamen to sail the ship. The rest held positions which detailed their roles in hunting whales, for example as harpooneers, 'boatsteerers' (who piloted the whaleboats), or 'line managers' (who managed the lines attached to the harpoons).[17] We might assume that everyone but the Able-Bodied and Ordinary Seamen would, like Walker and his friends, wait passively for the ship to reach the hunting-grounds. But a closer reading of the 'Articles' signed by the crew suggests otherwise. Those signing on with the *Esquimaux* agreed that notwithstanding the capacities [i.e., contracted working positions] set opposite the signatures hereto, each member of the crew engages to assist in any duty to the best of his ability for the general good.[18]

In other words, while a seafarer's wages and general rank was defined by their listed capacity, he was also contractually obliged to perform any task given by the captain. This clause, or versions of it, appears in virtually every whaling and sealing Agreement written during the second half of the nineteenth century.[19]

The plurality of roles was so common that the details of what crewmembers actually did while at sea were rarely recorded. Evidence emerges, however, if we look through ships' logs for entries about irregularities. There is not space in this chapter to develop an extensive reading of logbooks, but one example should give an indication of the potential that such sources hold. The *Esquimaux's* log for 1872 records the death of boatsteerer Edward Mays. Midway through the vessel's return to Scotland, Captain Charles Yule recorded that

> Edward Mays Boatsteerer while assisting to furl the upper fore Topsail during a gale fell from the yard on deck & sustained fracture of the skull and other internal injuries[;] he was immediately conveyed below and every means [was] adopted by the surgeon to restore consciousness but without effect[20]

The unfortunate Mays had been contracted to steer a whaleboat, but he died while climbing in the rigging with the sailing crew while crossing the North Atlantic. Flexibility in terms of duties and expectations was a basic feature of life for those engaged in Arctic whaling and sealing. Further exploration of the irregularities recorded in logbooks such as this would reveal more details about the multiple roles and activities undertaken by crewmembers.

The entries in Andrew Barclay Walker's *Cruise of the Esquimaux* also gesture towards the constant state of change aboard ship. These form our second example. Although Walker primarily depicted the *Esquimaux* as a hunting station, the whaling ship was first and foremost an ocean-going vessel. The barque had necessarily undergone structural alterations at different stages of its voyage, along the lines described above. In 1899, when the *Esquimaux* had left St John's and headed for Greenland, the ship reefed (shortened) many of the sails and transferred coal by the hundredweight from tanks below decks to the engine's bunkers.[21] Walker observed glibly in his diary, 'Taking advantage of the lovely weather, all the boats' crews were engaged preparing for the fray'. On deck, routine sailing duties were replaced with radically different tasks, including blacksmithing, splicing lines of rope, testing harpoon guns, and kitting out the whaleboats.[22]

Hunting normally did not take place from the decks of the *Esquimaux* itself, but the vessel was nevertheless dramatically transformed by its effort. Walker, his compatriots and most of the crew used the whaleboats to pursue seals, walrus, and narwhal; they also went out on the ice and ashore to hunt polar bear, Arctic fox, hares, and all manner of fowl. After a kill, the hunted prize would be brought back to the ship, and the gleaming deck and holds metamorphosed into a macabre scene of processed gore. Walker described returning after a successful whale hunt:

> The blubber was hoisted on deck, cut up into pieces about 1 foot square, and relieved of all bits of flesh .... It was then taken to the harpooneers, who are stationed on each side of the vessel, and was placed on the 'clash' or iron stanchions, with five spikes, fixed into a socket in the deck. The harpooneers cut off the skin ... and pass it on to the boat steerers, who ... stand behind the 'speck trough', which is 18 feet long by 2 feet wide, with a hole in the middle. Attached to this hole is a large canvas shoot, which goes into the blubber tank below, where the skeaman is on duty. The boat steerers have in front of them their cutting blocks, which are pieces of the whale's tail. ... [The blubber] then falls into the trough, and is shoved down the shoot into the tanks. ... The result of operations was about 2½ cwt. of whalebone and 6 tons of blubber.[23]

Once in the Arctic, then, the ship became a charnel house in which dead animals were processed and their valued products stored.

Walker's descriptions portray a practice that was common among all whaling ships. The constraints of space and the circumstance of decomposition meant that a catch had to be processed immediately. When birds or reindeer were shot for food, Walker noted that the fresh meat, 'was hung in the mast and fore tops ... which speaks well for this climate as a larder'.[24] But whalers, including the *Esquimaux*, also became home to opportunistic, somewhat haphazard, collections of living animals that were captured and brought back alive. Walker described how

> We arrived on board at 9 p.m. with a very odd cargo, consisting of one seal, three narwhal horns, six long tail ducks [all dead] ... also, ten young tern, two sand pipits and one sand piper, all alive ... R. [McKenzie] killed a young [polar] bear from the ship while I was away. This would not have occurred had I been there, as he ought to have been taken alive.[25]

The scene evoked here, in which an odd assortment of Arctic creatures scampered and hopped around the deck beneath dead animals suspended from the rigging, was typical for whaling vessels. It was relatively common for crewmembers to collect live creatures either for sale to menageries and zoos, or for personal pleasure.[26] Aboard the decks of Arctic whalers, death on a staggering scale was juxtaposed starkly against ebullient life.

Taken together, the sources examined above uphold an emphasis on the malleability of shipboard space. They also underline the importance of deploying a diversity of materials to understand how that space was inhabited and used. If considered only at face value, the lists of personnel in the

Crew Agreements might suggest that the people hired to hunt whales would spend the majority of the journey simply waiting for the ship to reach the hunting grounds. A closer reading of the Articles, and a consideration of the implications of Walker's journal entries, has shown, however, that this was simply not the case. Life at sea in the North Atlantic was monotonous only to those who remained in the cabins – and the only people to do so were passengers like Walker. The continually changing climate and state of the sea, and the arrival of new passengers (dead or alive) required crewmembers to swap in and out of roles and to constantly reconfigure the ship, often at great speed and at great personal risk.

Apparent opposites were constantly inverted or brought alongside, creating, in Foucault's terms, a heterotopic space. The ship was both a navigable vessel and a hunting platform; the mariners – regardless of their official capacity – variously adopted the roles of sailors, hunters, butchers, or even zookeepers. In continuously adapting the space of the ship to these changing conditions, the crew (who, we remember, were both part of and distinct from the physical vessel), likewise reconfigured their own roles and positions aboard. Malleability characterized the sailors' lives and defined their relationship with the vessel that they conducted.

Ships as administrative spaces

The mobility and malleability of merchant ships engendered cultural responses that ranged from evocative re-imaginings of life at sea through to more pedestrian administrative documentation. Merchant shipping records were not 'cultural' in the sense of artistic creativity, but they were manifestations of a central feature of nineteenth-century culture: an expanding bureaucracy. Indeed, the growing obligation to produce administrative documents began to encroach significantly upon shipboard life. In turn, this resulted in the creation of what are now substantial historical archives.[27]

It can be all too easy to assume that documents have been saved simply for the convenience of posterity, but the development of archives is a historical phenomenon in its own right. It is important, then, to consider where our evidence has come from. The wealth of administrative sources produced about 19th-century mercantile shipping derives from the British government's larger aim to facilitate economic and imperial dominance. From the 17th until the early 19th centuries, various kinds of vessel within Britain's mercantile fleet, including whalers, had been supported financially by a government wishing to foster a 'nursery of seamen' from which the Royal Navy could draw personnel.[28] The Navy's insatiable demands for manpower over the long 18th century, however, depleted its merchant shipping. In the decades following the Napoleonic Wars, nearly every aspect of maritime commerce was then brought under governmental purview, yet fears arose that the decaying condition and inefficiency of Britain's merchant fleet would eventually injure its economy and security. Confronting these issues, however, the government struggled with two internal problems. First, maritime concerns were spread over a bewildering array of governmental departments resulting in confusion, turf wars, and inefficiency. Trinity House supervised the administration of pilots and lighthouses, but the Admiralty took responsibility for maritime rules of navigation and the Treasury governed the registration of ships.[29] Second, the government lacked appropriate decision-making information with which it could identify areas of concern and take remedial action. In sum, the government required an administrative structure to support its broader aspirations.

Effective administrative controls came about through the steady consolidation of power within the Board of Trade.[30] In 1834, the government created a departmental library within the Board of Trade for the collection of commercial information and its use by government officials.[31] By 1850, the Mercantile Marine Act transferred all Admiralty functions relating to merchant seamen to the Board of Trade. It also established Shipping Offices (later renamed Mercantile Marine Offices) and required all shipmasters to maintain official log books. Subsequent acts further consolidated the Board of Trade's reach; the Merchant Shipping Act of 1854 declared that 'The Board of Trade is to be the department to superintend all matters relating to Merchant Shipping'.[32]

Prior to this surge in bureaucratic surveillance, the behaviour of officers and crew aboard British merchant ships had been controlled by customary rules and discipline. The 19th-century Crew Agreements simply codified these understandings about behaviour, authority, and personal rights. In 1872, the sailors on the *Esquimaux* thus agreed:

> to conduct themselves in an orderly, faithful, honest, and sober manner, and to be at all times diligent in their respective duties, and to be obedient to the lawful commands of the said Master, or of any Person who shall lawfully succeed him, and their superior Officers.

The Agreement reified in contractual form existing maritime standards regarding conduct and obedience. It also reaffirmed the hierarchical authority of the ship's officers and underlined seafarers' traditional rights to seek redress for any grievances against them:

> if any member of the Crew considers himself to be aggrieved by any breach of the Agreement or otherwise, he shall represent the same to the Master ... in a quiet and orderly manner, who shall thereupon take such steps as the case may require.[33]

With each new act, an increasing array of consolidated administrative measures was filtered down to the masters and crews through the form of Crew Agreements, official ship logs, and Official Notices.

The RGSS's steady demand for documentation at sea affected the way in which officers and crews inhabited shipboard space. Two such consequences were material; another related directly to the perception of space. First, requirements for documentation reinforced physical divisions within ships.

Officers, especially the master, needed berths with desks and living space that afforded enough stability and seclusion for keeping records. Even on merchant vessels the existence of such segregated places amounted to conspicuous and enduring statements of authority and privilege.

The increasing implementation of administrative measures meant government-imposed (i.e., landlubber) regulations now affected life and the usage of space aboard ship. The physical space allocated to each sailor and its condition were now monitored and reported. The 'Official Notice No. 2 – Accommodation for Seamen on Board Merchant Ships' (appended to the *Esquimaux*'s 1894 Agreement), reminded masters and shipowners of the standards established by the Merchant Shipping Act of 1867. It stated:

> Every place in any ship occupied by seamen, or apprentices, and appropriated to their use, shall have ... a space of not less than seventy-two cubic feet ... [it] shall be securely constructed, properly lighted and ventilated, properly protected from weather and sea ... Every such place shall be kept free from store or goods of any kind, not being the personal property of the crew in use during the voyage.

The notice goes on to apportion responsibility and consequences for failure to meet these standards. For structural faults, owners were held accountable and could incur a penalty 'not exceeding twenty pounds'. If masters failed to keep accommodations free of goods and stores, then they were liable to 'pay to each seaman lodged in such place the sum of one shilling a day for each day after the complaint'.[34]

The codification and enforcement of these standards thus led to changes in the way shipboard space was inhabited. Any master or shipowner who was found to have failed to comply was punished with steep fines. But the new administration also altered the way shipboard space was perceived in a less tangible way. Masters and shipowners, who of course had to uphold the government's stipulations, used the expanding bureaucratic framework to impose demands upon the crew, and they paid particular attention to regulating behaviour. The Board of Trade's published 'Regulations for Maintaining Discipline' sanctioned a transition in the way that misconduct was punished: mariners moved from a visible system of violent punishments (such as lashing) to one that deployed imperceptible forms of chastisement, such as fines and poor references. Ships began to leave a considerable paper-trail in their wake.[35]

In the *Esquimaux*'s 1882 Agreement and Account of Crew, the owners' portion of the contract stipulated that '[No] member of the Crew [shall] convey ... spirituous liquors on board the vessel, nor shall any Rifles, guns or Ammunition of any description be allowed on board except those belonging to the vessel'.[36] The Agreement then specified the scale of punishments to be meted out:

> Any one infringing these rules shall forfeit one months [*sic*] pay .... Any member of the Crew who is not in a fit state to perform his duties ... shall forfeit his wages until such time as he commences duty.[37]

The use of the lash, which continued to be common practice in some parts of the American merchant navy (and which is commonly depicted in stereotyped portrayals of discipline – or tyranny – aboard naval and merchant vessels), does not appear in British whalers' official ship logs from the second half of the 19th century.[38] Instead, the new administration developed a system of deferred fines. Misdeeds were recorded, and fiscal punishments were meted out in the form of forfeited pay. Such was the case in 1892, when Captain Jeffrey Phillips held back 12 days' pay for James Anderson having been on shore without leave in St John's for two days.[39] More egregious crimes resulted in lock-up until the ship reached port, where the accused was remanded into the custody of the MMO Superintendent.

The documents we consulted suggest that merchant mariners gained a very different incentive for regulating their own behaviour while at sea compared to the violent bodily discipline meted out in previous centuries. Repercussions for poor behaviour now resulted in financial loss and a record that followed a sailor's employment. Each sailor's conduct and ability were annotated in Crew Agreements, official ship logs, and discharge certificates. The latter quite literally followed a sailor from ship to ship, as they had to be produced prior to signing articles for a new voyage.[40] Thus, even illiterate crewmembers became increasingly aware that their own comportment was monitored and recorded. Each 19th-century merchant ship sailed within an authoritative hierarchy that stretched far beyond its physical confines or even its owners' purview.[41]

Beyond regulations touching the crew and ships, British maritime customs and legal rules also existed for the admission of non-mariners aboard ship. Since the 18th century, ordinances had prevented British whalers from taking aboard passengers who had not signed articles and were not, therefore, members of the crew. For this reason, women or small children rarely boarded whalers, and even stowaways, once discovered, were typically required to sign on as crew.[42] In 1874, the Captain of the *Esquimaux* recorded that four young men 'Having stowed themselves away have this day made there [*sic*] appearance on deck and I have put them on the articles as Ordinary Seamen'.[43] In 1899, Andrew Barclay Walker's friend Cecil Hammond skirted regulations by nominally signing on as the ship's 'purser', while another companion, John Collins, more appropriately took the title of valet.[44]

Maritime customs and legal rules of physical and administrative space, however, became much more ambiguous with regard to the whalers' interactions with the Arctic's indigenous populations. Every year, whaling vessels from Scotland travelled along the coasts of Greenland and Baffin Island often meeting Greenlanders and Inuit ashore or hosting them aboard the ship. Men and women, young and old, clambered aboard, and according to Walker, 'wander[ed] about the ship ad libitum'.[45] Whaling crews and Inuit shared food and entertained each other with music, dancing and games. On Monday 11 September 1899, Walker's journal recorded that the whaler *Esquimaux* had arrived at Dexterity Bay on Baffin Island, and he went ashore:

[I] made the acquaintance of Alnick, the head of the tribe here…There were some forty people in the settlement. Our boat was seized and filled with men, women, and children [whom we took to the ship …] the afternoon and evening were spent in playing poker, exhibiting the polyphone, and endeavouring to speak Esquimaux to an admiring crowd.[46]

In an instant, shipboard hierarchical restrictions within these strongly masculine spaces were relaxed or disappeared entirely, as children roamed the decks, men 'trokked' or bartered skins for knives, ammunition, beads, and tobacco, and women grouped together to make or repair highly valued 'kummings' or sealskin boots for exchange.[47] These encounters are indicative of the multitude of ways in which shipboard space and conventions were upended.

The spectres haunting administrative space

As with other sources discussed earlier, maritime administrative documents present limitations when considered in isolation. One of the most telling examples of these lies with their remarkable obliviousness to the Inuit. Like so many other whalers' documents submitted to the Dundee MMO, the *Esquimaux*'s 1899 official log book makes no references to the crew's interactions with the Inuit. The whaler *Esquimaux*, however, engaged with numerous Inuit groups.

The anecdotes in Andrew Barclay Walker's journal make very clear both the extent and nature of those interactions. Indeed, Walker himself became actively involved: He concluded his 11 September journal entry, quoted above, by noting that 'I am trying to take a half-caste boy back [to Dundee] on the ship. He is an orphan, and I fear is none to [sic] well treated on shore'.[48] Three days later, he wrote, 'after a bit of haggling with Alick [sic] we secured the native boy for an old pack of cards. He is delighted to come'.[49] No mention was made of the boy's passage in any of the *Esquimaux*'s official records. Instead, further evidence comes from a small news article in the *Aberdeen Press and Journal* mentioning Kidlaw, 'An Esquimaux Boy in Dundee'.[50] Similar decisions to take Inuit back to Dundee had in fact been made a number of times in previous years.[51] In every instance where Agreements are available, official records fail to reflect that an Inuk joined the ship. Occasionally a whaling captain would write a personal letter to the MMO Superintendent describing a situation.[52] Historians, therefore, must ask themselves: Was the omission of Kidlaw due to restrictions imposed by the format of the official document or did this reflect the will of the whalemen?

We suggest that the answer was a mixture of both. Crew lists were structured to elicit data pertinent to wage-paid crew members. It is not surprising that Kidlaw did not appear on the crew list, because the Inuit did not operate in a cash-based economy. Unusual situations cropped up from time to time, and ships' masters would annotate the official logbook accordingly, since anyone joining or leaving the vessel ought to have been accounted for there. It is important to note, too (in the context of 19th-century scientific debates about the development of the human species), that the Scots whalers did consider the Inuit as rough equals. Although contemporary journals and news articles often employed the term 'savage', the Scots in the Arctic and the Inuit mostly demonstrated a gritty but real mutual respect for one another. Oral histories told by the Inuit today still speak of relationships with the Scottish whalers, and many Inuit now claim a mixed Scottish heritage.[53] Kidlaw's omission from the official log book – he does not even appear as an annotation – was not an implicit commentary on any lack of human status. His absence reflects an active choice made by the captain.

It is difficult to fathom whether the captain was colluding with his crew and Walker to mislead the MMO Superintendent, or whether he had his own motivations to do so. What is clear, however, is that this episode in the *Esquimaux*'s history is suggestive of another kind of inversion, this time of the terrestrial administration's attempt to impose its authority over a ship operating at a distance.

In spite of appearances, then, official ship logs and other documentation did not necessarily record activities that deviated from the status quo. The new administrative space created by the RGSS and Board of Trade resulted in a wonderful set of historical resources. It turns out, however, that the seafarers' efforts to circumvent the rules and regulations imposed by terrestrial authorities mean that these sources occlude as much as they reveal. The archive is haunted by spectres such as young Kidlaw, who appear only if we read carefully across multiple genres of sources, from official records to diary entries and newspaper articles.

Conclusion

This chapter has examined how Foucault's notion of heterotopia might help us to think about the social significance of space. In doing so it has opened up several avenues for further research. Rather than offering a comprehensive exploration of how ships functioned as spatial heterotopia, we sought to establish how the notion of heterotopia might be understood in the context of 19th-century merchant shipping, and then to address the possibilities and limitations offered by sources that pertain to the lives of crewmembers.

Although mobility is fundamentally a ship's raison d'être, this characteristic alone does not make a ship a heterotopia 'par excellence'. Rather than focusing solely on a ship's mobility (normally represented as 'mundane' or simply as a list of destinations and way-points), we considered the physical and cultural conditions that brought about inversions of conventional spatial relations aboard ship. One of these was malleability, which characterized all kinds of shipboard life. Aboard any ship, the explicitly demarcated internal space had obvious consequences for daily life and ordinary interactions, but it might be altered in response to changing conditions. We then explored some of the specific forms of flexibility expected of 19th-century whaling ships, such as their transformation from vessel to hunting platform, as well as the juxtaposition of life and death aboard the vessel.

The second condition that we explored was the development of a perceived 'administrative space' relating to merchant shipping. Noting that the number of bureaucratic documents available in the archive increases greatly over the course of the 19th century, we first situated these documents within the cultural context of the expanding maritime bureaucracy. We then paid close attention to the way in which behaviour was monitored and recorded and identified how a transition in shipboard discipline came about as part of – and was chronicled through – these documents. Our concluding example underlined how activities that inverted or disrupted the status quo might *not* be recorded. In doing so, our analysis gestured towards the limited extent to which land-based bureaucratic measures were able to influence the use of, and behaviour within, merchant vessels operating in the furthest reaches of the Arctic.

Finally, the broader argument that we develop here is about the importance of using a wide range of sources for maritime social history. Reading multiple sources against each other is essential for gaining a more nuanced understanding of the lives of people such as ships' crews and Inuit, who did not produce their own written records but who, it turns out, very much inhabited shipboard spaces.

Notes

1. A special thanks to Professor Valerie Burton (Memorial University of Newfoundland), whose advice has been especially helpful for considering the administrative nature of maritime archives.
2. See David Livingstone, *Putting Science in Its Place: Geographies of Scientific Knowledge*, Chicago and London: University of Chicago Press, 2003, p. 7.
3. Michel Foucault, 'Of Other Spaces', trans. Jay Miskowiec, *Diacritics* 16, 1986, 22–7.
4. Ibid., 24.
5. Ibid., 27.
6. Levi G. Chafe, *Report of the Newfoundland Seal Fishery from 1863 to 1905*, St John's: Evening Telegram Job Print, 1905, p. 24–6. A 'swiler' is a Newfoundland seal hunter often armed with a gaff and rope.
7. Commonly dubbed 'crew lists' or 'Articles', we use the term 'Crew Agreements' in accordance with the Maritime History Archive (MHA) at Memorial University of Newfoundland. The MHA holds over 70 per cent of all British Crew Agreements from 1857 to 1938 and 1951 to 1976. For a detailed breakdown of principal British shipping records and where they are kept, see HTTP: https://www.mun.ca/mha/research/principalrecords.php (accessed 20 September 2021).
8. For the best introduction to British merchant marine Crew Agreements, see Valerie Burton, *More Than a List of Crew* (*MTLC*). Available HTTP: https://www.mun.ca/mha/mlc/ (accessed 20 September 2021) hosted by the Maritime History Archive, Memorial University of Newfoundland. *MTLC* exists to inform digital users about Crew Agreements and encourage new generations of scholars to use these documents.
9. Valerie Burton and Robert C.H. Sweeny, 'Realizing the Democratic Potential of Online Sources in the Classroom', *Digital Scholarship in the Humanities* 30, 2015, 177–84.

10 A few excellent examples from a large and productive historiography include: Stephen R. Berry, *A Path in the Mighty Waters: Shipboard Life & Atlantic Crossings to the New World*, New Haven: Yale University Press, 2015; Katherine Foxhall, *Health, Medicine, and the Sea: Australian Voyages, c. 1815–1860*, Manchester: Manchester University Press, 2012; and Joan Druett, *Hen Frigates: Wives of Merchant Captains Under Sail*, New York: Simon and Schuster, 1998.

11 Andrew Barclay Walker, *The Cruise of the Esquimaux, Steam Whaler, to the Davis Straits and Baffin Bay, April–October 1899*, Liverpool: Liverpool Printing and Stationery Company, 1909.

12 For a good example of such an approach (albeit with a different analytical purpose), see Nathan Perl-Rosenthal, *Citizen Sailors: Becoming American in the Age of Revolution*, Cambridge, Mass.: Harvard University Press, 2015.

13 Foucault, 'Of Other Spaces', 27.

14 Richard Henry Dana's (1815–1882) narrative account of his experiences as an Ordinary Seaman makes very apparent the extent to which the mariners' lives and that of the material ship itself were perceived as being intertwined. Richard Henry Dana, Jr., in Thomas L. Philbrick (ed.), *Two Years before the Mast and Other Voyages*, New York: Literary Classics of the United States, Inc., 2005, p. 36.

15 Walker, *Cruise*, pp. 15–16.

16 Berry, *A Path in the Mighty Waters*, p. 175. Many merchant vessels used a 'dog-watch' system, but Dundee whalers frequently employed three shifts of eight hours. Of course, everything was subject to change during a storm.

17 MHA, *Agreement and Account of Crew and Official Logbooks for British Empire Vessels*, S.S. *Esquimaux*, Official Number (hereafter O.N.) 52562, 1872.

18 MHA, *Esquimaux*, O.N. 52562, 1872.

19 The earliest whaling Crew Agreements held at MHA date back to 1863.

20 MHA, *Esquimaux*, O.N. 52562, 1872.

21 Walker, *Cruise*, p. 13.

22 Ibid., p. 18.

23 Ibid., p. 47.

24 Ibid., p. 33.

25 Ibid., p. 70; see also pp. 37, 69.

26 One news article highlighted a particularly infamous incident, 'Escape of a Polar Bear in Dundee', *Dundee Evening Telegraph*, 7 November 1878.

27 For a summary of some of these archives, see note 7.

28 Adam Smith, *An Inquiry into the Nature and Causes of the Wealth of Nations, Vol. II*, London: William Strahan, 1776; London, Ward, Lock, and Company, 1910, p. 18.

29 Roger Prouty, *The Transformation of the Board of Trade, 1830–1855*, London: William Heinemann, 1957, pp. 87–8.

30 For an accessible introduction to the Board of Trade's maritime oversight through its documents, see 'A History of the Agreements' at HTTP: https://www.mun.ca/mha/mlc/toolkit/history/ (accessed 20 September 2021). Other introductions to the Board of Trade at large include Susan Foreman, *Shoes and Ships and Sealing Wax: An Illustrated History of the Board of Trade 1786–1986*, London: HMSO, 1986; and Hubert Smith, *The Board of Trade*, London: G.P. Putnam's Sons, 1928.

31 Alistair Black and Christopher Murphy, 'Information, Intelligence, and Trade: The Library and the Commercial Intelligence Branch of the British Board of Trade, 1834–1914', *Library & Information History* 28, 2012, 186–201.

32 See 'Merchant Shipping Act, 1854', *The Mercantile Marine Magazine and Nautical Record*, 1854, 1, 333–41.

33 MHA, *Esquimaux*, O.N. 52562, 1872.
34 MHA, *Esquimaux*, O.N. 52562, 1894.
35 For an example, see: MHA, *Esquimaux*, O.N. 52562, 1879.
36 MHA, *Esquimaux*, O.N. 52562, 1882.
37 MHA, *Esquimaux*, O.N. 52562, 1882.
38 Known as 'hellships', American whaling vessels operating in the Alaskan Arctic, routinely carried out 'ship's discipline' with severe and demeaning physical brutality. See A. Basil Lubbock, *Round the Horn Before the Mast*, London: John Murray, 1903, pp. 33–4.
39 MHA, *Esquimaux*, O.N. 52562, 1892.
40 For example, the Shetland Museum and Archives holds four Certificates of Discharge from Scottish whalers for Daniel Henderson of Lerwick: SEA65639(a), S.S. *Esquimaux*, 1877; SEA65639(b), S.S. *Windward*, 1888; SEA65639(c), S.S. *Hope*, 1891; and SEA65639(d), S.S. *Polar Star*, 1892.
41 These measures amounted to the maritime extension of a culture of governmentality. For more on this form of self-discipline, see Graham Burchill, Colin Gordon and Peter Miller (eds.), *The Foucault Effect. Studies in Governmentality*, Chicago and London: University of Chicago Press, 1991.
42 For a commentary discussing the few known instances when British whaling captains' wives voyaged to the Arctic, see W. Gillies Ross, *This Distant and Unsurveyed Country: A Woman's Winter at Baffin Island, 1857–1858*, Montreal: McGill-Queen's University Press, 1997, pp. xxv–xxvi.
43 MHA, *Esquimaux*, O.N. 52562, 1874.
44 MHA, *Esquimaux*, O.N. 52562, 1899.
45 Walker, *Cruise*, p. 36. British whalemen referred to indigenous people in the Arctic as 'Eskimo', 'Esquimaux', 'Yakkie', or 'Greenlander' – the last of these terms was faithful to the word used by the indigenous people of western Greenland to refer to themselves. Today, eastern Arctic tribes prefer the name 'Inuit', while an individual is an 'Inuk'.
46 Walker, *Cruise*, p. 83.
47 Ibid., pp. 25, 35–37, and 83.
48 Ibid., p. 83.
49 Ibid., p. 84.
50 'An Esquimaux Boy in Dundee', *Aberdeen Press and Journal*, 1 November 1899. Walker never gave the boy's name. The paper's attempt is a phonetic spelling.
51 For more information on 'Eskimos in Dundee', see University of Dundee Archive Services, MS 254/3/1/3-4, 'David Henderson Collection'.
52 MHA, *Arctic*, O.N. 72543, 1876, inset letter. In his letter to the superintendent, Captain William Adams of the whaler *Arctic* begins, '[Sir] I beg to inform you that on my last voyage from Davis Straits I brought home with me to this Country … an "Esquimaux" named Alnack … In the circumstances I was at a loss to know whether or not to put him on the articles…' Alnack [also Olnick or Alnick] was the Inuit leader with whom Walker bartered for Kidlaw in 1899.
53 For an Inuit perspective, see Dorothy Eber, *When the Whalers Were Up North*, Montreal: McGill-Queen's University Press, 1989. For the Dundee perspective, see Malcolm Archibald, *Ancestors in the Arctic: A Photographic History of Dundee Whaling*, Edinburgh: Black & White Publishing, 2013.

7 Bars

Kate Ferris

The 'everyday': a temporal designator, a spatial designator

The 'everyday' is, of course, a temporal designator, connected very obviously to the chronology of human experience. Thus, at first glance, it might seem paradoxical to seek to understand the everyday lives, experiences, and practices of past historical actors using a spatial historical approach. However, whilst initial appearances might suggest a contradiction in examining 'everyday life history' through a spatial lens, in reality scholars labelled 'everyday life' historians (or *Alltagsgeschichte* historians in the West German context in which the approach first arose in the 1980s) have very often placed 'space' centre-stage in their analyses.[1] Certainly, the early *Alltagsgeschichte* historians who pioneered the approach in their studies of Nazi Germany, such as Alf Lüdtke and Detlev Peukert, were less overt in their spatial framing than more recent everyday life histories, often termed a 'second chapter' for the approach, have been in works published in the past decade or so.[2] Lüdtke, for example, opened his introduction to the collection of essays that first presented and delineated *The History of Everyday Life* approach by indicating the people who were the object of his scholarly interest – the 'everyday, ordinary people' or *kleine Leute* in the original German – but followed this swiftly with an indication of his spatial frame; these were 'individuals [who] emerge as actors on a social stage'.[3] For Lüdtke the 'social stage' that most occupied him, at least in that volume, was the factory floor in the early years of the Nazi rise to power and rule, although the public squares of German cities and towns where, for example, May Day celebrations were enacted, re-enacted, and appropriated also featured in his analysis.[4] Sheila Fitzpatrick who, without the *Alltagsgeschichte* label, explored everyday subjectivities and lived experiences in Soviet Russia under Stalin's rule, focussed particular attention on the spaces in which Soviet citizens interacted and exchanged with one another and with the Soviet party and state: kolkhoz markets, dachas, department stores, the Moscow metro, the communal dormitories of new industrial cities, as well as the ubiquitous snaking queues in which Soviet citizens, especially women and children, spent considerable portions of their days.[5]

DOI: 10.4324/9780429291739-10

More recent everyday historians of 20th-century dictatorship, including fascist Italy, which is explored in the empirical case study for this chapter, have retained the 'first chapter' historians' commitment to exploring questions of subjectivity, practice, enactment, and agency, but in addition are more consciously interested in the spaces in and through which lived experience is effectively enacted. In part, this has been articulated as a shift in focus, away from classic seats and sites of formal projections of power and towards those 'everyday spaces' in which the 'unofficial relations of power' were articulated and negotiated.[6] In the case of fascist Italy, this has meant recognizing that dictatorial authority, ideology, and policy were

> constructed not only in Palazzo Venezia, in state ministries, provincial prefectures and party headquarters [but] just as crucially, [were] enacted in the places inhabited and traversed by Italians in their day-to-day lives: in markets, streets, squares, bars, trains and train stations, factories, homes, shops, parish churches, and so on.[7]

However, exploring everyday life history spatially does not just entail a shift of venue. Crucially, it also requires the recognition of spatial agency and of the mutual interdependence and interplay between space/place and practice. 'Everyday spaces' are not simply passive venues in which the kinds of social/political/cultural/economic interactions and exchanges and agency-laden acts that interest everyday life historians took place. On the contrary, past individuals – historical actors – interacted with and were to some extent conditioned by the everyday spaces they traversed, visited, and inhabited; and vice versa, individual and societal interactions and practices shaped those spaces in both real and imagined terms. In this way, and following the example of Michel de Certeau's exhortations in *The Practice of Everyday Life*, everyday life historians must seek to understand how past spaces were *used*, recognizing space as both the product *and* shaper of social, political, cultural, economic exchanges, relations, and practices.[8] As such, we are primed to recognize the multivalent and flexible ways in which everyday spaces were used and experienced, given that 'common stops in the daily routine of local inhabitants [...could] hold a variety of meanings that are often in play simultaneously'.[9] In addition, we must note significant variations in the frequency and depth in the experience of, and engagement with, everyday spaces: some places may be traversed or visited only once or occasionally; others might be encountered fleetingly but frequently; others still would be regularly inhabited day-by-day, perhaps for several hours at a time. All still have the potential to shape, and to be shaped by, lived experiences and practices.

In this way, it should no longer seem so paradoxical to explore the everyday lives of past historical actors using a spatial history approach. Certainly, the 'everyday' is an extremely slippery category to define; it is perhaps best understood as an apparatus or heuristic frame through which to examine past practices and questions of subjective experience and agency rather than as a rigid, temporally defined category that pertains only to the study of routine, repetitive acts.[10] The spatial dimension of the 'everyday' – how everyday spaces were *used* and *experienced*, how they shaped the interactions and practices conducted within their bounds and how they were in turn produced and shaped by this usage – thus becomes a crucial constituent part of this analytical frame.

Our questions then turn to how we might begin to explore and to understand the spatial dynamics of everyday life: in particular, what kinds of historical sources might we examine and, more importantly, how should we 'read' and use them? Following the view that 'spatial history' is, in large part, a 'way of seeing' or reading sources, the following sections of this chapter discuss some of the problems and possibilities presented by the availability of sources and ways of seeing/reading sources relating to the spaces of everyday life. To do so, it uses the empirical case study of Mussolini's fascist dictatorship in Italy (1922–43), and a specific set of everyday spaces that were (and remain) key venues and agents in the playing out of Italian (especially male) social, cultural, political, and economic life, namely, bars.[11] Of course, the context of a fascist dictatorship and the choice of bars as a case-study space have certain particularities, some of which are discussed below. However, they also afford an opportunity to explore the range of possible source material and how we might use these to understand the spatial dynamics of everyday life in ways that are relevant to scholars grappling with similar questions in historical settings that may be far removed from the specific time, place, and context of bars in fascist Italy.

The sources of everyday spaces (in a dictatorship)

Inevitably, with almost any aspect of 'everyday life history', historians face the challenge of working with scarce, fragmented, and problematic source material. The daily experiences of 'ordinary' people very often pass unrecorded; certainly such records do not often find their way into the archives. These difficulties, which are presented to any historian who seeks to explore the past lives of 'ordinary' individuals, in any setting, are exacerbated by the dictatorial context under examination in this case. The dismantling of the apparatus of parliamentary democracy, the presence of regime censorship, and restrictions on reporting, the criminalization of oppositional political action and thought, increased state surveillance, and the fostering of a climate of violence, coercion, and fear all contributed to 'induc[ing] a form of daily self-censorship' and a retreat into the home, family, and self that was hardly conducive to the candid recording for posterity of one's genuinely held everyday thoughts, practices, and experiences.[12]

Despite these difficulties, there remain many types of sources that everyday life historians might use. Those prized most highly are the historical sources produced by the 'ordinary' individuals who are the object of study themselves, often labelled 'ego-documents'.[13] Such 'ego-documents' – diaries, letters, memoirs, oral testimonies, etc. – are not without issue for historians. Whilst they are particularly valued for the direct access they afford to the subjective felt experience and individual human-scale view of the very historical actors whose practices, lived experience, and agency the everyday life historian wishes to better understand, they remain filtered accounts written or recorded with a particular audience – even if only one's imagined/idealized self – in mind. In addition, these ego-documents are relatively rare.

In light of the relative scarcity of ego-documents – and given that academic best practice requires drawing evidence from a variety of source-types – everyday life historians must look to other source bases: contemporary published works and unpublished archival material, including official records produced by state representatives (and thus inevitably shaped by state ideals and policies). Such officially produced sources in fascist Italy include legislative, executive, and civil service records, the reports of national institutes, syndicates, and associations, and police and trial records. Given the depth of state intervention and surveillance of the media, sources such as newspapers, journals, magazines, newsreels, and radio transmissions in fascist Italy must also be considered 'official' (albeit occasionally deviant) productions. Officially produced source material must be handled with caution; official records convey a perspective on people and events shaped by the organization and/or individual that produces them. In addition, such sources foreground those moments and aspects of people's lives when they came into contact (and often conflict) with the dictatorial state.

At the same time, and with due awareness of their status as mediated refractions of the concerns and intentions of their producers, officially produced sources are also highly valuable to everyday life historians. Firstly, they help us to set the contours of how the regime and those (for example, journalists) affiliated with or bound to the regime prescribed and scrutinized given everyday spaces. If we are to understand how ordinary Italians *used* everyday spaces like bars, we need also to understand how fascist officials, nationally and locally, sought to control practices and modes of behaviour in and around these spaces, as well as the wider official rhetoric related to these. Secondly and perhaps more importantly for the purposes at hand, officially produced sources also provide access (however indirect and filtered) to the words and writings of those 'ordinary' Italians whose experiences and practices are the everyday life historian's ultimate subject of inquiry. Police records, for example, relating to incidents of political violence and conflict that occurred in Italian bars, often contain witness statements, intercepted letters, verbatim transcriptions of graffiti or 'speech-acts', all of which record the very words uttered or written by individuals somehow caught up in the events being investigated. In this way, these sources offer everyday life historians some access to their subjects' 'authentic voices' at times of their lives when these individuals directly encountered and interacted with agents of the fascist state.

The following section aims to show how this can be done. It makes use of varied source material – travel guides, diaries, police records – that taken together allow us to explore the interplay between space, agency, and practice; how bars were used as everyday spaces of political sociability and interaction within the fascist dictatorship; the impact that these spaces had on political interactions and vice versa.

Case study: bars as spaces of everyday political sociability and interaction in fascist Italy

Esercizi pubblici, places licensed to serve alcohol in Italy for consumption on the premises, such as bars, *osterie*, *trattorie*, and cafés, were, in the early decades of the 20th century (and, indeed, remain to this day) key spaces of everyday, predominantly adult sociability and interaction. Bars and other spaces where alcohol was consumed played host to and facilitated various social and economic functions, as places where (mostly but not exclusively male) friendships could be fostered away from domestic settings (largely imagined as female-dominated spaces) and where economic exchange and personal advancement could be transacted.[14] In addition, bars also operated as political spaces. Whilst there are many questions that historians might ask about the everyday usage of bars – for example in relation to questions of gender identity and practices (certainly, bars as everyday spaces were far less gender-segregated than might be supposed) or with respect to the projection of national, regional, and local identities or to questions of how notions of 'modernity' were imagined and effected in the social and cultural practices of everyday consumers in bars – this case study focuses its questions on bars as political spaces. Specifically it asks: how, if at all, did the arrival to power of Mussolini's fascist dictatorship in 1922 affect the everyday political interactions that took place within bars and, to reframe the question, how did bars as political spaces help shape the political encounters and interactions – including episodes of violence – that took place within them during the dictatorship?

Long before Mussolini's black-shirts entered through their doors, a key function of Italian bars had been to host political interactions among those who drank and worked therein. Individual bars were often known for having particular political affiliations (in the same way that many were also known for harbouring particular sporting affiliations) dictated by the views of the host and/or their regular frequenters. Others were known as places where people of different – often opposing – viewpoints (and, not unrelated, different socio-economic status) could meet to talk politics.

The particular political affiliations associated with individual bars, and their functions as venues for exchanges between people of shared as well as divergent political viewpoints, are attested to in a variety of sources, including published guide-books to Italian *osterie* and bars. Numerous such publications were written both by Italians and foreigners in the early 20th century, adapting the already-established guide-book format to the relatively new fashionability of *osterie* as places imagined to be imbued with authentic *italianità* (Italian-ness): according to one guide-book author, these were 'the places that were the most humble, characteristic and frequented by ordinary folk', where one could drink local wines, eat local dishes, and converse with 'real' Italians.[15] This came from the most famous of the genre, Hans Barth's *Osteria: Guida spirituale delle osterie italiane da Verona a Capri* (1909), originally written in German (Barth was a German journalist and long-time resident of Italy), then translated into Italian. It went through at least three editions, the latest including a preface written by the Italian 'warrior-poet' and friend of fascism, Gabriele D'Annunzio. The *osteria* guide-books produced by Italians during the fascist dictatorship, sometimes by a collective of drinking companions as in the case of the multi-authored *Osterie Romane* (published in 1937 with a preface by then fascist minister for education, Giuseppe Bottai), other times by single authors such as Elio Zorzi's *Osterie veneziane* (1928) or Chino Ermacora's *Vino all'ombra* (1935), explicitly referenced Barth's work. They also followed Barth's modus operandi of depicting not only the physical setting, menu, and humour of the host of each *osteria* described, but of also offering detailed anecdotal vignettes reporting particular events, personal conversations, and interactions that the author(s) had witnessed or participated in within the spaces of each *osteria*. Of course, the everyday life historian has to be aware that at least some of these vignettes may be (partly) products of the author's imagination, but these sources remain important resources for establishing the connections between bars and politics and for setting out the possible parameters of political interactions and practices to which they could play host.

From the drinking companion-authors of *Osterie Romane*, for example, we learn of the political-social disruptions that could emerge among previously companionable regulars in response to contentious political debates. In 1914, 'the first conflicts among the clientele' erupted at the *Osteria all'Tempio d'Agrippa* in Rome, a bar whose mix of regulars included socialists, radicals, and Catholics as well as senators and journalists, as they divided into camps of advocates of intervention or neutrality 'in the period before the Great War'.[16] In his 'sentimental guide' to the bars and *osterie* of North-eastern Italy, Chino Ermacora depicted these as key spaces of local-national identity-construction and display, explicitly connecting them to the political struggles and conflicts around irredentism and nationalism across the late-19th to early-20thtwentieth century. As such, he revealed how one of the young *padroncine* of the *Trattoria della Ghiacciaia* in Udine made the inn a centre for Italian 'resistance' following the city's occupation by Austrian troops after Italy's defeat at Caporetto in November 1917. The *Trattoria del Monte*, also in Udine, was presented as a noted 'refuge' for irredentism (the political belief and movement for the incorporation into newly unified Italy of territory then ruled by Austria-Hungary to the north and north-east of Italy's pre-war border) and also for Great War interventionism.[17] The *Monte*, as Ermarcora called it, went on to become the 'newsroom' for (reportedly) the first interventionist national newspaper, *Ora o mai!* and, later, the place where 'the first [local] Mussolinians gathered'.[18]

Bars were spaces where not only 'Mussolinians' but also socialists, anarchists, and communists gathered. As such, bars constituted key venues for the playing out of political violence between fascists and socialists during the *biennio rosso* – the two 'red years' from 1919 to 1921 marked by factory occupations, strikes, and violence between socialists (and, from 1921, communists) and fascists - and in the runup to the fascist take-over of power in 1922.

To a significant extent, the political violence of the *biennio rosso* and period leading up to the fascists' March on Rome (October 1922) can be read as a contestation for control over the everyday spaces of Italian political life: *Case del Popolo* – the neighbourhood centres that were the seats of local socialist associational life – were assaulted, as were the local HQs of the *fasci di combattimento* (fascist fighting squads), conventional political party headquarters, and newspaper offices. Factories were occupied, and Italian squares, streets, and homes became contested sites of left- and right-wing demonstrations, intimidation, and violence, in which several thousands died and many more were wounded and/or forced into exile.[19] Contemporary newspaper reports and also the memoirs and diaries of those who participated in the political violence of the years before October 1922 clearly indicate that bars were important spaces in which political power was contested between fascists and anti-fascists.[20]

The diaries and memoirs of participants in the political violence in the years before and after Mussolini's accession to power can help us to understand the frequency with which bars were the sites for violent encounters between fascists and socialists. They also demonstrate that bars were spaces visited specifically for the purpose of enacting violence. To take an example from Venice, the diary of Raffaele Vicentini, an early convert to the fascist black-shirt, noted several occasions on which the city's bars hosted violent political encounters with socialists. His entry for the 8 April 1921, for example, recorded that a fellow fascist was assaulted by 'a group of subversives led by none other than the bar-owner' as he sat in *Bar Roma* in *calle dei Fabbri* reading *Italia Nuova* (a local fascist newspaper). In reprisal, fascist *squadristi* later 'entered the locale, approached a group of subversives and, without speaking, took the copy of *L'Eco dei Soviet* [a pro-Soviet newspaper] out of the hands of one of them and tore it up'. The 'fierce battle' that ensued resulted in gunshot injuries to one fascist and one communist.[21] On 20 May 1921, he wrote that two fascists, 'as they were sitting quietly' in the Camozza brothers' bar – wearing the fascist *distintivo* (badge) – were assaulted, prompting the local fascist squad to later 'devastate' the bar.[22] On 6 June 1922, a 'quiet' game of cards amongst a group of fascists and *Cavalieri della morte* (a paramilitary organization) in a bar on Via Garibaldi in the working-class district of Castello was interrupted when members of the royal public security guard sought to arrest one of the company, reputedly for having played the harmonica at a late hour. In the course of resisting arrest, one of the *Cavalieri* was shot by a guard and died, 'provok[ing] very lively agitation in fascist circles'.[23]

Whilst some of these political clashes were depicted as chance encounters, it is evident also that those involved recognized bars and other locales where men congregated to drink alcohol as spaces conducive to engaging in politically violent activity. Vicentini acknowledged that he and his fellow *squadristi* would frequent the working-class bars of Via Garibaldi and Campo Santa Margherita in order 'to drink and sing war and patriotic songs' with the express hope of inciting a (violent) response from the workers drinking there.[24]

The next step is to ask: What impact did the arrival to power of fascism and (from the mid-1920s) the subsequent dismantling of Mussolini's pseudo-parliamentary rule and consolidation of a fascist dictatorship have on the political interactions that took place within Italian bars? Here, it is important to note a number of broader contextualizing factors which effectively left Italian bars in a somewhat ambiguous position vis-à-vis the regime during and following the transition towards (a would-be totalitarian) dictatorship in the mid-1920s.

Firstly, Italian fascism had a somewhat ambivalent relationship with alcohol. On the one hand, in line with the projected image of Mussolini as a model of health and virile masculinity, it was claimed that he absteemed from alcohol. In step with the claims made by Italian temperance and prohibition campaigners that excessive alcohol consumption damaged the physical and moral health of the nation, in the first decade of its rule the fascist regime frequently espoused temperance rhetoric and passed legislation in 1923, 1925, and 1927 that aimed to curb alcohol consumption, especially among working-class and rural communities, for example by reducing the number of bars per capita and making changes to prescribed opening hours, and by increasing sales tax on wine. On the other hand, wine production was a vital facet of regional economies in Italy, with a powerful and active lobby – self-styled the 'defenders of wine' – to support and protect the interests of wine producers, distributors, and merchants. This lobby emphasized wine as a symbol of *italianità*, with an illustrious heritage stretching back to antiquity and the capacity to aid processes of nation-building in contemporary, fascist Italy. In the 'battle for wine' that ensued during the 1920s and into the 1930s between temperance advocates and wine industry lobbyists, it was the latter who claimed victory in 1937 via their journal *Enotria*.[25] Despite the legislation of the 1920s, which aimed to reduce the number of *esercizi pubblici* from one per every 500 inhabitants to one per every 1,000 inhabitants, the numbers of bars and other establishments licensed to sell alcohol did not decrease.

Secondly, the dismantling of the apparatus of parliamentary democracy entailed dissolving the long-standing spaces of working-class and rural (political) sociability, including the *Case del Popolo*, *Camere del Lavoro* (labour union centres), political parties and their headquarters, left-wing newspapers and their printing presses. At the same time, the regime's after-work organisation, the *Opera Nazionale Dopolavoro*, created in 1925, set about replacing or subsuming all non-fascist leisure-time associations.[26] OND centres, housed in local neighbourhood and city headquarters, as well as in larger factories and other sites of major employment, often contained their own *spaccio* or bar, as did the network of newly constructed neighbourhood *Case del fascio* (local PNF headquarters that effectively aped the functions of the *Case del popolo*).[27]

The effect of these developments was to drastically reduce the amount of unregulated, or even semi-regulated, space and scope for working-class and rural Italian political interactions. Arguably, in this environment, the political functions of bars actually took on increased significance during the fascist dictatorship as one of relatively few remaining non-official spaces for (political) sociability still available to ordinary Italians in their day-to-day life.

Of course (and herein the ambiguity), designating bars 'non-official' spaces in fascist Italy does not mean that we should view them as entirely unregulated spaces that somehow fell outside the gaze of the dictatorship. As has been noted, bars were subjected to a legislative and policing 'crackdown' over the 1920s.[28] Moreover, the regime itself recognized that bars were politically ambivalent spaces which continued to house and shape (potentially suspect) political interactions: as places where people both congregated and transited through day-by-day, and aided by the apparent ability of alcohol to act as a lubricant for political discussion, bars were key zones of operation for the regime informants who supplied fascist prefects and OVRA (secret police) chiefs with reports based on the gossip, rumours, and 'public mood' they overheard and detected.[29] Indeed, bar owners often became informants themselves.[30] In contrast, bars that had or developed reputations for harbouring 'subversive' people and behaviours could find themselves under surveillance and even closed down.[31] The bar licensing system was controlled by the police.[32]

Thus, bars were politically ambivalent spaces: places where regime informants and spies could try to gauge the public mood and into which representatives of the regime – black-shirted militia, party leaders, and Carabinieri and police officers – could enter in order to police everyday practices and utterances and, ultimately, to shore up the dictatorship. However, they were also suspect spaces in which Mussolini and his dictatorship could be ridiculed, criticized, and – in ways that were surely relatively microscopic and ephemeral but not without significance or impact – 'unmade'.[33]

The final set of questions examined briefly here explores this ambivalence surrounding bars as spaces of political interactions that might work to shore up or 'unmake' the dictatorship. How did people *use* bars as political spaces during the dictatorship? Did political practices and interactions in bars change over the course of the dictatorship? How did bars spatially shape those practices? Here, the evidence available in the archives of the fascist Public Security Police is invaluable. These record detailed accounts of events investigated as pertaining to the political crimes of 'subversive acts' and 'offences against the head of government'. The files contain material (investigative reports, telegrams, letters, court documents) written by police officers, court and prefecture officials but also include, as witness statements, letters from the accused/convicted, etc., source material that emanates from those Italians accused of having committed these political crimes. The 'General and Reserve Affairs' (AGR) files of the Public Security Police held in the Archivio Centrale dello Stato in Rome are arranged by year, province, and nature of political crime and number several hundred. Given the number of files, and that each file contains detailed evidence relating to multiple individual cases/incidents, it would be impossible to examine the full run of archival records. Instead, a significant sample (100 files in total) has been consulted, chosen for their geographical and chronological spread to ensure coverage of the country and empire, as well as the full span of the dictatorship.[34]

Table 7.1 Bergamo: incidents recorded in the 'subversive activity' files by location

Year	Bar etc.	Street/ piazza	Private home	Fascio/ OND/ official	Other	Location not specified	Total	% of total in bars
1924	10	14	3	0	0	7	30	33
1927	5	2	0	0	0	1	8	63
1932	9	4	0	0	1	1	13	69
1935	6	6	0	1	1	2	17	35
1937	18	8	1	0	4	2	33	55
1939	7	3	0	2	1	1	14	50
1941	5	1	0	1	3	2	11	45

Examined from a quantitative perspective, these sources can tell us much about the continuities and changes in the ways bars were used as spaces for political interactions and exchanges over the 20 years of fascist rule. In the first place, they allow us to track how frequently bars provided the venue for political interactions and conflict that went reported and resulted in police investigations relative to other spaces of political exchange. Here, all episodes reported and investigated as 'subversive acts' in the province of Bergamo in the industrial north of the country have been surveyed at intervals between 1924 and 1941 (see Table 7.1).

Whilst the total number of incidents recorded fluctuated over time quite considerably, the figures indicate that bars remained consistently the most prevalent location for low level political 'crimes' in the province, with the notable exception of 1924, when the dictatorial apparatus had not yet been installed (for example, political parties were not yet banned). The percentage of low-level political crimes that took place in the province's streets and piazzas increased again relative to bars around the mid-1930s, possibly connected to the Ethiopian War (1935–6), but it should be noted that all the incidents recorded in 1935 as taking place in streets and squares were cases that took place on the journey home from a bar (and in which the accused was described in police reports as 'drunk'). Of course, these are records only of those political altercations that went noted, and investigated, by the fascist authorities. We might speculate that private homes were spaces that hosted and facilitated politically 'subversive' discussions, actions and practices more often than is indicated by these figures. Nevertheless, the figures illustrate the persistence of bars' functioning as venues for the expression of unsanctioned and potentially regime-hostile views and beliefs through both the consolidation of the dictatorship (1925–9) and the regime's so-called imperial-racist-totalitarian turn of the later 1930s.[35]

In addition, using the police records both quantitatively and qualitatively, we can delve deeper to observe not only that bars continued to function as spaces for political exchange and conflict, but to identify continuities and changes in the particular ways in which bars were used and experienced, spatially, and in the ways in which these multivalent spaces framed and shaped such political encounters. It is clear that in the early years of the fascist regime (1922–6), the political encounters recorded in the police records as taking place in bars tended to involve multiple actors. For example, in the province of Bergamo, in 1924, the police files record three incidents in the province's bars investigated as potential 'subversive acts' involving 5 or more actors. In the same year in Arezzo, the only incident investigated involved 12 individuals; in Turin of the two incidents recorded as taking place in bars, one involved 4 'socialists', the other 19 'communists'.[36]

In addition, these early-years political encounters in bars were more likely than those of later years to involve organized political groups and to be premeditated in the sense that these spaces appear to have been consciously chosen and pre-arranged as venues for the discussion of politics or the seeking out of political opponents. The incident in the province of Arezzo mentioned above, recorded in late December 1924 in the village of Montemarciano, actually resulted from a pre-arranged 'punishment expedition' embarked upon by 'around 15' fascist black-shirts who travelled to Montemarciano 'with the intention of putting in their place some of the local subversives'. The black-shirts entered the *osteria* 'of a certain Antonio Gozzi, situated in the middle of the village and frequented by subversive elements'. Inside, the fascist black-shirt Federico Coppi slapped Vasco Bindi, presumed 'responsible for subversive propaganda', inciting a scuffle in which gunshots were fired just outside the *osteria* door.[37]

No serious injuries resulted from the 'punishment expedition' to the *osteria* in Montemarciano. However, the use of physical violence in this case is typical of the political encounters in bars during the early years of the dictatorship. Such encounters were significantly more likely than encounters in later years to involve physical violence, including serious violence, sometimes resulting in death. In Bergamo, for example, in 1924, eight violent political encounters were recorded as having taken place in bars or *osterie* or following evenings spent drinking in one, four of which resulted in loss of life.[38]

By the end of the 1920s, however, and through the 1930s, the political encounters involving bars and alcohol in the police records had changed significantly. After 1927, it is rare to read a case involving more than five actors and by the 1930s the cases of political conflict in bars involving large groups of actors, evident pre-meditation, and serious physical violence all but disappear. Instead, the commission of 'subversive acts' and, from the turn of the decade, the newly distinct crime of 'offending the head of government' were much more likely to involve individuals or pairs, usually family members or friends, and to constitute 'micro-acts' that appeared to be – or certainly were later presented as – spontaneous actions: the singing of 'The Red Flag' whilst drunk; an off-colour joke at the expense of Mussolini, the King, local fascist leaders, or militia members.[39]

These changes in the use of bars as political spaces resulted from the combination of multiple factors. The consolidation of the regime's repressive network after 1925–6 and the extension of what historian Paul Corner described as Italians' recognition and acceptance of, and negotiation within, the 'limits imposed on [their] behaviour' as the fascist regime entered its more 'total' – at least in its intent – phase of rule in the 1930s help account for this evolution from political encounters in bars that were pre-meditated, politically organized, and relatively large-scale to interactions that were mostly (apparently) spontaneous outbursts involving individuals or small groups more likely to be family or friends than political allies.[40] The risk to life and livelihood of hosting political meetings affiliated to now-banned parties was high, as many bar owners who found their businesses either temporarily or permanently closed as a result, discovered to their cost.[41] Both bar owners and their clients adapted to the changing regulation of their everyday spaces of operation: still, bars evidently continued to be spaces for the enactment of political sociability; political discussions and interactions evidently continued to take place within their walls – no doubt at least partially thanks to the often disinhibiting intoxicating products sold and consumed therein.

Conclusion

What emerges from the evidence contained in the files of the public security police is that, despite the hostile legislation of the 1920s, the acknowledged use of bars as key stomping grounds by regime spies and informants, and the consolidation of the wider apparatus of dictatorship during the 1930s, bars continued to be key spaces, and alcohol a key lubricant, that facilitated public political discussion among friends, family, and acquaintances, and fellow drinkers. The insights gleaned into the functioning of bars in fascist Italy adds to our understanding of the functioning of other everyday spaces – markets, colonial settler villages, the piazza, the home – as well as other facets of the spatial history of Italian fascism reached by scholars.[42] The fascist regime's creation and domination of national and local ceremonial or 'monumental space' in order to display its aestheticized politics and to mobilize Italians, as well as its colonization of space, both within Italy (for example, through the dredging of the Pontine Marshes and creation of five 'internal colony' new towns) and through fascism's violent imperialist expansion in the Mediterranean and North and East Africa, have been and continue to be documented.[43] The spatial re-designation of what the regime considered 'private' and 'public' space, physically and metaphorically, has also been explored.[44] This case-study, and other examinations of everyday spaces, add to this bigger picture of the spatial history of Italian fascism by seeking to show how, and how far, the dictatorship moved in – and out – of Italians' everyday worlds but also – and crucially – how Italians' use of everyday spaces, and the practices and interactions (with the regime and with each other) that took place therein, themselves conditioned and were conditioned by space.

More broadly, the case study points to some of the possibilities in terms of how historians might approach and use various source materials to uncover the functioning of everyday spaces in a dictatorial context (and also in a non-dictatorial setting). As a 'way of seeing' the past and of reading sources, it follows that there is not necessarily a corpus of distinct sources that belong (exclusively) to 'spatial history'. Certainly, some source types might particularly or obviously lend themselves to a spatial history approach such as those that provide visual representations or depictions of space like maps, architectural plans, photographs, films, and newsreels, or those material objects that actually existed in past spaces and which remain physically present for historians to materially examine, or those that deal specifically with travel across space such as travel guides and motoring brochures. In its exploration of the everyday spaces of bars in fascist Italy, this case study has made recourse to only one source base that would fit the above source types: *osterie* guides, a sub-genre of travel guides. Undoubtedly, the examination of, for example, photographs and architectural plans or depictions in contemporary fiction of interwar Italian bars could only enrich this study. However, this study has wished to demonstrate how one might also use other kinds of (textual) sources that are in no way peculiar or necessarily particularly suited to spatial history – diaries and police records – to understand how everyday spaces were used by ordinary people in a dictatorial setting, how those spaces were shaped by, and themselves influenced, the myriad interactions and relationships that were enacted within them. Certainly these source types have proved invaluable to everyday life historians who, as we've noted, have increasingly taken more overt interest in the spatial dynamics of the lived experiences of past 'ordinary' individuals. As we have also seen, whilst these source materials (even ego-documents) have to be used with due care and attention to the mediated and framed access they offer to past historical actors' everyday thoughts, actions, and practices, they have much to tell us about how spaces of everyday life were created and *used* through experience and practice and how these interacted and overlapped with other spatial (and non-spatial) frames to form the plurality of spaces that comprise and shape historical lived experience.

Notes

1 Paul Steege, Andrew Stuart Bergerson, Maureen Healy and Pamela E. Swett, 'The History of Everyday Life: A Second Chapter', *Journal of Modern History* 80, 2008, 358–78, here 363–8.
2 Alf Lüdtke (ed.), *The History of Everyday Life: Reconstructing Historical Experiences and Ways of Life*, trans. William Templer, Princeton: Princeton University Press, 1995; Detlev Peukert, *Inside Nazi Germany: Conformity, Opposition and Racism in Everyday Life*, trans. Richard Deveson, Harmondsworth: Penguin, 1989. The 'second chapter' assertion is made in Steege et al., 'The History of Everyday Life'. Examples of more recent works that are informed by *Alltagsgeschichte* include Kate Ferris *Everyday Life in Fascist Venice, 1929–1940*, Basingstoke: Palgrave Macmillan, 2012; Andrew Stuart Bergerson, *Ordinary Germans in Extraordinary Times: The Nazi Revolution in Hildesheim*, Bloomington: Indiana University Press, 2004; Shannon Lee Fogg, *The Politics of Everyday Life in Vichy France: Foreigners, Undesirables, and Strangers*, Cambridge: Cambridge University Press, 2009.

3 Alf Lüdtke, 'Introduction: What Is the History of Everyday Life and Who Are Its Practitioners?', in id. (ed.), *History of Everyday Life*, p. 4.
4 Alf Lüdtke, 'What Happened to the "Fiery Red Glow"?: Workers' Experiences and German Fascism', in id. (ed.), *History of Everyday Life*, pp. 198–251.
5 Sheila Fitzpatrick, *Everyday Stalinism: Ordinary Life in Extraordinary Times – Soviet Russia in the 1930s*, New York: Oxford University Press, 1999.
6 Belinda J. Davis, *Home Fires Burning: Food, Politics, and Everyday Life in World War I Berlin*, Chapel Hill: University of North Carolina Press, 2000, p. 5; Steege et al., 'The History of Everyday Life', 361.
7 Joshua Arthurs, Michael Ebner and Kate Ferris, 'Introduction', in id. (eds.), *The Politics of Everyday Life in Fascist Italy: Outside the State?*, New York: Palgrave Macmillan, 2017, p. 8.
8 Michel de Certeau, *The Practice of Everyday Life*, trans. Steven Rendell, Berkeley: University of California Press, 1988, pp. xi–xxiv, 15–42.
9 Steege et al., 'The History of Everyday Life', 364.
10 Arthurs, Ebner and Ferris, 'Introduction', p. 6.
11 Included in this group are all forms of *esercizi pubblici* (licensed premises) in Italy, which alongside classic bars include also cafés, *osterie, trattorie, spaccie* (bottle shops), and *bettole* (dives). For ease, the chapter will refer simply to either *esercizi pubblici* or bars, intended as a shorthand to include all licensed venues in which alcohol was sold and also consumed on the premises.
12 Philip Morgan, '"The Years of Consent"? Popular Attitudes and Forms of Resistance to Fascism in Italy, 1925–1940', in Tim Kirk and Anthony McElligott (eds.), *Opposing Fascism: Community, Authority and Resistance in Europe*, Cambridge: Cambridge University Press, 2004, pp. 163–79 (169 for the quotation).
13 Rudolf Dekker, 'Introduction', in id. (ed.), *Egodocuments and History: Autobiographical Writing in Its Social Context since the Middle Ages*, Hilversum: Verloren, 2002, p. 7. In relation to Fascist Italy, these sources were recently used to excellent effect by the late Christopher Duggan in *Fascist Voices: An Intimate History of Mussolini's Italy*, New York: Random House, 2012.
14 On how these multiple functions operated in another dictatorial context see David Gilmore, 'The Role of the Bar in Andalusian Rural Society: Observations on Political Culture under Franco', *Journal of Anthropological Research* 41, 1985, 263–77.
15 Hans Barth, *Osteria: Guida spirituale delle osterie italiane da Verona a Capri*, trans. Govanni Bestolfi, Rome: Enrico Voghera editore, 1926, p. xiv. It should be noted that these *osteria* guidebooks had intended readerships, including wealthy foreign travellers to Italy and the authors' Italian friends and wider fellow middle- to-upper-class Italians.
16 *Osterie Romane* 64–5.
17 Chino Ermacora, *Vino all'ombra*, Le Tre Venezie, 1942, p. 104.
18 Ibid. 107.
19 Michael Ebner, *Ordinary Violence in Mussolini's Italy*, Cambridge: Cambridge University Press, 2011, p. 9.
20 See Giulia Albanese, *La marcia su Roma*, Rome: Laterza, 2006. The prevalence of bars as loci for political violence is noted by Michael Ebner in *Ordinary Violence*.
21 Raffaele Vicentini, *Il movimento fascista veneto attraverso il diario di uno squadrista*, Venice: Soc. Acc. Stamperia Zanetti, 1935, p. 109.
22 Ibid., p. 125.
23 Ibid., p. 223.

24 Alesandro Casellato, 'I sestieri popolari', in Mario Isnenghi and Stuart Woolf (eds.), *Storia di Venezia: L'Ottocento e il Novecento*, vol. II, Rome: Istituto della Enciclopedia Italiana, p. 1583.
25 *Enotria*, January 1937 pp. 36–7.
26 Victoria de Grazia, *The Culture of Consent: Mass Organisation of Leisure in Fascist Italy*, Cambridge: Cambridge University Press, 1981. The leisure activities of upper-middle- and upper-class Italians, which predominantly took place within private and commercial spaces, were far less regulated and scrutinized than were working-class and rural peasant leisure activities.
27 Lucy Maulsby, 'Case del fascio and the Making of Modern Italy', *Journal of Modern Italian Studies* 20, 2015, 666–8.
28 Victoria de Grazia, *How Fascism Ruled Women: Italy, 1922–1945*, Berkeley: University of California Press, 1992, p. 202.
29 This is clear from the informants' reports held in the Archivio Centrale dello Stato (henceforth ACS) in Rome. See for example: MIN CUL POP gab. b.163 f.1052; MI DGPS OVRA b. 7 f. III zona; MI DGPS OVRA b. 7 Sardegna.
30 Ebner, *Ordinary Violence*, p. 57.
31 See, for example, MIN INT DGPS AGR 1924 b.58 Arezzo and Bergamo.
32 Ebner, *Ordinary Violence*, p. 57.
33 'Unmade' is a paraphrase of Ruth Ben-Ghiat's observations about the 'making' and 'unmaking' of the fascist project in *Fascist Modernities: Italy, 1922–1945*, Berkeley: University of California Press, 2001, p. 15.
34 Material relating to the following years were examined: 1924; 1927; 1931–2 (filed together); 1935; 1937; 1939; 1941).
35 Alexander De Grand, 'Mussolini's Follies: Fascism in Its Imperial and Racist Phase', *Contemporary European History* 13, 2004, 127–47.
36 MI DGPS AGR 1924 b. 58 Bergamo; MI DGPS AGR 1924 b. 58 Arezzo.
37 MI DGPS AGR 1924 b. 58 Arezzo.
38 MI DGPS AGR 1924 b. 58 Bergamo.
39 See, for example, MI DGPS AGR 1932 sec 1a b.12 Como.
40 Paul Corner, 'Collaboration, Complicity and Evasion under Italian Fascism', in Alf Lüdtke (ed.), *Everyday Life in Mass Dictatorship: Collusion and Evasion*, London: Palgrave Macmillan, 2016, p. 79; see also de Grand, 'Mussolini's Follies'.
41 See, for example; MI DGPS AGR 1924 b. 58 Bergamo.
42 Kate Ferris, 'Consumption', in Arthurs, Ebner and Ferris (eds.), *Politics of Everyday Life*, pp. 123–49; Roberta Pergher, 'Between Colony and Nation on Italy's "Fourth Shore"', in Jacqueline Andall and Derek Duncan (eds.), *National Belongings: Hybridity in Italian Colonial and Postcolonial Cultures*, New York: Peter Lang, 2010, pp. 89–106; Mario Isnenghi, *L'Italia in piazza: I luoghi della vita pubblica dal 1848 ai giorni nostri*, Bologna: il Mulino, 2004; de Grazia, *How Fascism Ruled Women*.
43 Mabel Berezin, *Making the Fascist Self: The Political Culture of Interwar Italy*, Ithaca: Cornell University Press, 1997; Ruth Ben-Ghiat and Mia Fuller (eds.), *Italian Colonialism*, New York: Palgrave Macmillan, 2005; Mia Fuller, *Moderns Abroad: Architecture, Cities, and Italian Imperialism*, London: Routledge, 2007.
44 Luisa Passerini, *Fascism in Popular Memory: The Cultural Experience of the Turin Working Class*, trans. Robert Lumley and Jude Bloomfield, Cambridge: Cambridge University Press, 1987.

8 Rivers

Mark Harris

In this chapter, I ask what kind of historical space is a river and explore the documentary sources that can be used to answer that question in the Amazon region of South America. River spaces offer an intriguing perspective on the history of human activities because of water's shifting character. Rivers can erase the past through floods and deposit their load to cover old land. The way people learn to live and work with a river offers crucial insights into our relationships with the environment. With the modern age we have sought to control, like no other human society before, the riverine space with dams, embankments, channelling, and dredging. The effect has been to transform completely the riverine character of this space, affecting not just the water itself but animals, fish, insects, the banks, floodplains, and neighbouring lands. In short, some rivers have had the life taken out of them. Rivers are interesting to the spatial historian because the way people live around rivers – human spatial arrangements which include what they do to the waters – is connected to their social values and political structures.

More literally, rivers define, join, and divide human spaces. Some rivers, such as the Thames or the Nile, come to stand for a whole country, and others cut across many, such as the Rhine or the Danube.[1] They mark boundaries between nations and meander through cities giving distinct identities to either side. Yet whether humans use riverine spaces for symbolic and ritual purposes, or to harness their powerful energy, the irrepressible fact is that once in a while rivers will fight back. Flash floods happen – and can be made worse by human management of waters. This force of nature has encouraged environmental historians to give rivers individual attention, their own history, and be treated as subjects in their own right.[2] For example, Mark Cioc has written a history of the Rhine with the subtitle 'an eco-biography', a life story of an ecological entity.[3]

With inspiration from this approach we come to the Amazon, which has resisted much of the modernization seen along other major rivers of the world – the Mississippi, for example. Even though hydroelectric dams have been built on some of the Amazon's tributaries, the region has escaped large-scale manipulation of its flowing waters. It is simply too powerful. In line with the more subjective approach, the Amazon provides a case study to explore how unfettered riverine space has been part of human life. There is

DOI: 10.4324/9780429291739-11

no single Amazon. People have lived there for at least 12,000 years in quite distinct societies evident in the archaeological record, and it was colonized by Europeans in the 17th century. Different societies have interacted with the rivers of the Amazon in quite diverse ways. Although it has changed, the Amazon River is a constant through history. How can these social and natural changes and continuities be combined in a life story of the Amazon?

I try to answer this question through my own mix of sources. I have conducted research with floodplain dwellers in communities along the Amazon River and some of its tributaries in the state of Pará, Brazil.[4] I have also investigated archival collections in Europe and South and North America to examine the historical emergence of Amazonian riverine cultures that threaded together diverse kinds of people: Indians and Africans of several nations and poor Europeans.[5] Often survivors of traumatic episodes in the colonial and national periods (c. 1650–1870), these individuals created enduring riverbank societies as the Portuguese established an insecure domination. To reconstruct the spatial histories here I have had to make my own imaginative leaps for the documents contain only fragments of information. These reconstructions are based in familiarity with the setting and experience of living along the rivers and the changing height of the river following the wet and dry seasons in the basin.

Some relevant background on rivers

Recent scholarly writing on rivers has tended to the environmental and the political.[6] Although these are hard to separate in the life of a river, scholars who emphasize the political dimension in their studies have always gone back to a key figure, Karl Wittfogel. He was a German historian of China who managed to reach the United States in the 1930s after his internment by the Nazis. His book, *Oriental Despotism*,[7] was the first attempt to make water central to the theoretical understanding of the development of society. He argued that in arid and semi-arid regions of the world the large-scale irrigation of water led to the rise of an elite state bureaucracy. To serve cities and fields for agriculture, water had to be led there deploying many people in the construction of channels. Directing the flow of water gave elites great control over labour and a centralized state. A hydraulic society then was one where water control was critical to its successful functioning. Although Wittfogel's main example was imperial China, he considered many others such as the Andean mountains and Inca irrigation. In the Amazon, water control was, and is, supposedly unnecessary though there is archaeological and contemporary evidence of water manipulation, if not for agriculture.[8] This raises the question whether ancient (and contemporary) Amazonian societies did not develop into states because there was no need to control water.[9]

Wittfogel's ideas have spawned much debate. Simon Schama, an art historian who wrote a famous book on landscape and memory that became a television series, divides ancient hydraulic societies into two, circular and linear. In the former, which includes the ancient Egyptians' relationship to the Nile, the emphasis is on the importance of the source of the river for fertility and a mythical return there. The Romans, by contrast, sought to straighten their rivers, giving them road-like forms.[10] This distinction is relevant because I argue that ancient Amazonia had a circular attachment to moving water. Rome was the centre of the vast Roman Empire. All roads led there. Archaeological evidence has revealed lowland south America, which includes the Amazon, was sewn together by river and roads in a polycentric pattern that traversed natural frontiers such as watersheds and ecological zones. It has also shown that the chiefdom societies had distinct cycles of growth and decline, which meant that core areas continually shifted and there was no unilinear move to greater centralization and the emergence of a state. Outside these concentrations of human settlement, sites of special power were marked by petroglyphs (rock art), such as rocks near running water, waterfalls, rapids, semi-submerged stones in a river, suggesting their mythological magnitude. These places, many of them in or near rivers, can be seen as spaces of encounter, intercultural positions outside of towns or cities. I see these spaces as hubs for gatherings, large and small, although it is impossible to know with certainty given our current knowledge. The cyclical movement between them and centres of population through the ritual – and practical – role of flowing water is what kept life going in ancient Amazonia – and if you like, kept the state from coming into existence.[11]

This rotating model is quite dissimilar to Wittfogel's one. Nevertheless, his line of thinking has endured and has recently been updated by the historian of the US, Donald Worster. He studied the push to the American West in the late-19th century, where rivers were transformed to support the frontier society in the desert-like environment, which became the 'hydraulic west'. This political drive for agricultural development came at great cost to the natural world.[12] In a similar vein, another American historian, Richard White, looks at the Columbia River of the Pacific Northwest but he reaches different conclusions. The river is seen as an 'organic machine', echoing the notion of hydraulics. But human history for White is not there to suppress nature.[13] Instead different societies live with the river in different modes – and on several scales. Nature is always there even if dams are built, and fewer kinds of fish, animals, and insects make their home on or near the river. Societies work with the river in their specific ways, creating their own mixtures of social and natural history.

Here I would like to introduce the notion of scale, which is critical to spatial history. Scale has various meanings but here I use it to refer to the range, or reach, of an activity over an area. Thus, fishing for salmon in a river is on larger scale given the need for salmon to spend most of their lives in the sea and only come upriver to breed, which makes them easier to catch. But fishing for pike occurs on a much smaller scale. Depending on the activity the river space can extend to the ocean or be quite localized. The notion of scale then can be applied to all sorts of practices and indeed to the way a community or society uses a river. Critical to the notion of scale is the ability to move along a continuum from face-to-face encounters to transnational connections along fluvial pathways. We have to recognize that rivers lend themselves to multi-scaling and do not confine us to one single scale, however that may be defined. But scale is not only about space or geography; it concerns timescales – hence the powerful phrase spatial history.

This chapter will now proceed to sketch a spatial history of the Amazon. I start with the period that preceded European arrival in the early 16th century, consider the impact of European presence, then concentrate on one river by analyzing a document written by a Jesuit in the 1750s, and end with the present-day reality on that river. To reconstruct the ancient Amazon I draw on archaeological studies that make reference to ceramics, their styles and distribution, human made soils (known as dark earths because of their high charcoal content), and landscape modifications. Most archaeologists of the Amazon see the place as transformed by human work. There is no pristine Amazon, they say. This makes the discussion about spatial history even more relevant and valuable for it takes us into a 'deep history'.[14] We can compare and contrast different ways of living in the Amazon and understand how the rivers are a constant feature flowing through them all.

The Amazon River, and its peoples

We know that the ancient people of the Amazon were different to contemporary indigenous societies. From at least 600 CE, they lived alongside large navigable rivers on bluffs and floodplains but also higher land near water sources. Their level of social and economic development was integrated with riverine resources, but by no means did they only live along the main river trunk itself. They traded in precious objects such as green stones, gold, and shells across the northern part of the continent up to the Caribbean. The societies of early to mid-16th century lowland South America were large and complex in social and political organization. Some sites must have been city-like with many thousands of people (perhaps in the tens of thousands). Although they had agriculture and were sedentary, there was significant dependence on extractive activities such as fishing, hunting, and collecting from a semi-cultivated forest.[15]

The challenge that scholars have had is to understand the political and social character of these societies on the eve of Europe's advance to the Americas. The Amazon was quite different from the state-like highland society of the Inca, their neighbours in continental terms. Power, for the lowlanders, was hierarchically organized with specialized roles for priests or shamans and chiefs, but these were not centralized societies, which used tax collecting, labour obligations, coercion, and violence to enforce and legitimate authority. Rule was instead worked by persuasion through leaders, or a council of elders. Disputes led to a dissatisfied group moving away and fragmentation, making for the refounding of settlements elsewhere. On a larger scale this breaking-up may account for the rise and fall of Amazonian polities before the European arrival.

The first European descent of the Amazon, in 1541–42, left a remarkable piece of writing by an impressionable Spanish Dominican, Gaspar de Carvajal.[16] He observed long lines of massive wooden and adobe buildings with plazas on the riverbanks but also big gaps in between these townships. His indigenous informants told him that these neighbours were at war, so these spaces have been considered a buffer zone by archaeologists. They might also have been frontiers that instead of dividing people, wove together groups in diverse ways over time – marriage, rituals, feasting, and war, for example. When the whites made landfall, the crew noticed well-laid tracks extending for many miles into the hinterland, like a 'royal road' in one instance. Once, they followed a path for half a day but gave up because they did not seem to be reaching a destination. Carvajal was told there were even larger population centres far away from the main Amazon riverbanks. Most famously the Amazon women were said to live in a city with huge coloured temples many weeks' journey to the north, roughly locating them in the Amazon watershed region with the Orinoco.[17]

The combination of the archaeological evidence and the historical sources indicate these Amerindian societies were spread out evenly over the whole region, rather than concentrated on the main riverbank, which we might assume from reading Carvajal. This arrangement is just as one might expect from well-established patterns of social development and expansion. But it is very different from the principally river-focused Amazon that the Portuguese sought to create in the 17th century, and dominates our understanding of the region today. Read carefully nowadays, Carvajal's relation helps reconnect indigenous societies with their spatial histories. The scale of the Amerindian was continental, composed from an organic series of many centred networks and linked by rivers and roads, which traversed key hubs in between these core areas.

One further element of riverine space that was noticed by the Europeans was the seasonality of food availability and residential patterns. The first descent of the Amazon took place near the peak of the wet season. Houses on stilts were clearly visible from the river. Some of these were likely temporary structures for use when procuring fish and preparing them for consumption elsewhere and trade for other foodstuffs. As above, riverine space was part of a wider spatial matrix of livelihood making and exchange networks. In other words, human settlement in the aquatic environment was integrated in a regional network formed along rivers, flooded and otherwise. The scale of these regional systems depended on the item being traded (very large-scale for precious items and small-scale for food). This regional interdependence and articulation with different scales of interaction linked to different products and activities is paralleled by the example of Cahokia, the centre for the whole of the Mississippi basin in the early part of the last millennium.[18] In the Amazon there were multiple centres of population and meeting points for seasonal trading and work and ritual observances, many of which have not been located by archaeologists. Put crudely, much of social life was taking place outside of the towns and city like places in temporary habitations and while journeying.

The story of El Dorado, a huge lake where gold apparently emerged in abundance according to early European chroniclers, could be reconsidered from the point of view of the seasonality of social life. The lake was said to lie somewhere in the watershed mountains between the Orinoco and Amazon basins and almost certainly was seasonally inundated and the location for regular indigenous meetings and trade in precious goods.[19] Its promise of treasure enchanted one of the favourites in the Elizabethan court, Sir Walter Ralegh in the early 1580s. The text that resulted from Ralegh's first expedition, *The*

Discoverie, is clearly a product of its time but it can be read critically now for its portrait of native societies, amongst other themes.[20] Like Orellana in Carvajal's account, Ralegh sought to dominate the river not by conquering it but by travelling along it with ease. This aim was impossible. The ships were too big for the shallow portage at the mouth, so the crew had to move into smaller canoes that had to be paddled. They could not find enough food and feared the strange beasts beneath the river's surface. Ralegh met Topiawari, a powerful chief, at the mouth of the Caroni River (about 200 miles up the Orinoco) and was given information about a golden lake much further upstream. On they went, remarking that the Indians became more numerous and nobler looking. But the river became so fast flowing and the rains so hard, they turned back downriver, pockets empty. The river had tamed the ambitions of the whites. Had they continued up the Orinoco they would have reached its headwaters, where the Casiquiare canal provides a navigable channel to the Amazon via the Negro River. Early colonial chroniclers knew about this route because indigenous people used to trade European goods across it (see Figure 8.1).

El Dorado was a lake, a city, and also a man.[21] Its meaning and location shifted according to indigenous informants.[22] Even so, all the indications (mythical or real) are that it was perceived to be in the headwaters of one of the rivers, which come down from granite highlands, that mark the watershed

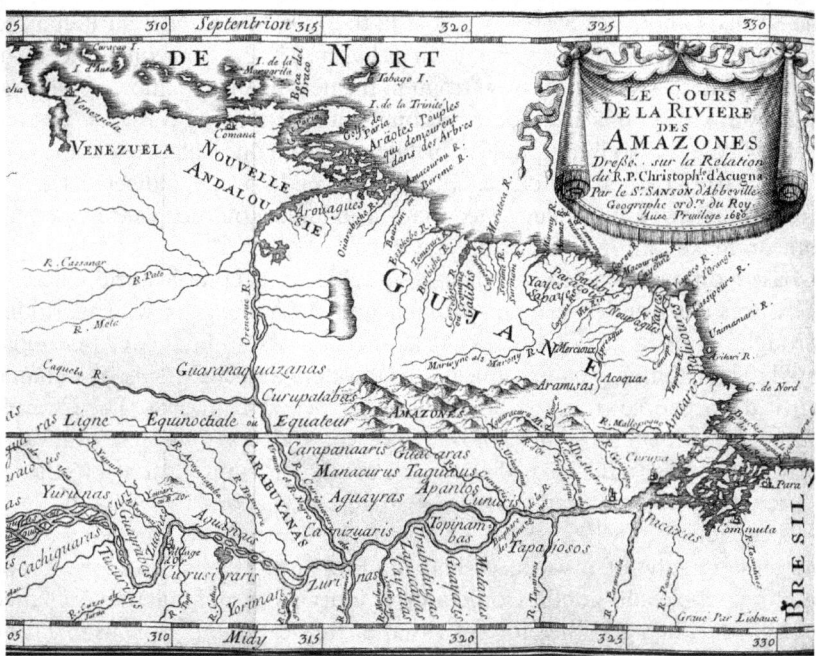

Figure 8.1 Map of the Amazon area of Guiana. It belongs to the work "Relation de la Riviere des Amazones", Paris, 1680.

Written by Cristobal de Acuna (1597–1676), Spanish Jesuit missionary. © Album/Alamy Stock Photo.

with the Amazon River in the south. From colonial times this region has been marginal to societies that were dominated by the Atlantic coast. Yet in these stories, it was a source of wonder and wealth. For sure, no abundant source of gold has been found but that is the nature of stories and myths: they don't need to be true in the way the Europeans put them into texts. They simply need to hold cultural power. The headwaters are the beginnings of life and source of fertility. Hence their ritual and practical significance in a society whose motif was circulation.

Returning to indigenous spaces and their clash with European conquest

If El Dorado is read alongside the story about the women-only society – the Amazons from Carvajal amongst others – situated again in the headwaters of a north bank Amazon tributary (such as the Branco, Trombetas, or Nhamundá) then we have indications of two sources of power and authority in the middle of northern South America. These places articulate the Orinoco, Amazon, and the smaller basins of rivers that flow into the Atlantic where the Dutch and French established colonial presence. Collectively these places divide into Guiana and the Amazon. But what is the nature of the frontier between these regions and whose is it? There is a second known connection, in addition to the Casiquiare canal, that was well used in the early colonial period. This is a portage between the headwaters of the Essequibo and Rupununi in Guiana, and the Branco (whose headwaters are close to the Trombetas, and is a tributary to the Negro and on some maps is known as Parima) in the Amazon. Canoes can be carried across from one river system to the next in an area known as Pirara. In the wet season this area floods, becoming a lake, making travel across even easier. There was a vibrant trade in metal tools from one watershed to the next in the early colonial period, as Europeans tried to buy indigenous allies. Archaeologists have also suggested that Carib indigenous people traversed the route to make war on enemies in the Amazon.

Based on the archaeological, cartographic, and ethnohistoric evidence there was no frontier between Guiana and Amazonia, as we have today through nation-state boundaries. At least there was no boundary that separated. This frontier was natural, the headwaters and highlands of Guiana, but it drew people from various downriver societies upwards. Their travels embroidered them together as they exchanged products of high cultural value from one region to the next. If they were conducting important trade, then it is likely they were observing religious rituals together and feasting.

These activities, when put together with the stories above, present a native moral geography of a world of back-and-forth movement along rivers, of rotating people and goods. Could all this journeying and material exchange relate to fertility, and therefore of ritual birth, death, and replenishment be like the Ancient Egyptians' venerated Osiris? This image of a wheel with the hub in the hilly and boggy lands either side of the watershed is my own.[23] It can be visualised in a series of late 16th century and 17th century maps which were largely based on Ralegh's and Carvajal's accounts, such as in Figure 8.2.

Figure 8.2 German map of the Guianas based on Ralegh's account, 1599 (John Carter Brown, J590 B915v GVL8.1 / 2-SIZE, reproduced with the kind permission of the John Carter Brown Library).

Note the lake in the middle and the figure of the woman striking a pose with a long bow in her hand. © British Library Board. All Rights Reserved/Bridgeman Images.

Amerindian affairs stretched across the whole land mass. Key meeting places included watersheds, rapids, seasonal lakes, and interfluves and can be seen as nodes of scalar articulation. Divisions between regions grew with greater European presence, as well as the catastrophic population decline of indigenous people. No longer were these long-distance journeys viable for so few people. By the end of the colonial period in the early 19th century the north part of Lowland South American had been split up. A series of much smaller regional societies became established along tributaries and the interfluve areas between them. In the following section we will consider one of them – the Tapajós/Madeira complex. In ancient times this was the southern tip of the Guiana/Amazon space.

What is striking in this map is the combination of story-like figures and geographical features. The representation of the new world by the old world mixed them into one understanding.

Colonial realignments and indigenous spaces

From early on the Portuguese understood their ability to conquer the region depended on their control of the riverways, rather than colonizing the land and enslaving Indians in large numbers. They sought to achieve this aim by establishing forts on bluffs, at the mouths of major rivers or in strategic areas where the river narrows. They designed specially outfitted large canoes with gunwales, ensured they had strong native paddlers, and trained soldiers for fighting on the river. In collaboration with the missionaries, these soldiers and traders went up each tributary seeking to 'pacify the riverbanks'. Hardly penetrating the interfluve forests of the Amazon by the early 18th century, a network of private interests and state institutions had become present over most of the Amazon basin. The Treaty of Madrid in 1750 would give the Portuguese formal possession of this space. From this time, the Portuguese overseas council explicitly pursued a policy that would put the rivers 'at the service of the colony ... to establish civilisation'.[24] In practice, there was much more of a mutual relationship between nature and society: people worked with the rivers, and the flowing waters gave people work and life-giving resources. These novel social patterns gave rise to a new spatial arrangement, an extensive network of hamlets and forts on the riverbanks and floodplains. If you like, a new scale was born, one that encompassed the 'Amazons', all the rivers in the water-basin.

A cartographic representation from 1753 of this aquatic assemblage in a new regional society is provided in Figure 8.3. The main Amazon river and its villages are the focus of the representation. The tributaries and the interior

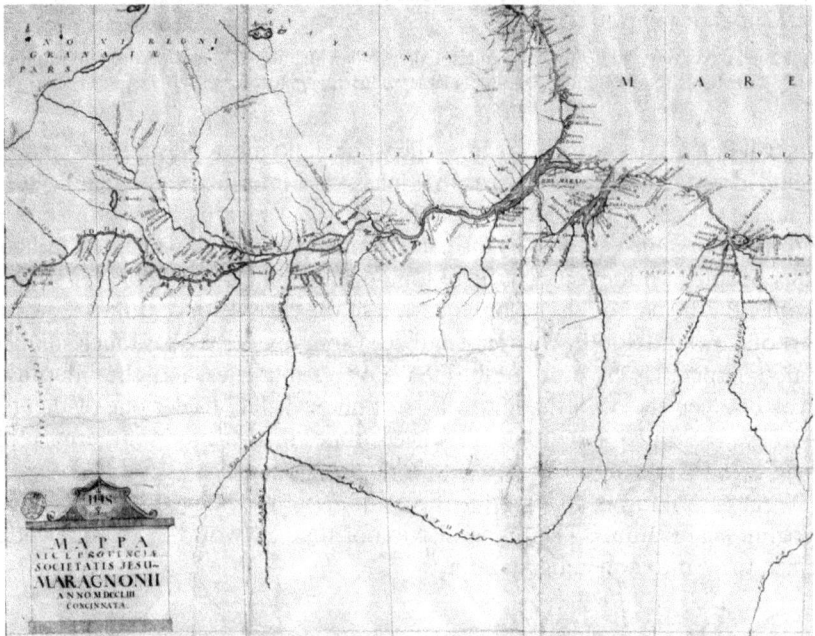

Figure 8.3 Part of a map of the Amazon showing riverbank Jesuit missions and other centres of the population, 1753 (Biblioteca Pública de Evora, GAV 4 N25).

Notice that all settlements have a dual position, inward and outward facing: these places are like portholes, stop-off points along a water highway.

of northern South America have been stripped of their stories (as conveyed in Figure 8.2). A lake, no longer El Dorado or Parime, is still indicated, but is much reduced. The lands and rivers have been emptied of significance by the people who drew the map. But – as we will see in the following section – this was not the complete picture. If we look at particular rivers through rarely read documents, a messier situation is revealed. Indigenous people continued to live, and grow, in these remoter regions (as seen from the white viewpoint) and influence the course of colonial society and its policy towards the region.

So far, I have traced in broad terms the transformation of the physical meaning of riverine spaces in the early modern period for northern South America. Amerindians and Europeans perceived the rivers and their capacities in various ways. Amerindians had built up a series of political, economic, and ritual associations that emphasized spatial rotation, up and down the rivers, and polycentricity. Europeans, having suffered an initial taming by the rivers, through their dependence on native labour and knowledge for navigation, developed their own mastery of the riverine space. This was not born of control of water movement, as in other 'hydraulic societies', but in linear spatial extension, a much simpler – but very effective – form of movement along the rivers. Its scale was transatlantic and consolidated by the riverbank settlements that were established to reconstitute the river as a concourse for Portuguese values and interests, which had their own connections, imaginary and material, to Europe.[25]

Having outlined two basic and large-scale spatial patterns, it is now time to shift focus and zoom in to one river. Here we will see how the various scales articulate and the need for a multi-scalar analysis. Up to the European arrival the Tapajós region was part of the continental network outlined above with strong connections to the north. This reach was reduced over the colonial period, the consequences of which we can now look at in the following section.

A riverscape in the Amazon

The Tapajós River is a significant tributary of the lower course of the Amazon River. Its headwaters start in central Brazil and flow northwards for about 1,900 km, passing over the Brazilian shield, an ancient and weathered area of sedimentary rock. This means the water is clear, almost greenish, carrying little in its load; and when it meets the creamy coffee Amazon, there is an eye-pleasing mix of colours. Some 500 km to the east is the Xingu River. On the western side is the Madeira River, whose sources are in current day Bolivia in the foothills of the Andes some 3,200 km away. All these rivers were important regions of Amerindian settlement before the Europeans came with their own significant regional centres. As we saw in previous sections, over the colonial period older connections and rivalries gave way to a new way of relating to the riverine spaces and the forests in between. In the Tapajós, there are many reasons for this transformation. Principal among these are the impact of disease and Portuguese slaving expeditions between 1630 and 1750 and the Jesuit missionization in the Lower Tapajós from 1660 and 1750.

The Portuguese themselves did not always carry out their own search for slaves. Instead, indigenous allies often did this work, trading with the Portuguese for iron tools, cloth, and alcohol. Indian groups could become stronger or weaker as a result of their alliance with the Portuguese. Similarly, those who were raided for slaves might have joined forces with the enemies of the whites to better resist these unwanted incursions. These relations transformed over time as political leaders changed and opportunities developed. In terms of the waterscape, this meant Amerindian activity avoided the river spaces, for this was where the Portuguese and their allies travelled and installed themselves. The hinterland, that is the interfluve areas, became the main location for these reconfigurations of Indian friendships and for their conduct of war and revenge attacks. The deeper the slavers and missionaries went to find 'recruits', the more colonial pressures became part of Amerindian affairs. As more Indians were brought to settle by force or voluntary relocation along the main riverbanks at strategic locations – mouths of major tributaries, prominent bluffs, near good soils or hunting grounds, away from insects – new communities and a regionally interlinked society came into being. Rivers were the connection; all rivers led to the capital, Belém, at the mouth of the Amazon.

From archaeological and ethnohistorical studies, we know there was a large settlement of near city-like proportions at the eastern side of the mouth of the Tapajós, one of the most populous along the Amazon riverbank and the centre of a large-scale regional network that involved the trade of green stones.[26] By the end of the 17th century the powerful Tapajó people, after whom the river is named, were much reduced through slaving, infighting, and disease. The Jesuits had managed to establish a mission on the site of the ancient city. From this base, they supported slaving expeditions that went up the Tapajós and built a network of missions, exchanging people and goods. Soon that network had grown to about 8,500 Indians in 11 missions overseen by the Jesuits on the Xingu, Tapajós, and Madeira, the principal rivers of the south bank of the Amazon.[27] This number was a fraction of those living outside the missions further upriver. It was the idea of retrieving these people, as well as tropical products, that drove the missionaries and the soldiers to explore such a vast region.

However, it was not just the 'red gold' of Indians themselves that the Portuguese desired to labour in their fields and homes. In the early 1740s gold miners from central Brazil discovered gold and diamond deposits near the headwaters of the Tapajós. In May 1742, a small group of Brazilian miners, Indians, and blacks left the Puxacazes River in Matto Grosso and then travelled downriver all the way to the Tapajós village and mission. Their progress was hampered by the existence of stony rapids and dangerous whirlpools along the way, not to mention the need to procure food. They set out in small canoes but managed to obtain larger ones from the Guariteres people. While the better, more substantial canoes helped the crew on the river, the men had to carry them over the rapids, which was extremely hard work. The voyage took 106 days. The details of this trip were written down by Jesuit missionary

Manuel Ferreira in the late 1740s. He talked to the miners and relayed their information in a chronicle-type document, which included other knowledge from his Jesuit colleagues. Accompanying this 'news' was a map that he intended to cover the entire length of the miner's expedition, but he failed to confirm some details. So, he only drew as far as the Jesuits were working, about a quarter of the length (see Figure 8.4). Nevertheless, the visual representation, along with the report, offers fascinating insights into this riverine world that stretched from central to north Brazil, concerning what kinds of people lived where and their relations. I suspect Ferreira wanted to write 'a discovery of the Tapajós' that would enter the annals of exploration literature; yet what is of interest in the present context is its inability to tame its subject. It is as though all the knowledge he sought to convey was bubbling up and spilling over the edges. His report was never published but remains available for consultation in the Public Archive of Evorá, Portugal, one of the key archives for Jesuit Portuguese history.[28] The document may never have circulated outside the Jesuits because the gold and diamond find ran dry quickly – and the Portuguese decided in the 1750s that the Madeira River was a much better way of consolidating territorial gains in the west than the Tapajós route to central Brazil. All rivers were not equal in the service of empire.

The main part of Ferreira's document follows the Tapajós from the mouth upriver, countering the downward flow of the miner's voyage, with which he opens the document. The river can be split into three sections, (1) the lower Tapajós riverbank villages (mission zone), (2) the Amerindian sphere indirectly and directly affected by outside influence, and (3) the unmapped upper Tapajós and its headwaters. Nonetheless there is much movement between each of these sections. Moreover, if outsiders could travel this course without fear and injury there was an absence of aggression from Indians. The Amerindian sphere in the headwaters might appear remote from the missions at the mouth but it was proximate to other colonial areas, such as the mines in central Brazil. Indeed, Ferreira ends his report with the comment that the Tapajós headwaters connect to the Rivers Paraguay and Plate in the south.

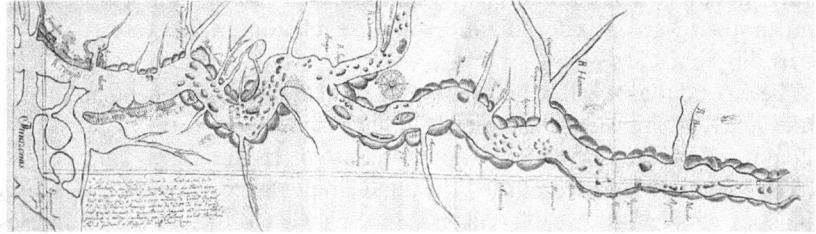

Figure 8.4 Map by Manuel Ferreira of the lower and middle parts of Tapajós River as far as Jesuits have worked.

Biblioteca Pública de Evorá, Códice CXV 2-15 P7A.

Tapajós space: natural marks, seasonality, and residence

Two themes relating to activities in the riverscape emerge from the report that are relevant to the discussion of spatial history: the first concerns the environmental features along the river course, and the second are the spatial and residential arrangements. How do physical features shape social life? Rapids are of critical importance because they pose an obstacle to movement, especially in the dry season. Archaeological sources for the Tapajós and elsewhere in the Amazon, such as different styles of ceramics, indicate that rapids, rather than riverbanks, acted as a meeting point for settlements in the interfluve areas in the pre-colonial period.[29] For the Europeans, cataracts impeded progress upriver if they lacked the knowledge and labour to continue, so they acted as barrier. Indeed, the map represents the last of the missions on the western bank as the point the rapids begin, as one goes upriver. This kind of break is found on other rivers in South America in the colonial period. The report devotes much attention to the space around the rapids and comments on the line of paths that the crew took. The miners had made a boundary-smashing expedition after all: it was possible to travel safely up and down this river with the correct knowledge.

A further feature was the location of seasonal huts on the riverbank belonging to the Apencuria and Periquito people in the upper Tapajós. These temporary homes were for the preparation of fish and other river animals for consumption elsewhere. Indeed, one of the most striking aspects of the reports is the absence of descriptions of Amerindian villages on the riverbank. These were all situated away from the riverbanks on smaller affluents in what were referred to as kingdoms or lands in the report. After 200 years of (direct and indirect) harmful white activity, this relocation is not surprising. However, it is not clear in this report if indigenes became more mobile and took to moving seasonally as a recent adaptation to avoiding whites.

A final way in which the report indicates how rivers related to social life is the outlining of groups living on neighbouring tributaries to the Tapajó River: some were at war and others at peace. This suggests that in this space a river was identified with a kind of people, all related, and spread out in villages along the riverbanks, perhaps even according to a hierarchy of sibling groups, as has been documented in the contemporary North-West Amazon.[30] Marriages and other ritual exchanges took place between one river's 'nation' and another; war was a continuation of these relations, rather than a qualitatively different state.

The second theme is the distribution of residential arrangements in the riverine space. I have mentioned already the city-like place at the mouth at the Tapajós, where at least a thousand people lived. In the upper Tapajós, just before it split into two tributaries, the miners reported the existence of a large kingdom with many sizeable villages. In each village there were 80 houses, each one with 30 people. Even if the numbers are not exactly reliable, it is obvious there was a significant regional society in the upper Tapajós that matches ones found elsewhere in the Amazon, such as the neighbouring upper

Xingu, that has not been found by archaeologists (as mentioned above). At this time, then, the European and the Amerindian were parallel worlds each with their own relation to the riverine space – and connected through various kinds of mediators, such as traders, missionaries, and miners.

The settlements on the lower Tapajós grew in size and importance, and acted as meeting points for these two spatial patterns. Indeed, if we were to focus on these missions, as they were at the time of Ferreira's report, what micro patterns – other scales – would we find? Recent research has shown the frequency of contact for trading (and raiding) purposes but also marriage, which implies visits and ritual celebrations.[31] By the late colonial period, the villages had their own integrity and were not simply waystations in between two worlds for they developed their own links with other semi-colonial places. Their scale was given by the circulation of people and goods in this network. Analyzing this movement makes evident other spatial patterns such as family relations, labour exchange, and even the seeking of ritual cures from reputable shamans. Depending on the scale different relations come into view that direct our understanding. This is a shift in perspective rather than hierarchical levels since different forces fix and form each scale.

Moving to the present

Over the colonial period and after independence in 1822, these worlds grew further apart. Amerindians sought less interference and more autonomy, even though the Portuguese and Brazilians sought to penetrate further into these remote spaces of the tributaries and headwaters. Meanwhile, a new Amazonian society developed along lower parts of the main rivers as well as in the riverbank villages, the missions, and the farms. The rivers threaded together diverse individuals and their families, giving them new bonds, but drawing on older expertise to survive and make a livelihood. In effect, these diverse people became the new owners of the Amazon River. They made the riverine environment through their work. Their labour involved new technology, such as metal hooks for fishing and new vessels for navigating the currents. While the rivers enabled colonial domination by facilitating a network over the basin, they also shaped cultural and material life in a manner that privileged those with regional knowledge and skills.

The cities and towns along the Brazilian Amazon and its tributaries in contemporary times grew out of the struggles outlined above. Until the middle of the 20th century this riverine way of life was dominant. And as a result, the Brazilian part of the Amazon has always been a remote, a poor, and what economists see as an undeveloped one. Since the Second World War, governments have pursued policies to integrate the Amazon into the nation. This has involved encouraging colonists from outside the Amazon to move there, cultivate land in what they are told is an empty territory. This has meant road building, which has led to land conflict with indigenous people and existing small farmers and environmental destruction. Land said to be empty by the government was certainly not empty for those living there, even if they make use of the place very differently. Their spatial histories were invisible. This shift has also meant the rivers have become less central to the region's economic and political functioning. For the first time, large spaces of forested areas have been occupied by non-indigenous people with huge farms of soy beans, for example.

Conclusion

In this chapter I have examined four different scales of relations with the riverscape of the Amazon: (1) the Amerindian view of a circulating river that was in place before and around European arrival in northern South America; (2) these ancient Amerindian spatial practices which were disaggregated and transformed by colonial intervention, leading to a series of fragmented expressions in tributaries, one of which was examined in the Tapajós River; (3) the Portuguese view of a colonial world dominated by a political and economic capital at the mouth of a huge river system focused on Lisbon across the Atlantic but bearing the marks of its dependence on Amerindian cultures; (4) finally a new regional culture that emerged from the broken up Amerindian societies and who made new communities along the main riverbanks, circulating amongst themselves. Despite their differences, the people who participated in these histories all contended with the rivers. In this liquid history, the waters have dissolved these pasts, not in the sense of obliterating them, but in the creation of a belonging that is continually renewed. This ownership by the river is what has integrated apparent disparities and divergences.

Let me end by returning to scale. Each of the schemes above creates its own scale, so it might be more helpful to use the verb of scale – scaling – and bind it up with the work and energy of humans and rivers. The activities on and around rivers produce different levels and depths of engagements: some intense and close up, while others make use of the immeasurable surfaces for long-distance movement. Scaling on the river, then, is a composite technique for knowing about spatial extension, which does not confine itself to national boundaries or watersheds. This approach helps shift our terracentric notions towards a more liquid appreciation of human life. Embracing a multi-scalar perspective, I have sought to capture the ongoing tension among scales treated as diverse perspectives and composed of different relations and spatial patterns.

Notes

1 This notion of a river as sewing (*couture* in French) together places and people was central to Lucien Febvre's study of the Rhine, the first academic study of a river. See Albert Demangeon and Lucien Febvre, *Le Rhin: Problèmes d'histoire et d'économie*, Paris: Colin, 1935.
2 Christof Mauch and Thomas Zeller (eds.), *Rivers in History: Perspectives on Waterways in Europe and North America*, Pittsburgh: University of Pittsburgh, 2008, p. 2.
3 Mark Cioc, *The Rhine: An Eco-Biography, 1815–2000*, Seattle: University of Washington Press, 2002.

4 Anthropologists call this research 'fieldwork', involving long periods of time, e.g., a year, participating in daily life and talking with people on a regular basis. See Mark Harris, *Life on the Amazon: The Anthropology of a Brazilian Peasant Village*, Oxford: Oxford University Press, 2000.
5 Mark Harris, *Rebellion on the Amazon: Race, Popular Culture and the Cabanagem in the North of Brazil, 1798–1840*, Cambridge: Cambridge University Press, 2010.
6 Mauch and Zeller, *Rivers in History*, p. 7.
7 Karl Wittfogel, *Oriental Despotism: A Comparative Study in Total Power*, New Haven: Yale University Press, 1957.
8 Hugh Raffles, *In Amazonia: A Natural History*, Princeton: Princeton University Press, 2002.
9 Some scholars have argued that Amerindian societies were against the state because of their appetite for war, which prevented social stability.
10 Simon Schama, *Landscape and Memory*, London: Fontana, 1996, pp. 257–62.
11 Stephen Rostain, *Islands in the Rainforest: Landscape Management in Pre-Columbian Amazonia*, Walnut Creek: Left Coast Press, 2013. The classic discussion about the non-evolution of the state in the Amazon is Robert Carneiro, 'A Theory of the Origin of the State', *Science* 169, 1970, 733–8.
12 Donald Worster, *Rivers of Empire: Water, Aridity and the Growth of the American West*, New York: Oxford University Press, 1985, p. 7. A European equivalent can be seen in the mastery of water in Germany, starting with the Prussian draining projects in the Oderbruch. See David Blackbourn, *The Conquest of Nature: Water, Landscape, and the Making of Modern Germany*, New York: W.W. Norton, 2006.
13 Richard White, *The Organic Machine: The Remaking of the Columbia River*, New York: Hill and Wang, 1995, p. 112.
14 An excellent discussion about scale that incorporates deep history can be found in Sebouh David Aslanian, Joyce Chaplin, Ann McGrath and Kirstin Mann, 'How Size Matters: The Question of Scale in History', *American Historical Review* 118, 2013, 1431–72. Deep history, according to McGrath, is a reminder to probe beyond superficial readings of the past and to avoid a linear understanding of time. There are starts, stops, leaps and returns in the making of social life that make for temporal unpredictability (ibid., 1436).
15 Denise Schaan, *Sacred Geographies of Ancient Amazonia: Historical Ecology of Social Complexity*, Walnut Creek: Left Coast Press, 2013, p. 15.
16 José Toribio Medina (ed.), *The Discovery of the Amazon*, New York: Dover, 1988 (originally published 1934).
17 Medina (ed.), *Discovery*, pp. 202–13.
18 Thomas Pauketat, *Cahokia: Ancient America's Great City on the Mississippi*, Harmondsworth: Penguin Books, 2010.
19 Neil Whitehead, 'The Mazarui Pectoral: A Golden Artefact Discovered in Guyana and the Historical Sources Concerning Native Metallurgy in the Caribbean, Orinoco and Northern Amazonia', *Journal of the Walter Roth Museum of Anthropology* 7, 1990, 124.
20 Neil Whitehead (ed.), *The Discoverie of the Large, Rich and Bewtiful Empire of Guiana by Sir Walter Ralegh* (1596), Norman: University of Oklahoma Press, 1997; and Joyce Lorimer, *English and Irish Settlement on the River Amazon, 1550–1646*, London: Hakluyt Society, 1989.
21 John Hemming, *The Search for El Dorado*, London: Michael, 1978.
22 Whitehead (ed.), *Discoverie*, p. 87.

23 It nevertheless emerges from discussions of the location of El Dorado as near lake Macu, and the Pirara portage point in between the two basins, Whitehead (ed.), *Discoverie*, p. 87.
24 Maria Gnerre, *Roteiro do Maranhão a Goiaz pela Capitania do Piauhy*, PhD Thesis, University of São Paulo, 2006, p. 151.
25 The word 'concourse' is from Diarmid Finnegan's comments on a draft of this chapter, for which I thank him.
26 Schaan, *Sacred Geographies*, pp. 126–9, Bruna Rocha, *Ipi Ocemumuge: A Regional Archaeology of the Upper Tapajós River*, PhD thesis, University of London, 2017.
27 ARSI (Archivum Romanum Societatis Iesu, Rome), *Brasiliana 10* (II), f. 338–338v, 1730.
28 BPE (Biblioteca Publica de Evora), Manuel Ferreira, 'Breve noticia do Rio Topajoz', CXV/2-15 a n.° 7, f. 51r–54r, 1753.
29 Bruna Rocha and Vinicius Honrato, 'Floresta Virgem? O Longo Passado Humano da Bacia do Tapajós', in Daniela Alarcon, Brent Millakan and Mauricio Torres (eds.), *Ocekadi: Hidrelétricas, Conflitos Socioambientais e Resistência na Bacia do Tapajós*, Brasília: International Rivers, 2016.
30 Stephen Hugh-Jones, 'Clear Descent or Ambiguous Houses? A Re-Examination of Tukanoan Social Organisation', *L'Homme* 33, 1993, 95–120.
31 Barbara Sommer, 'Colony of the Sertão: Amazonian Expeditions and the Indian Slave Trade', *The Americas* 61, 2005, 401–28; and Heather Roller, *Amazonian Routes: Indigenous Mobility and Colonial Communities in Northern Brazil*, Stanford: Stanford University Press, 2014.

9 Infrastructures

Frithjof Benjamin Schenk

Infrastructure and the production of space

Social life in modern societies is shaped by a multitude of infrastructures. Infrastructures have been conceptualized as 'media of social integration of the first order'[1] and as 'the physical components of interrelated systems providing commodities and services essential to enable, sustain, or enhance societal living conditions'.[2] In contemporary usage, the term denotes a large variety of networks and services, such as airports and railway stations, roads and bridges, grids of telecommunication and public transportation, and electric power and water supply systems. As a seemingly neutral media of public welfare, 'infrastructure mediates between political rule and everyday life, becoming part of both'.[3] Originally the term 'infrastructure', first used by French railwaymen in the 1870s, referred to stationary constructions facilitating geographical mobility. After the Second World War, the notion found its way into military debates on logistics and political discourse on economic integration. The inflationary use of the term in discussions about development policy, or the alleged obligation of public authorities to provide the economy and society with a set of 'infrastructures', led to an increasing dilution of the term's meaning.[4] Accordingly, the field of infrastructure history, which has taken shape as a distinct historical sub-discipline over the last two decades, has focussed on a large variety of historical phenomena, processes, and problems.[5]

This chapter discusses the history of technical infrastructure as a specific approach in the field of spatial history. I will restrict myself to one particular aspect, namely, the history of railroads, which has been described as the 'infrastructural leading medium [*infrastrukturelles Leitmedium*] of the nineteenth century'.[6] Technical infrastructures in general, and railroads in particular, may be regarded as ideal objects of enquiry in the field of spatial history. If we treat space not as a 'given', but as a phenomenon that is shaped by human imagination, perception, regulation, and appropriation, the history of infrastructure offers an excellent lens through which to understand the 'production of space' in its various dimensions. As Henri Lefebvre has argued, the process of 'spatialization', that is, the transformation of 'natural space' into 'social space', results from the interplay between human practices, perceptions (*le perçu*), and representations of space (*le conçu*).[7] All three

DOI: 10.4324/9780429291739-12

dimensions become relevant when we approach the history of infrastructure from a spatial point of view.[8] If we take the railways as an example, we may analyze the construction of bridges, tunnels and railway tracks, the mobility of railway passengers, or the use of trains for military purposes as social *practices* that shape social space in a specific way. At the same time, the development of a railway network, the construction of carriages for passengers, or the design of railway stations always depend on certain *visions* and theories of how social space should be organized and structured. Moreover, the various modes of spatial *perceptions*, whether the impressions of railway passengers during their journeys or the authority's view of the capability and weaknesses of a country's railway system, have an impact on the production of space in the modern era. Finally, the analysis of the development and usages of technical infrastructures like railroads enables us to study the 'production of space' both on a macro- and a micro-level. Whereas, for example, discourses on the development of an imperial railway network relate to the former, the study of social practices and the mutual perceptions of passengers in a poky train compartment shed light on the latter.[9]

It is commonly understood that the invention, spread, and use of railways have dramatically altered traditional concepts of time and space since the early-19th century.[10] Proponents of the steam-driven means of transportation promised nothing less than a victory of humankind over time and space. Advocates of the construction of railroads in the 1820s and 1830s dreamt of abolishing geographical distance with the help of accelerated movement through space.[11] Apart from the steam engine, it was the locomotive which became the most potent symbol of technological progress in the advent of high modernity.[12] The few critics who perceived the railroads as a device of the devil and as a threat to the traditional order were soon silenced. Political discourse in many European countries, and on the American continent, was increasingly dominated by engineers who believed in the unlimited possibilities of technical progress.[13] By connecting geographically distant places, regions, and people, these optimists were convinced that railroads would not only contribute to an integration of markets and trigger economic development, but that they would simultaneously overcome political boundaries and eventually transform the world into a prosperous, civilized, and peaceful place.[14]

The optimistic and unquestionably utopian tone of this discourse had a strong impact on traditional narratives of railway history. Echoing the basic assumptions of modernization theory, many classic studies of railway history treat the new means of transportation as an undisputed harbinger of progress, as a power of spatial (territorial) integration, and as a driving force of technological and economic development. In the meantime, railway historians have long treated their topic primarily in the context of economic history, as a key factor in industrialization and the emergence and consequence of modern capitalism. These narratives focused predominantly on the heroic engineers who constructed steam engines, rolling stock, railway stations, steel bridges, and vast networks of railway lines which helped national economies grow at previously unthinkable rates and contributed (together with the telegraph and steam ships) to a rapid integration of national and global markets. As a side effect, the rise and political impact of new social groups such as railway workers and railway entrepreneurs became subjects of scholarly interest.[15]

These economics-centred and overwhelmingly optimistic narratives of traditional railway history have been challenged by new approaches since the late 1970s. Wolfgang Schivelbusch's formative study on the history of the railway journey (first published in German in 1977) shifted the focus of analysis, making railway passengers and their modes of behaviour, cultural practices, and spatial perception objects of scholarly enquiry. Schivelbusch declared the railway station, with its specific architectural shape and outlook, the design of railway compartments, the cultural practice of reading during a train ride, and the 'panoramic travel' as promoted in handbooks for railway passengers to be new and promising objects of historical analysis. Schivelbusch argued that railways, by connecting geographically distant places, both created and annihilated space. While new rail connections undoubtedly broadened travellers' geographical mobility, the space 'in between' two cities, i.e. the distance between their respective railway stations, 'vanished' from the mindset of an increasingly mobile population. In the railway age, train passengers were conveyed from one spot to the other with unprecedented speed. They found themselves confined in their train compartment, reading books and viewing the landscape through the frame of the window as part of a picturesque 'panorama'.[16]

Schivelbusch's book, translated into English in 1979, had a strong impact on the emergence of a *New Railway History* in Great Britain in the 1990s and early 2000s. Promoted by, among others, Michael Freeman, Matthew Beaumont, Colin Divall, and Ian Carter, it moved the 'world the railways made' (Nicolas Faith) into the focus of cultural history.[17] Whereas traditional railway histories had been primarily based on economic statistics, administrative data, and technical drawings by railway engineers, these new studies added railway passengers' travelogues and handbooks, timetables, architectural drawings, and 'objects' such as railway stations and train carriages to the list of valid and fruitful historical sources.

Moreover, new approaches in railway history challenged traditional narratives of the spatial integration, territorialization, and social consolidation brought about by the construction and use of the steam-driven means of transportation. Inspired by debates about 'critical infrastructures', scholars have highlighted, for example, that the new means of transportation increased the vulnerability of modern societies and states which had become increasingly dependent on these networks and services.[18] Other scholars have reminded us that the decision to connect one city to a national railway network usually implied that other spots on the map had to wait, often in vain, for this act of spatial integration.[19] Finally, we have learned that the allegedly 'democratic' railway system also deepened social cleavages in space by providing first-class passengers, for instance, with more space, comfort, and speedy transportation than customers with lesser financial means. The following case study on the 'production of space' in the Russian Empire before the First World War is inspired both by the cultural turn in railway history

174 *Frithjof Benjamin Schenk*

and by the critical stance on the railways' potential to integrate, both socially and spatially, in the era of modernity.

Railroads and the idea of territorial integration: the Russian case

According to one of the most important narratives in railway history, the construction and use of railroads in the 19th and early-20th centuries significantly contributed to the economic and political integration and spatial cohesion of both nation-states and multinational empires.[20] As early as 1833, the famous spokesman of liberalism in Germany, Friedrich List, dreamt of creating a united German nation-state with the help of a railway network. Part of List's published plea was a sketch of a map, depicting an anticipated network of railway lines both connecting and spatially integrating the separated German principalities. By showing the 'German' territory as a white plain without any internal administrative boundaries and structured only by arteries of transportation (rivers and railroads), this image (Figure 9.1) offered an

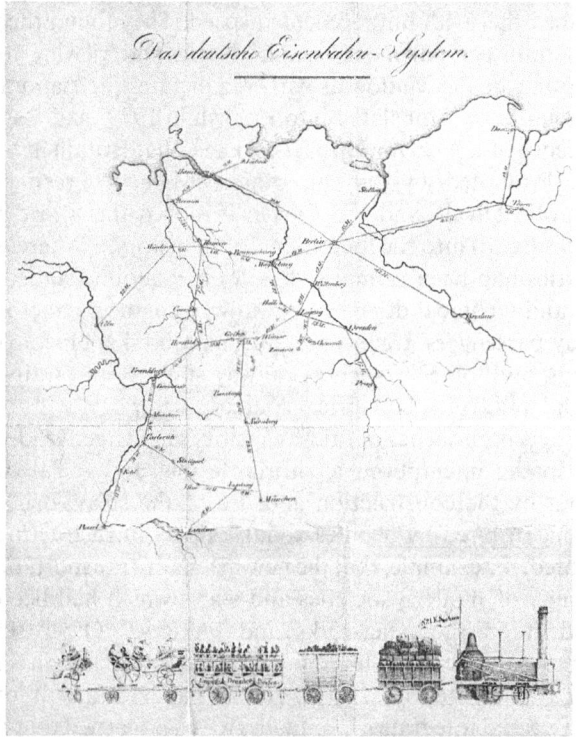

Figure 9.1 Friedrich List's *Das deutsche Eisenbahn-System*, 1833.

Source: Friedrich List, *Über ein sächsisches Eisenbahn-System als Grundlage eines allgemeinen deutschen Eisenbahn-Systems und insbesondere über die Anlegung einer Eisenbahn von Leipzig nach Dresden*, Leipzig: Liebeskind, 1833. Available HTTP: https://digital.slub-dresden.de/werkansicht/dlf/93171/6 (accessed 14 May 2021).

appealing political message and promise; once the envisioned network of railroads was realized, the creation of a single German nation-state would be easy to achieve and even a 'natural' outcome.[21] Echoing List's vision of 1833, the influential Russian publicist Mikhail Katkov, an admirer of German railway policy, proclaimed in 1883 in his newspaper *Moskovskie vedomosti* (*Moscow News*) that 'after the bayonet, it is the railways that consummate national cohesion'.[22] To this day, an iron network of infrastructure is still recognized as the backbone or skeleton which enables territorial integration and stabilizes territorial integrity.[23] The historian Charles Maier, for example, has argued repeatedly that railroads played a decisive role in what he called 'territorialisation' during the 19th and 20th centuries in Europe and beyond.[24] In recent historiography, however, it has been argued that modern infrastructure had an ambivalent impact on the integration of territories and societies. While, on the one hand, it encouraged spatial and social integration, it also provided 'effective' means for spatial *dis*integration and destruction, especially in times of war.[25]

From the very beginning, proponents of the construction of railroads in Russia perceived the new means of transportation as a tool of territorial integration and political rule. Tsar Nicholas I, who approved the construction of Russia's first railroads in the 1830s and 1840s, imagined that the iron horse (*chugunka*) would help him overcome the legendary Russian problem of 'roadlessness' (*bezdorozh'e*) and diminish the country's vast geographical distances. At the same time, he additionally hoped that the railroad would enable the imperial administration to better communicate with the peripheries and to react more swiftly and effectively to upheavals in the imperial borderlands. With the help of trains, the Tsar was convinced, loyal troops could be sent more quickly to regions in turmoil. Some of his advisors even anticipated that in the future, Tsarist troops need only be concentrated in a small number of geographical nodes from which they would be speedily deployed via the railways to regions which needed to be 'pacified'.[26]

At the end of the 19th century, when the construction of transcontinental railway lines became an issue of growing importance for the Tsarist administration, the new means of transportation was ascribed an important role in consolidating and culturally homogenizing the vast imperial space. By using railways – in particular the Trans-Siberian Railroad – as a tool for the resettlement of millions of Russian peasants from the European to the Asian part of the empire, Tsarist administrations at the turn of the 20th century aimed at strengthening the link between the centre and the periphery, transforming the multinational empire into a *Russian* imperial realm and averting the danger of separatist movements in the borderlands.

These (and other) spatial imaginations found their reflection in various artefacts presented at the World Exhibition of 1900 in Paris. This international showcase was of the utmost political and economic importance for the Tsarist administration. At this point, the Trans-Siberian Railroad was not only the longest, but also the most expensive single railway line ever built. The Russian government had declined to involve private investors, but lacked sufficient capital of its own to realize this huge infrastructural enterprise. The Russian Minister of Finance planned to borrow mainly from the French capital market. Western investors had to be persuaded, therefore, that the construction of the Trans-Siberian Railroad would be an economically successful and politically harmless project and worth an investment of almost 800 million roubles.[27]

In order to spread this message at the World Exhibition in Paris, the Russian Minister of Finance, Sergei Vitte, issued an impressive, large-scale *Guide[book] to the Trans-Siberian Railroad* in Russian, French, German, and English, which praised Siberia as an integral part of the Russian Empire and a 'land of the future' which promised fabulous economic potential.[28] Vitte dreamt of attracting the flow of trade between Asia and Europe with the help of a transcontinental railway corridor across Russia. A map attached to the *Guide to the Trans-Siberian Railroad* showed the envisioned railway line as part of an uninterrupted rail connection between the Atlantic in the West and the Pacific in the East (Figure 9.2). This red line on the map represented a Jules Verne–like dream of spatial integration, uniting not only the European and the Asian parts of the Russian Empire but also establishing Russia as a firm link between the European and the Asian continents. Visitors to the *Pavilion of Empress Marie* in Paris could experience this vision through a virtual train journey from 'Moscow to Peking' in four luxurious railway carriages provided by the *Compagnie International Waggons Lits* (CIWL). Artists of the Paris opera house had produced a multi-layered canvas, depicting a typical Siberian landscape which was moved alongside the motionless train to give the 'passengers' the impression of travelling the huge

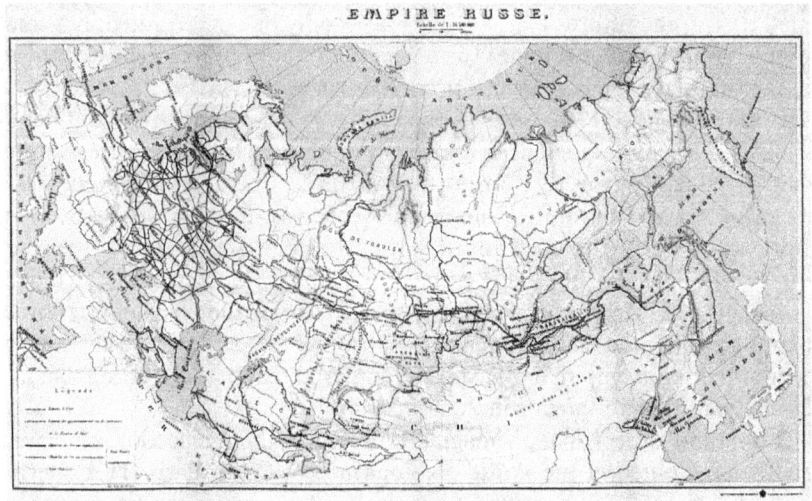

Figure 9.2 Map of the Russian Empire, 1900.

Source: *Die Große Sibirische Eisenbahn: Pariser Weltausstellung des Jahres 1900*, ed. by Kanzlei des Ministerkomitees, Saint Petersburg 1900.

geographical distance between the Russian and the Chinese capitals in only 45 minutes. At the 'final destination' of their railway 'journey', the guests were welcomed by employees in Chinese dress. The illusion was almost perfect and a huge attraction for the visitors.[29] It did not matter that the Trans-Siberian Railroad was still under construction in 1900, that the 'Boxer rebellion' in China was blurring hopes of a peaceful connection between Europe and Asia, and that a train ride from Moscow to Peking was possible only in the passengers' imagination. Despite these problems, both the producers and the audiences of these visions of territorial integration still wanted to believe in the almighty transformative power of modern infrastructure.

When it came to the spatial integration and social transformation of their own country, the adherents of railway construction in Russia imagined – like their contemporaries in other countries – that the modern means of transportation would help the westernized elite to civilize the whole country and transform its population into a well-ordered society of loyal subjects. In 1900, Russia already possessed a railway network of almost 52,000 km, the second longest in the world after the United States. Before the outbreak of the First World War, more than 68,000 km of railway track were in operation in the Tsarist Empire. The railway, with its elaborate structure of train schedules, technical regulations, and educated and uniformed personnel, was imagined as an important agent of civilization in both the European and the Asian parts of the empire. By travelling on the railroad and respecting its rules of behaviour, optimistic reformers imagined that rude Russian peasants and 'uncivilized' non-Russian subjects would be gradually transformed into human-beings with refined manners; the whole Russian empire would become, as it were, a community of first-class railway passengers.

These utopian visions found their representation both in the material outlook of the modern urban railway station and in detailed instructions for Russian passengers on how to properly behave at these 'places of modernity'.[30] As in other European or American cities – one thinks of the central station in Antwerp, Saint Pancras in London, or the Grand Central Station in New York City – Russian urban railway stations were not just buildings where trains arrived and departed. Designed as modern city gates with historicizing facades, railway stations optically mediated between the steel-and-glass construction of the platform building and the architectural body of the respective city-scape.[31] Thus, railway stations soon became outstanding symbols of representation. They symbolized a belief in technological progress and the wealth of the respective railway company. The new building of the railway station of the private *Moscow-Vindava-Rybinsk Railway Company* (MVRRC) in Saint Petersburg (today the city's '*Vitebskii vokzal*'), erected between 1900 and 1904, serves as a good example.[32] The impressive Art Nouveau building, designed by architect Stanislav Brzhosovskii, miraculously survived two World Wars and was at the moment of its inauguration in 1904 one of the most modern railway stations in Russia (Figure 9.3). First- and second-class passengers entered the building through an impressive hall decorated with a statue of Tsar Nicholas I and paintings depicting the ports

Figure 9.3 Building of the Moscow-Vindavo-Rybinsk Railway Company in Saint Petersburg (today *Vitebskii vokzal*), erected 1900–1904.

Source: Alex 'Florstein' Fedorov, CC BY-SA 4.0, https://commons.wikimedia.org/w/index.php?curid=34440243.

of Saint Petersburg and Odessa. Together with the coat of arms of the Belorussian city of Vitebsk fixed on the façade of the building, these images provided the cornerstones of a mental map representing the geographical territory served by the railway company.

As in other countries, the Russian railways were in principle a 'democratic' means of public transportation accessible to people from all social strata. In reality, the spatial design of both railway stations and train carriages were strongly shaped by the clear-cut hierarchy of Russia's social class system. Architectural drawings, technical sketches by railway engineers, and the physical shape of railway buildings give ample evidence of this fact. Whereas passengers in first and second class were provided 149 square-*sazhen* (approximately 680 square meters) of waiting space, including restaurants, in MVRRC's newly erected railway station in Saint Petersburg, third-class travellers were squeezed into waiting rooms covering only 140 square-*sazhen*.[33] The discrepancy becomes more obvious when we consider that less than one-tenth of the station's 20,000–24,000 daily passengers could afford first- or second-class tickets. Nevertheless, the planners of the Russian railway system strongly believed in the 'civilizing' force of the modern means of transportation. An 1898 guidebook of the state-run *Southwestern Railway Company* reads, for example, that

> [e]very *versta* [1.07 km] of completed railway track symbolises the achievement of a new level of [our] country's cultural development.

Each new railway station forms a new centre from which enlightenment and the light of knowledge will spread into the depths of deserted no-man's land.[34]

Russian railway stations became sparkling objects of an imagined well-ordered and modern society, where trains would depart on time, where passengers would refrain from drinking and gambling, and where civilized railway employees in uniforms would treat customers politely and refuse all bribes.[35]

Railroads and the emergence of new threats to the spatial order of the empire

Reality proved most of these optimistic visions ill-founded. On the one hand, the adherents of railway construction in Tsarist Russia simply overestimated the impact of the technological system on patterns of its users' social behaviour. On the other, they failed to appreciate that the new means of transportation would not only enhance order and security in, and territorial integration of, the multi-national empire, but that the construction and use of railroads posed new challenges to its political and social order.

From the very beginning, debate raged over whether, and how, the new means of transportation would threaten traditional concepts of political stability and imperial rule. In early years, conservative bureaucrats perceived an augmented level of geographical mobility among the population as a threat to the traditional order. The railways, Georg von Kankrin, the Russian Minister of Finance, argued in the early 1840s, would make 'unstable people even more unstable', contribute to a 'levelling of social classes', and spread harmful democratic ideas among the population.[36] By the turn of the 20th century, this debate was already history and Russia experienced an unprecedented increase in the geographical mobility of its population. Whereas in 1894, Russian railway companies sold only 55.6 million train tickets, the 1912 yearbook of the statistical bureau of the Ministry of Ways of Communication (MPS) registered 231.4 million railway passengers, of which 207.3 million had purchased either third- or fourth-class tickets.[37] This development reflected various socio-economic trends of the late-19th and early-20th centuries: the growth of the empire's population, processes of urbanization and industrialization, peasant colonization in the Asian borderlands, the differentiation of railway tariffs in the 1890s, and the liberalization of the Tsarist passport regime in 1906.

The success of the Russian railroads in attracting more and more passengers and thus contributing to increased geographical mobility among almost all social classes caused great difficulties in managing and monitoring the flow of people. The authorities' growing awareness of the problem can be discerned from debates on the Hajj of Muslim pilgrims from Turkestan in Central Asia to Mecca and Medina, and on the issue of 'wild [i.e. uncontrolled] migration' of Russian peasants travelling the transcontinental railroads to Asia in search of new arable lands.[38] Whether the significant increase of geographical mobility actually contributed to the process of territorial integration in late Tsarist Russia is still an open question, however. Russian railway passengers who traversed the empire, leaving a large number of travelogues, perceived their motherland both as an integrated political space and as a highly fragmented and culturally diverse spatial entity. While some authors were impressed by the vastness of the newly integrated imperial realm and the uniform outlook of its transport infrastructure, others were shocked and frightened by the empire's ethnographic differences and cultural fragmentation.[39]

The unequal division of space in Russian railway carriages provoked an interesting political debate among representatives of the new scholarly discipline of social hygiene in the 1870s. From the mid-19th century onward, Russian scientists, like their counterparts across Europe, were increasingly concerned by the harmful impact of the railroads on public health. The Russian scholar Vladimir Porai-Koshits published an elaborative article on 'The railroads from the perspective of social medicine' in 1870 and 1871.[40] In this text, he tackles, among other things, the quality of breathing air in different classes of Russian railway carriages. Referring to the newest state of medical knowledge, he argues that an adult person in an enclosed space physically requires approximately 60 m^3 of breathing air per hour. According to his analysis, however, a Russian passenger travelling first class in a full train carriage of the *General Society of Russian Railroads* could only expect an average of 6.37 m^3 of breathing air. In third class, which only offered 2.14 m^3 of breathing air, the situation was far worse. From a medical standpoint, Porai-Koshits argued, train carriages with a volume of 90–120 m^3 should be equipped with a well-functioning system of ventilation and should carry no more than 12 to 16 passengers each. Interestingly, he made this argument both for passengers in first and third class. Going further, he even argued that representatives of the working class – who, as a rule, travelled third class – needed *more* breathing air than the nobles and bureaucrats found in first- and the second-class train carriages:

> There can be no better proof of the human desire to exploit others. The air seems to be a common good that belongs to all living creatures on earth, but here it serves [as a means] to realise profit. In the meantime, the laws of physiognomy are spumed recklessly [on the Russian railways]. In fact, third-class passengers should be provided with more breathing air because here you mostly find workers whose physiognomy and lungs are stronger and accordingly need more oxygen [than passengers in first class].[41]

Astonishingly, this almost revolutionary thesis was later debated among Tsarist bureaucrats, but with no real effect on the conditions of third-class railway carriages, which, in fact, did not improve but dramatically worsened over the following decades.

The Tsarist administration was confronted with yet another, much more visible political challenge in the railway age. In the early 1860s, when the first Russian network of railroads became operational, Polish rebels, who took up arms during the January uprising of 1863 to fight for their national cause, identified the tracks of the Saint Petersburg-Warsaw railroad as the despised Tsarist regime's Achilles heel. They burnt railway bridges, attacked military trains, and cut the wires of the railway telegraph. They could not, of course, completely stop the relocation of loyal troops to the rebellious provinces, but their actions caused severe problems for the Russian army's attempts at suppressing the rebellion.

A few years later, also in the Russian heartland, political activists started attacking trains and railway tracks to achieve their political goals. When the Executive Committee of the terrorist organisation *Narodnaia volia* (*People's Will*) decided to kill Tsar Alexander II in the summer of 1879, they deliberately chose the emperor's train as the first target. Although this attempt was not 'successful', the emperor's security agencies subsequently paid more attention to the new issue of railway terrorism.[42] After the first Russian Revolution of 1905, freight trains increasingly became targets of criminal, separatist, and revolutionary activities; in 1908, the Revolutionary Fraction of the Polish Socialist Party, under the leadership of Jósef Piłsudski, attacked a postal train not far from the city of Vilna in Lithuania, looting an impressive sum of 200,000 roubles. During these years, armed raids of postal and cargo trains also became an issue of growing concern for the authorities in the provinces of the Northern and Southern Caucasus.[43]

Last but not least, the huge number of railway workers, representing the largest professional group in Tsarist Russia, moved into the focus of authorities at the turn of the century. Railwaymen significantly contributed to the outburst and spread of anti-Jewish pogroms in the early 1880s and at the beginning of the Russo-Japanese War.[44] At the same time, Russian railwaymen were soon targeted by political agitation of socialist and anarchist groups. In fact, the first Russian Revolution and the general strike in October 1905, which compelled Nicholas II to transform Russia into a constitutional monarchy, would not have been possible without the collaboration and support of the Russian railwaymen. Securing the loyalty of this powerful pressure group, which controlled the arteries of imperial communication, became a task of increasing importance for the Tsarist administration. All of these incidents perfectly illustrate that the railroads, originally envisioned by the Tsarist authorities as a mighty state weapon for the promotion of imperial cohesion, could be easily seized by the government's opponents and used in their fight against it, thereby threatening not only the Tsarist regime, but also the empire's territorial integrity.

How to preserve political stability and territorial integrity in the railway age?

There was no coherent political debate in late Tsarist Russia on the specific challenges posed by the railroads to the imperial order. Nevertheless, the authorities were deeply concerned by the problems mentioned above. One issue of concern was the increase of geographical mobility in the imperial borderlands. As the January uprising in Poland and the western borderlands had shown, the railways were not only a tool of imperial rule but an aid to insurgents pursuing their revolutionary goals. This experience had a visible impact on political debates over the construction of new railway lines in the imperial borderlands. In 1864, when different agencies were discussing the itinerary of the so-called Southern line, some warned that railway construction would foster the national movement in Ukraine.[45] Others even dreamt of bypassing Poland altogether when new railway lines were to be planned. In the end, however, security concerns were often outweighed by economic reasoning. The authorities simply had to acknowledge that Poland, the Caucasus, and the Ukraine, despite posing security problems, represented very dynamic and productive economic centres of the empire. Even if the government decided not to build railroads in these politically sensitive regions, there were still private entrepreneurs keen to invest in their development.

Being unable to slow down the construction of railroads in politically unreliable borderlands, the Tsarist administration had to concentrate on policing both passengers and the personnel running the imperial infrastructure. In fact, at the end of the 19th century, the Russian railway gendarmerie had grown to become the biggest single police unit under the control of the imperial Ministry of the Interior. As far as the control of railwaymen was concerned, certain social and ethnic groups regarded as politically unreliable were excluded from service. In the first place, this regulation affected Jews who were not allowed to work either for the railroads or the telegraph.[46] After the January uprising in 1863, some officials additionally argued for the exclusion of Polish subjects from service. This idea, however, could not be realized as there were simply not enough trained Orthodox experts to replace the Catholic railway employees.

From a historical perspective, it seems obvious that the railroads played a decisive, if ambivalent, role in Russia's (internal) political and social transformation in the late-19th and early-20th centuries. Contemporaries, by contrast, perceived and treated the modern means of transportation in the 1910s primarily in the context of debates on Russia's *economic* development and geopolitical strategies. In 1908, for example, the Russian Parliament (the third Duma) debated a plan of the Ministry of War encompassing the construction of new strategic railroads in the western borderlands, the construction of a railroad along the Amur River in the Far East, connecting the railway networks of Finland and Russia, and the construction of a second track of the Trans-Siberian Railroad.

In the meantime, the parliament once again established a special commission to investigate the (economic) performance of Russia's private and governmental railway companies.[47] Two years later, when the former Minister of Ways of Communication, Minister of Finance, and Prime Minister, Sergei Vitte, gave a talk at the Institute of Engineers of Ways of Communication in April 1910, he decided to focus on the problem of economic non-profitability in the empire's railway network.[48] During these years, the political elite was primarily concerned by two questions: first, why were the Russian railroads in such a desperate need of state subsidies (which in 1908 totalled 98 million roubles) and how was this situation to be remedied? Second, how could Russia's geostrategic position in the country's western and eastern borderlands be improved through new strategic railroads? Building and running railways in an empire of this size was both extremely expensive and difficult to manage. In an economically backward country like Russia, the decision to supply the country's Asian peripheries with railway connections meant that less state revenue remained for building strategic railways in the west. While debating these undoubtedly important issues, the political elite lost sight of the fact that the empire's railway system not only caused economic problems and failed to meet geopolitical targets, but that it also contributed to large-scale social and cultural transformations which challenged the empire's social, political, and territorial (or spatial) order in a completely different way.

When the Tsarist empire entered the railway age at the beginning of the 19th century, the proponents of technical modernity perceived the railways as a mighty tool of imperial power. After the defeat in the Crimean War, almost everybody in the Tsarist administration was convinced that Russia could only keep up with its rivals in Western Europe if it systematically modernized and developed its network of communications. In fact, Russia's imperial expansion during the second half of the 19th century – such as the conquest of Turkestan or the economic incorporation of Manchuria – was closely linked to the politics of railway imperialism.[49] The construction of railroads led not only to a previously unthinkable level of geographical mobility among the empire's population but also to new threats to the imperial order. After the first Russian Revolution in 1905, socialist railwaymen took control of large parts of the Trans-Siberian Railroad and even hoisted red flags in the Siberian cities of Chita and Krasnoiarsk. This episode can be interpreted as a prelude to the revolutionary events of 1917 when the railways – formerly a mighty tool of imperial rule – were turned into a weapon of its rivals. It is commonly acknowledged that Russian railwaymen played a decisive role during the events of the February Revolution.[50] It may not be an irony of history, therefore, that Tsar Nicholas II was forced to sign the instrument of abdication in March 1917 inside his own railway carriage.

Notes

1 'Gesellschaftliche Integrationsmedien erster Ordnung', quoted in Dirk van Laak, 'Infrastruktur-Geschichte', *Geschichte und Gesellschaft* 27, 2001, 367–93, here 368; see also his *Alles im Fluss: Die Lebensadern unserer Gesellschaft – Geschichte und Zukunft der Infrastruktur*, Frankfurt/Main: Fischer, 2018, especially pp. 21–30.
2 Jeffrey E. Fulmer, 'What in the World Is Infrastructure?', *Infrastructure Investor*, July/August 2009, 30–2.
3 Julia Obertreis, *Imperial Desert Dreams: Cotton Growing and Irrigation in Central Asia, 1860-1991*, Göttingen: Vandenhoeck & Ruprecht, 2017, p. 42.

4 The German term 'Infrastrukturen' corresponds with the English notions of both 'infrastructure', 'social overhead capital', and 'public utilities'. Dirk van Laak, 'Garanten der Beständigkeit: Infrastrukturen als Integrationsmedien des Raumes und der Zeit', in Anselm Doering-Manteuffel (ed.), *Strukturmerkmale der deutschen Geschichte des 20. Jahrhunderts*, Munich: De Gruyter Oldenbourg, 2006, pp. 167–80, here p. 169.

5 A good overview is Dirk van Laak, 'Infrastruktur-Geschichte'; id., *Das 'vergrabene Kapital' und seine Wiederentdeckung: Das neue Interesse an der Infrastruktur, Materialien der Interdisziplinären Arbeitsgruppe Globaler Wandel – Regionale Entwicklung*, vol. 11, Berlin: Berlin-Brandenburgische Akademie der Wissenschaften, 2010; Jens Ivo Engels and Gerrit Jasper Schenk, 'Infrastrukturen der Macht – Macht der Infrastrukturen', in Birte Förster and Martin Bauch (eds.), *Wasserinfrastrukturen und Macht: Von der Antike bis zur Gegenwart*, Munich: De Gruyter Oldenbourg, 2015 (*Historische Zeitschrift*, Beihefte, N.F. 63), pp. 22–58.

6 Dirk van Laak, *Imperiale Infrastruktur: Deutsche Planungen für eine Erschließung Afrikas, 1880 bis 1960*, Paderborn: Schöningh, 2004, p. 9.

7 Henri Lefebvre, *The Production of Space*, trans. Donald Nicholson-Smith, Oxford: Blackwell, 1991, French 1974.

8 Colin Divall and George Revill, 'Cultures of Transport: Representation, Practice and Technology', *Journal of Transport History* 26, 2005, 99–112, here 104–6.

9 Frithjof Benjamin Schenk, 'Die Produktion des imperialen Raumes: Konzeptionelle Überlegungen zu einer Sozial- und Kulturgeschichte der russischen Eisenbahn im 19. Jahrhundert', in Karl Schlögel (ed.), *Mastering Russian Spaces: Raum und Raumbewältigung in der russischen Geschichte*, Munich: Oldenbourg, 2011, pp. 109–27.

10 Peter Borscheid, *Das Tempo-Virus: Eine Kulturgeschichte der Beschleunigung*, Frankfurt/Main: Campus, 2004; Stephen Kern, *The Culture of Time and Space, 1880–1918*, Cambridge, Mass.: Harvard University Press, 1983.

11 Charles S. Maier, *Once within Borders: Territories of Power, Wealth and Belonging Since 1500*, Cambridge, Mass.: Harvard University Press, 2016, esp. ch. 5.

12 Richard White, *Railroaded: The Transcontinentals and the Making of Modern America*, New York: W.W. Norton, 2011, pp. xxii, 507.

13 Dirk van Laak, *Weiße Elefanten: Anspruch und Scheitern technischer Großprojekte im 20. Jahrhundert*, Stuttgart: Deutsche Verlags-Anstalt, 1999, pp. 53–97.

14 Per Högselius, Arne Kaijser and Erik van der Vleuten, *Europe's Infrastructure Transition: Economy, War, Nature*, London: Palgrave Macmillan, 2015, pp. 1–5; Maier, *Once within Borders*.

15 For the Russian case see, for example, Irina Michailovna Pushkareva, *Zheleznodorozhniki Rossii v burzhuazno-demokraticheskikh revoliutsiiakh*, Moscow: Nauka, 1975; Henry F. Reichman, *Railwaymen and Revolution: Russia 1905*, Berkeley: University of California Press, 1987; A. M. Solov'eva, 'Zheleznodorozhnye "koroli" Rossii. P. G. fon Derviz i S. S. Poliakov', in A. K. Sorokin (ed.), *Predprinimatel'stvo i predprinimateli Rossii ot istokov do nachala XX veka*, Moskva: Rosspen, 1997, pp. 266–85; Leonid Ivanovič Korenev, *Zheleznodorozhnye koroli Rossii*, Sankt Peterburg: Peterburgskii gos. universitet putei soobshcheniia, 1999.

16 Wolfgang Schivelbusch, *Geschichte der Eisenbahnreise: Zur Industrialisierung von Raum und Zeit im 19. Jahrhundert*, 4th ed., Frankfurt/Main: S. Fischer, 2007, first published 1977, pp. 35–45, 51–66, 67–83.

Infrastructures 185

17 Michael Freeman, *Railways and the Victorian Imagination*, New Haven: Yale University Press, 1999; id., 'The Railway as Cultural Metaphor: "What Kind of Railway History?" Revisited', *The Journal of Transport History* 20, 1999, 160–7; Ian Carter, *Railways and Culture in Britain: The Epitome of Modernity*, Manchester: Manchester University Press, 2001; Matthew Beaumont and Michael Freeman (eds.), *The Railway and Modernity: Time, Space, and the Machine Ensemble*, Bern: Peter Lang, 2007; Colin Divall and Hans-Liudger Dienel, 'Changing Histories of Transport and Mobility in Europe', in Ralf Roth and Karl Schlögel (eds.), *Neue Wege in ein Neues Europa: Geschichte und Verkehr im 20. Jahrhundert*, Frankfurt/Main: Campus, 2009, pp. 65–84. Apart from Schivelbusch's seminal book of 1977, the school of New Railway History was inspired by works like Michael Robbins, *The Railway Age*, London: Penguin, 1962; John R. Stilgoe, *Metropolitan Corridor: Railroads and the American Scene*, New Haven: Yale University Press, 1983; Nicholas Faith, *The World the Railways Made*, London: Pimlico, 1990.

18 Engels and Schenk, 'Infrastrukturen der Macht', p. 31; Erik van der Vleuten, 'Infrastructures and Societal Change: A View from the Large Technical Systems Field', *Technology Analysis & Strategic Management* 16, 2004, 395–414, here 402; Erik van der Vleuten and Arne Kaijser, 'Networking Europe', *History and Technology* 21, 2005, 21–48, here 22; van Laak, *Alles im Fluss*, pp. 221–41; Jens Ivo Engels (ed.), *Key Concepts for Critical Infrastructure Research*, Wiesbaden: Springer, 2018.

19 Frithjof Benjamin Schenk, *Russlands Fahrt in die Moderne: Mobilität und sozialer Raum im Eisenbahnzeitalter*, Stuttgart: Steiner, 2014, p. 122.

20 Maier, *Once within Borders*, ch. 5.

21 Friedrich List, 'Über ein sächsisches Eisenbahnsystem als Grundlage eines allgemeinen deutschen Eisenbahnsystems und insbesondere über die Anlegung einer Eisenbahn von Leipzig nach Dresden', 1833, in Erwin v. Beckerath and Otto Stühler, *Friedrich List: Werke, Bd. 3: Schriften zum Verkehrswesen. 1. Teil: Einleitung und Text*, Berlin: Hobbing, 1929, pp. 155–95.

22 Moskovskie vedomosti, 4.8.1883, quoted in Valentina A. Tvardovskaia, *Ideologiia poreformennogo samoderzhaviia. M. N. Katkov i ego izdaniia*, Moscow: Nauka, 1978, p. 79.

23 Jordi Marti-Henneberg, 'European Integration and National Models of Railway Networks (1840–2010)', *Journal of Transport Geography* 26, 2013, 126–38. On railway and road maps as primary sources, see Alexander Badenoch, 'Myths of the European Network: Constructions of Cohesion in Infrastructure Maps', in id. (ed.), *Materializing Europe. Transnational Infrastructure and the Project of Europe*, Basingstoke: Palgrave Macmillan, 2010, pp. 47–77.

24 Charles S. Maier, 'Consigning the Twentieth Century to History: Alternative Narratives for the Modern Era', *The American Historical Review* 105, 2000, 807–31, here esp. 819–21.

25 Högselius, Kaijser and van der Vleuten, *Europe's Infrastructure Transition*.

26 Schenk, *Russlands Fahrt in die Moderne*, pp. 37–50.

27 In the end the project proved to be almost twice as expensive. See Steven G. Marks, *Road to Power: The Trans-Siberian Railway and the Colonization of Asian Russia, 1850–1917*, Ithaca: Cornell University Press, 1991, p. 217.

28 Aleksandr Dmitriev-Mamonov, *Guide to the Trans-Siberian Railway*, London: David and Charles, 1972. Russian original: Ministerstvo Putei Soobshcheniia and A. I. Dmitriev-Mamonov (eds.), *Ot Volgi do Velikago okeana. Putevoditel'po*

Velikoi Sibirskoi zheleznoi doroge s opisaniem Shilko-Amurskago vodnago puti i Manchzhurii, Sankt Peterburg, 1900.
29 David C. Fisher, 'Exhibiting Russia at the World's Fairs, 1851–1900', PhD thesis, Indiana University, 2003, pp. 220–1. For personal accounts see, for example, Eugen Zabel, *Transsibirien: Mit der Bahn durch Rußland und China, 1903*, München: Frederking & Thaler, 2008; Bodo Thöns (ed.), *Transsiberien: Mit der Bahn durch Russland und China*, Darmstadt: Frederking & Thaler, 2003, pp. 49–51; Annette M.B. Meakin, *A Ribbon of Iron*, Westminster: Archibald Constable & Co., 1901, p. 12.
30 On the notion 'places of modernity' see Alexa Geisthövel and Habbo Knoch, (eds.), *Orte der Moderne: Erfahrungswelten des 19. und 20. Jahrhunderts*, Frankfurt/ Main: Campus, 2005.
31 Schivelbusch, *Geschichte der Eisenbahnreise*, ch. 11, pp. 152–7, Jürgen Osterhammel, *Die Verwandlung der Welt: Eine Geschichte des 19. Jahrhunderts*, Munich: C. H. Beck, 2009, pp. 437–40.
32 I. A. Bogdanov, *Vokzaly Sankt Peterburga*, Sankt Peterburg: Filologicheskii Fakul'tet S. Peterburgskogo Un-ta, 2004, pp. 47–52.
33 *Al'bom grazhdanskikh sooruzhenii Moskovsko-Vindavo-Rybinskoi zheleznoi dorogi*, o.O. 1908.
34 *'Illiustrirovannyi putevoditel' po iugozapadnym kazennym zheleznym dorogam*, Kiev, 1898, p. x.
35 'Pravila dlia passazhirov v poezdakh zheleznykh dorog' and 'Pravila pol'zovaniia passazhirskimi pomeshcheniiami zheleznodorozhnykh stantsii', 1891, quoted in Sergei Ivanovich Fedorov, *Spravochnaia knizhka dlia nizhnikh chinov zhandarmskikh politseiskikh upravlenii zheleznykh dorog*, Sankt Peterburg, 1903, 2nd ed., pp. 173–6.
36 Alexander Graf Keyserling, *Aus den Reisetagebüchern des Grafen Georg Kankrin, ehemaligen Kaiserlich Russischen Finanzministers in den Jahren 1840–1845. Mit einer Lebensskizze Kankrin's nebst zwei Beilagen. Erster Theil*, Braunschweig: Verlag der Hofbuchhandlung von Eduard Leibrock, 1865, p. 23.
37 *Statisticheskii sbornik Ministerstva Putei Soobshcheniia*, vyp. 131, vol. 2–3, 1915.
38 On the 'Russian hajj' see Eileen Kane, *Russian Hajj: Empire and the Pilgrimage to Mecca*, Ithaca: Cornell University Press, 2015.
39 Schenk, *Russlands Fahrt in die Moderne*, pp. 225–71.
40 Vladimir Ignat'evich Porai-Koshits, 'Zheleznyia dorogi v sudebno-meditsinskom i gigienicheskom otnosheniiakh', *Archiv sudebnoi meditsiny i obshchestvennoi gigieny*, 6, 1870, 1, pp. 122–48; 2, pp. 62–140; 3, pp. 48–84; 7, 1871, 1, pp. 59–81; 2, pp. 78–99.
41 Porai-Koshits, 'Zheleznyia dorogi', 7/2, 1871, pp. 87–8.
42 Frithjof Benjamin Schenk, 'Attacking the Empire's Achilles Heels: Railroads and Terrorism in Tsarist Russia', *Jahrbücher für Geschichte Osteuropas* 58, 2010, 232–53.
43 See, for example, 'Okhrannyia mery', in *Zheleznodorozhnoe delo* 1, 1908, p. 11.
44 Frithjof Benjamin Schenk, 'Travel, Railroads, and Identity Formation in the Russian Empire', in Eric Weitz and Omer Bartov (eds.), *Shatterzone of Empires: Coexistence and Violence in the German, Habsburg, Russian, and Ottoman Borderlands*, Bloomington: Indiana University Press, 2013, pp. 136–51.
45 Alfred Rieber, 'The Debate over the Southern Line: Economic Integration or National Security', in Serhii Plokhy and Frank Sysyn (eds.), *Synopsis: A Collection of Essays in Honour of Zenon E. Kohut*, Toronto: Canadian Institute of Ukrainian Studies Press, 2005, pp. 371–97.

46 'Tsirkuliar' tekhnichesko-inspektorskogo komiteta zheleznykh dorog', 4926, 22.8.1875, in *Sbornik ministerskikh postanovlenii* 2, 1877, p. 124; 'Tsirkuliar po eksploatatsionnomu otdelu', 21839–21845, 12.6.1896, in *Sistematicheskii sbornik uzakonenii i obshchikh rasporiazhenii, otnosiashchikhsia do postroiki i ėkspluatatsii zheleznykh dorog kaznoiu i posledovavshikh v period vremeni*, 1: s nachala 1881 g. po 31 maia 1898 g. vkljuchitel'no, Sankt Peterburg 1900, p. 650.

47 'Itogi raboty 3-i Gosudarstvennoi Dumy', in *Zheleznodorozhnoe delo*, 1908, 46, pp. 287–8.

48 Graf S. Iu. Vitte, 'Nekotoryia soobrazheniia o prichinakh defitsitnosti russkoi zheleznodorozhnoi seti', *Zheleznodorozhnoe delo*, 1910, 17–18, pp. 89–94.

49 Dietrich Geyer, *Der Russische Imperialismus: Studien über den Zusammenhang von innerer und auswärtiger Politik 1860–1914*, Göttingen: Vandenhoeck & Ruprecht, 1977, esp. pp. 22–3, 34–41, 76, 107–8, 141–8, 239–50; Edward R. Glatfelter, 'Russia, the Soviet Union, and the Chinese Eastern Railway', in Clarence B. Davis and Kenneth E. Wilburn (eds.), *Railway Imperialism*, Westport: Greenwood, 1991, pp. 137–54; Sören Urbansky, *Kolonialer Wettstreit: Russland, China, Japan und die Ostchinesische Eisenbahn*, Frankfurt/Main: Campus, 2008.

50 Roger Pethybridge, *The Spread of the Russian Revolution: Essays on 1917*, London: Palgrave Macmillan, 1972; Frithjof Benjamin Schenk, '"Fly, My Locomotive Fly…": Railway Tracks of Power during the Russian Revolution', in Deutsches Historisches Museum and Schweizerisches Nationalmuseum (eds.), *1917 Revolution: Russia and the Consequences*, Dresden: Sandstein, 2017, pp. 40–51.

10 Border zones

Lisa Hellman

As has become apparent in recent years, globalization neither precludes the making of borders nor attempts to exclude people and commodities from crossing them. Indeed, borders are drawn and redrawn all over the world. This was equally true a few hundred years ago when borders were created in parallel with increased economic, intellectual, and cultural integration, and in light of perceived threats posed by ethnic groups or genders. Nowhere is this more apparent than in harbours and port cities, the hubs of early modern globalization, and few analytical tools highlight these processes better than the concept of space. Analysis of the construction of two such harbour spaces reveals forced integration as well as separation, and demonstrates that borders are not simply lines drawn on a map, but the consequence of continuously fluctuating power relationships. Indeed, border-making is a story of everyday human practices and the hierarchies they create. Considering who borders were meant to keep apart, as well as who could cross them, illuminates acts of exclusion and permeability as well as of ethnic and gendered division.

This chapter compares the borders surrounding the foreign trade quarters in Japan under Tokugawa rule and in China during the Qing dynasty. The way foreign trade was regulated demonstrates contact between the two polities and the existence of sometimes contrasting and sometimes overlapping attitudes towards the control of foreign trade. In the ports of Japan and China, parallel processes of border-making produced spaces with striking similarities and differences. By closely inspecting the resultant border zones, as well as the policies of control, we can see how borders are the outcome not only of regulations but also of subversions and strategies in the face of these regulations.

In the 17th and 18th centuries, China and Japan acted as independent agents in the intra-Asian and European trade systems. Neither was a European colony nor under European control, and both limited their trade with Europe by restricting it to the ports of Canton and Nagasaki, respectively. A large body of research has shown these harbours to have been vibrant global hubs home to a mix of ethnic, cultural, and economic groups.[1] This does not mean, however, that trade was a free-for-all. Rather, just as in European ports at the time, Canton and Nagasaki were regulated in a

DOI: 10.4324/9780429291739-13

number of ways based on state ambitions to balance profit and control. Neither the population nor the borders of Tokugawa Japan and the Qing Empire seamlessly match those of present-day Japan and China. During the Qing dynasty, China was an expansive and multi-ethnic empire, while Japan expanded its territory northwards.

Scholars have long been interested in sites of intercultural encounter. A classic work in this field is by the literature scholar and theorist Mary-Louise Pratt, who coined the term 'contact zone' to describe 'social spaces where cultures met, clashed and grappled with each other'. She was primarily interested in spaces of 'highly asymmetrical relations of power', using examples of European colonial contacts.[2] However, the concept of 'contact zone' can also facilitate analysis of the foreign trading quarters in Tokugawa Japan and the Qing Empire, where the relationship between European and non-European was the inverse of that in the colonies. Postcolonial analyses and concepts, such as Pratt's, stress that encounters between cultures in the early modern period not only tied the world together, but also created hierarchical relations between peoples and regions. Where Pratt uses the concept of space to discuss diverse arenas and the relationships within them, other scholars focus on how these relationships create the spaces themselves. One influential theorist of space, Michel de Certeau, emphasizes how spaces are given meaning (and are thereby created) by the people living in, acting in, and understanding them. This constructivist understanding is clear in his separation between the concepts of place and space. For de Certeau, a *place* is a perceived stable order in which elements are distributed, while *space* is composed of intersections of mobile elements, that is, the practices and movements of people.[3] In short: space is a practised place, and, thus understood, is the focus of this chapter.

De Certeau's discussion of space includes ideas about borders; as his starting point he questions the idea that the nation-state functions as an absolute space. He therefore sees borders not as a line separating distinct spaces, but as constituting a substantive third space between them.[4] Seeing borders as a separate space consistently constructed by practices and discourse helps us deconstruct relations between states, their borders, as well as the resulting power relations within individual states. Each of these elements lies at the heart of studies on borderlands.[5]

De Certeau's theory of space has been met with criticism. The feminist geographer Doreen Massey, for example, finds his concept of place to be monolithic, static, unaffected by time, and liable to entail a deeply problematic 'taming of space'.[6] Part of this tension is a result of her focus on de Certeau's notion of place, rather than on his socially constructed notion of space and the diverse practices by which space was constructed. For de Certeau, there is a mutually constitutive relationship between space and discourse: storytelling, just as walking and mobile practices, also carves out agency.[7]

Massey's critique is motivated by her insight into the social and material dimensions of space, emphasizing the potential for social change in the process of constructing space. In contrast, the agency de Certeau ascribes to space is momentary – narratives can subvert power and place but cannot change place itself. For the historian, de Certeau offers an important reminder that analyses of space need to include both discourse *and* practices, that is, literary depictions as well as lived experiences. Massey's view of space offers a stress on processes of inclusion and exclusion and the resulting focus on the classed, ethnic, and gendered dimensions of spatial divisions.

For an analysis of borderlands, such an intersectional view is imperative; borders form where someone is or is not allowed to move and live. Practices of restricting and transgressing borders are intertwined with the construction of the groups enforcing the restrictions on the one hand and those transgressing them on the other.[8] This simultaneous creation of the groups and the spaces they were permitted – or even forced – to inhabit, is crucial for our understanding of intercultural relations in places like Nagasaki and Canton. For example, all foreign traders in Canton were forced to stay and work in the same limited area. That isolation, together with the restrictions on learning Chinese, led to the development of a language particular to the foreign quarters: Pidgin English.

There is a deep-rooted myth that Tokugawa Japan and the Qing Empire were 'closed' countries during the 17th and 18th centuries, a myth which is connected to narratives of Japanese 'national isolation', termed 'Sakoku', or of 'Eastern xenophobia' more generally. Such narratives present the borders surrounding the countries as impermeable, cutting Japan and China off from the rest of the world and setting them on a different historical trajectory from states that remained in contact with Europe. This interpretation, however, is no longer viable; social, cultural, and economic historians have shown that the countries were neither completely isolated nor that they intended to be.[9]

As they established their systems of trade regulation, Japan was recovering from a civil war, which had concluded with the establishment of the Tokugawa regime in 1603, while China had experienced the Manchu invasion, the toppling of the Ming rulers, and the foundation of the Qing dynasty in 1644. Both countries were fraught with ethnic, political, and religious tensions. Policies regarding trade should be seen in relation with attempts to establish overarching control rather than as an expression of some form of xenophobia or desire for national seclusion. Canton and Nagasaki offer windows into the simultaneous creation and policing of two early modern city borders, a process in which Europeans held very little sway.

This spatial analysis will be carried out in three steps, firstly by considering the material constitution of the harbour spaces, secondly by mapping who was kept in, and, finally, by discussing who was kept out and why. Analyzing the social as well as the physical borders surrounding Canton and Nagasaki on the basis of space takes us straight into the lived experience of intercultural power relations. Consideration of hierarchies of gender, ethnicity, and social and economic standing highlights not only the local domination of Europeans by non-Europeans, but also the multiplicity of groups living and working in the foreign quarters and the multi-layered structure of their interactions. The focus on the construction of these spaces, meanwhile, offers a way to make this study one of practices, and to trace inclusion, exclusion, and the simultaneous creation of groups. In short, the spaces help break down a binary view of intercultural contact, highlight the limits of national perspectives, and illuminate the different ways in which gender discourses and gendered practices impacted this contact. In the two cases presented here, it was European male traders who were subject to conscious policies of exclusion.

The location and layout of the harbours

It is first necessary to consider the spatial characteristics of international trade in China and Japan, starting with its restriction to one port, and the resulting process of border creation. China and Japan employed similar regulations of their maritime trade with Europe. These rules were not introduced overnight, but evolved gradually. In Japan, maritime foreign trade was increasingly regulated over the course of the 17th century. Similarly, in the Qing Empire, a set of rules evolved from the end of the 17th century, which by the mid-18th century was referred to as the 'Canton System'.[10] As a result, foreign maritime trade in the Qing Empire was restricted to the foreign quarters outside the city walls of Canton, and in Tokugawa Japan to trading quarters, first in Hirado and later in Nagasaki.

The two ports ultimately selected were situated in the south of each territory, far from the capitals of Beijing and Edo (modern-day Tokyo). While the Dutch and the English considered trade negotiations to be an intrinsic part of their diplomatic relations with foreign powers, the Tokugawa system of foreign relations differentiated between the countries it traded with and the countries with which it had diplomatic relations. A trader could thus not engage in the latter, while the Tokugawa leader, known as the shogun, would not negotiate with the former. The Dutch were, however, expected to undertake a journey to Edo to pay tribute to the shogun, just as the lords of the various feudal domains would. In China, imperial tribute was paid through intermediaries and few foreigners made the trip to Beijing themselves. In both contact zones, the distance between the port and the court enforced the theoretical division between commercial and diplomatic contacts.[11] The officials in China and Japan drew inspiration from each other regarding relationships with European and other foreign traders and the means to profit from trade without relinquishing political control.[12]

In addition, one should consider the material characteristics of the ports. The harbour of Canton lies at the heart of the Pearl River delta. Here, sediment forms mud banks that make the river shallow and precarious for large ocean-going vessels. As a result, all large ships, such as the East Indiamen of the European trading companies, depended on the services of Chinese towing boats and local pilots in their journey from the mouth of the river to their point of anchorage. If Chinese officials and European traders were in disagreement, it was easy for the officials to prevent traders from leaving by simply withholding the pilots and towing boats. The characteristics of the Pearl River itself thus constituted a control mechanism which Chinese officials could use against the foreign traders.[13] The control of mobility between and within spaces is a key instrument of power and one that spatial analysis is particularly well suited to illuminate.

In the case of Japan, the Dutch began their trade operations in the port of Hirado, close to Nagasaki, where they were based from 1609 to 1640. In the early sixteenth century, the Japanese government constructed the fan-shaped island of Deshima in Nagasaki, initially to house the Portuguese traders in Japan, who were treated with increasing suspicion because of their connection to the Christian missions. When the Portuguese were banned from Japan in 1639, it was the Dutch who were made to move to Deshima. Eventually the Chinese trading group also found themselves under increased controls as well, and they were housed in a separate quarter in Nagasaki after 1689.[14]

In Hirado, Dutch access to surrounding territory was restricted from about 1612 by a wall which was extended several times.[15] Despite this, the Dutch were able to move around with relative freedom. This changed after the move to Deshima. Nicholas Couckebacker, a Dutch trader, had gloatingly referred to Deshima as 'a prison for the Portuguese', without knowing that the Dutch would soon be the ones to live there.[16] The island of Deshima was walled-in and strictly supervised, but it was not the sole compound to be so controlled; the gates of the Chinese quarters, too, were locked from the outside.

By the late-18th century, Chinese merchants moved more freely around the city. Earlier on in the century, by contrast, a Chinese report from Nagasaki stressed how '[w]hen general trading merchants travel to that land, they are corralled in the city. There are high fences surrounding them [...]; it has a main gate and is guarded by heavily equipped soldiers'.[17] In Canton, the foreigners were not walled-in, but kept out. The foreign quarters had been placed outside the city wall, meaning that the wall stood between them and the bustling metropolis. They were, moreover, not free to walk as they pleased in the countryside or along the waterfront, but were restricted to the quarters designed for foreign trade. Walls were used to police urban environments throughout early modern Europe, as well as in China. Diverse ethnicities were separated within Europe; several cities had separate Jewish districts, and in Moscow and the eastern Mediterranean European traders were allocated houses in particular areas of the city. Walls were also built in European colonial and trade settlements in Asia and elsewhere. In Ayutthaya, Manila, Batavia, and Bengal, Europeans constructed walls to separate different trading and ethnic groups. The walling off of foreign traders in Japan and China was thus a typical example of urban control in the early modern world. The description of walls in Nagasaki in Canton were also classic reflections of the role narratives of mobility played in these two contact zones.

In Canton, as well as in Nagasaki, the foreign traders' own descriptions of their quarters, and thereby the discursive construction of the spaces, included recurring discussions of the walls separating them from the Asian countries beyond. The walls were a common topic in European travel writings and diaries. For example, Chaplain Gustav Fredrik Hjortberg, who visited Canton in 1748, wrote in detail about the construction, height, width, and length of the wall (1,856 steps he determined, after walking alongside it). The wall barred him from a Chinese world in which he was not welcome – one in which foreigners were 'regaled with stones, dogs, abuse and the throwing of sticks'.[18] Such descriptions of the experience of the foreign quarters also align with a spatial discourse within Europe, specifically that of the 'walled-off Asia', in which spatial restrictions were presented as proof for oriental despotism. The world beyond the wall – namely, what could not be seen – was imagined as representative for Asia as a whole and the walls as physical symbols of the Asian powers barring them from exploring it.[19]

The spaces of Canton and Nagasaki were constructed not just through text, but also in images. The Dutchman Isaac Titsingh, who worked as a trader in Japan for several years in the 1770s, depicted Nagasaki by painting the Dutch and Chinese districts (see Figure 10.1). Together, these two isolated spaces symbolized Nagasaki and its foreign trade. Numerous Japanese paintings of the two trading spaces were made and commonly sold as diptychs. While initially produced by court painters, such diptychs were also distributed as woodblock prints for a broader audience, showing their symbolic strength and exotic appeal.[21] Similarly, paintings of the foreign quarters of Canton became part of an industry of export art aimed at the European market, but they were also sold within China as symbols of exotic European trade. Depictions in text or art of both harbours mirrored trade restrictions and local power, but the Chinese quarters, the fan-shaped island of Deshima, and the factories of Canton became symbols of an exotic 'other' in Europe – a glimpse into their supposedly exclusive, and restrictive, nature.

Who was kept in

Next, we turn to the regulation of the ports and how that regulation was part of the continuous creation of borders around them. In the foreign quarters of Canton, all foreign traders lived and worked within a few streets. Their houses, called factories, were lined up along the waterfront and functioned as offices, warehouses, and homes. Notably, the houses were not owned by the foreign traders, but were rented from Chinese merchants. In time, the factories were rebuilt in a European style and the costs of their upkeep and refurbishment increasingly fell on the Europeans and, by the end of the 18th century, Americans. Most European trading groups rented one house each; eventually, certain buildings became so closely associated with that particular group that they could rent them out in turn.[22] The way in which the foreign traders lived, the style of their houses, and the division of the groups reinforced by the separate but neighbouring houses simultaneously worked to create a Chinese, as well as a European, notion of joint foreign trading groups in China. This group of foreign traders excluded the crew, but included all men charged with the trade of the respective European company and all private non-Chinese traders, who also lived in the foreign quarters. At the same time, however, the fact that the traders lived in different houses and worked for different trade ventures created divisions between them.

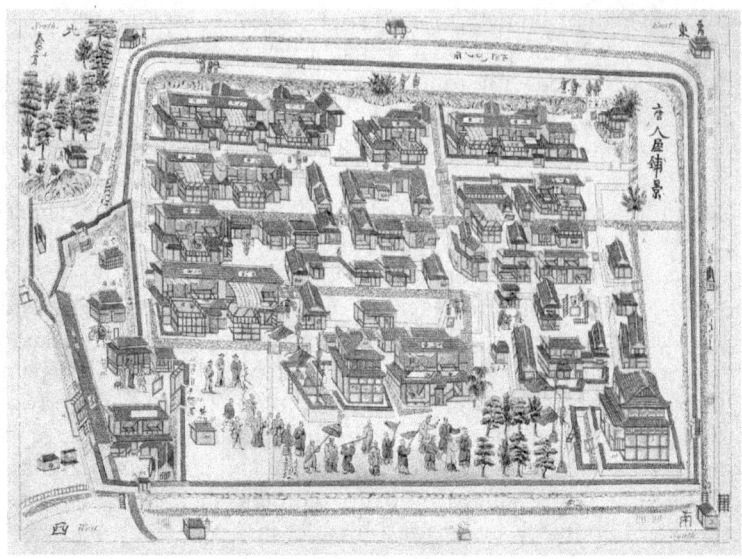

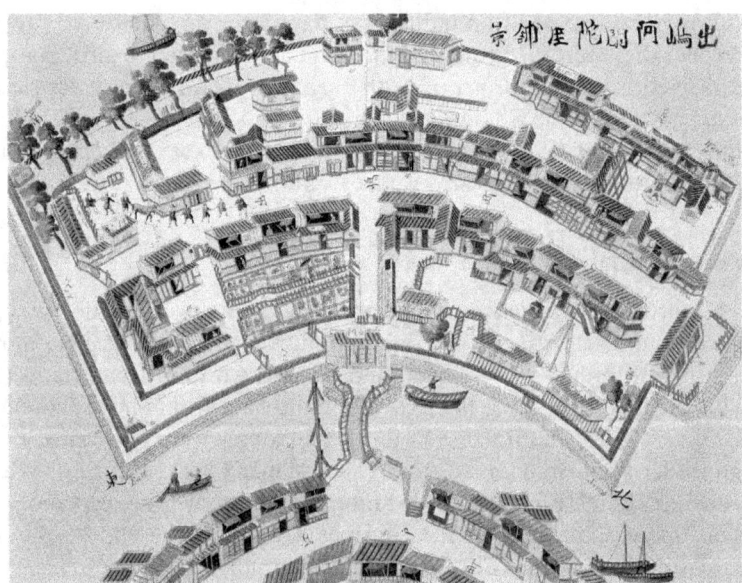

Figure 10.1 Isaac Titsingh's (c.1740–1812) depictions of the Chinese (above) and the Dutch (below) trading quarters.

Both © The Stapleton Collection/Bridgeman Images.[20]

While all the foreigners of Canton were crammed together in their quarters, in Nagasaki, Chinese and Dutch traders occupied separate areas. Just as in Canton, foreigners in Nagasaki were not allowed to own land and were initially forced to rent their houses. The Dutch lived in Japanese-style houses which, unlike foreign traders in Canton, they did not rebuild. The Chinese traders in Nagasaki, however, did refashion their rented houses in a Chinese style. These buildings were thus often depicted from the outside as an exotic and a distinct symbol of foreign trade. In the spatial representation of the Dutch quarters, meanwhile, painters had to rely on other visual signposts, such as the Dutch food bell that stood outside one of the houses, but which in depictions was often moved inside.[23] The spatial construction of Nagasaki and its visual representation both reflected the strong control exerted over foreign traders and the weakness of the Europeans' position.

The spatial construction of the foreign quarters was characterized not only by restrictions on housing, but also by the multiplicity of groups – of different genders and ethnicities – that lived and worked there. All foreign traders were welcome in Canton, but they were normally treated as one homogenous entity by the Chinese authorities. This contact zone was linked not only to European trade networks, but also to those of Jewish, Parsi, Armenian, and Muslim traders. The Japanese constitute an exception as they were forbidden to leave Japan by the 18th century.[24] Canton was thus a melting pot of interactions, and Nagasaki was little different.

In Japan, by far the largest and most influential foreign group of traders during the 18th century were the Chinese, who operated in a separate trading area in Nagasaki, where they lived and worked.[25] The second largest group was the Dutch, who occupied a similar walled-off area in another part of the city. Though representing a European power, the Dutch company's trade in Japan was more closely connected to the Asian trade sphere than to Europe. A ship which operated under a Dutch flag, moreover, did not mean that its crew was exclusively Dutch. The Dutch quarter in Nagasaki, close to that of the Chinese, was the aforementioned island of Deshima.[26] From 1613 to 1623, the British East India Company tried to establish itself in Japan and cooperated with the Chinese trading houses in Nagasaki, in an attempt to benefit from their experiences.[27] The main connection between the foreign trading groups in Japan, however, was not collaborative but competitive as Chinese merchants challenged their Dutch counterparts for access to the Japanese market.

The foreign trade quarters of Canton gathered multiple groups in one space, whereas the Nagasaki system had two foreign trading groups in two parallel spaces. Overall, the conditional living arrangements for foreigners in Canton and Nagasaki functioned as a constant cap on foreign ambitions; the traders were only permitted to trade there and could not establish settlements. In both ports, additionally, the isolation of the traders in one space created one joint trading group based on the basic division between foreigners and locals, rather than on the subgroups within the group of foreigners. Finally, these regulations were mirrored in visual and exoticizing representations of these same trading groups within China and Japan.

Who was kept out

In Canton, the only people who lived permanently in the foreign quarters were the handful of men overseeing the foreign trade and their servants.

Priests, doctors, and other officers moved between the ship and the foreign quarters. However, the largest group of foreigners, the sailors, spent most of their time in China in the harbour of Whampoa a few kilometres down the river. In Nagasaki, the Dutch and Chinese contingents included traders, their servants, and other administrative personnel, while the sailors stayed on board the ships at anchor in the harbour. In both harbours only a few Japanese and Chinese merchant houses were allowed to take part in the bulk of the foreign trade – as was common at this time, monopoly systems dominated on both the European and the Asian sides.[28] The trading spaces of Canton and Nagasaki were small; indeed, one could leisurely cross them in a matter of minutes, and it was here that the contracts were drawn up.

The limited size of the foreign quarters made it impossible to accommodate all sailors, which most probably was the intention of local Chinese and Japanese authorities, as well as of the foreign traders themselves. Making the sailors stay on board, away from the foreign quarters, was a way of keeping them in check and separate from sensitive trade negotiations. Fights often broke out between sailors, and their potential for rowdiness was considered particularly threatening to the maintenance of good relations with the Chinese. In 1830, the American David Abeel said, 'the conduct of sailors, permitted to enjoy "liberty" on these islands, has been highly offensive to the Chinese'.[29] In Deshima, traders voiced similar concerns as to how the behavior of drunken sailors would be perceived by the Japanese.[30]

Exploring space as the product of processes of exclusion and inclusion can help us understand the multi-layered power relationship in the ports and the flexible nature of the groups involved. To preserve the harmony of trade, the foreign traders willingly joined forces with Japanese and Chinese border authorities to control the mobility of the sailors.[31] Paying attention to who was excluded from the foreign quarters, furthermore, shows the importance of including the ship as part of the harbour environment – it constituted a home, as well as a space for the isolation of the crew, during the months in port.

In both China and Japan, the seclusion of the foreign spaces was also reinforced by religion. The officials regulating the contact zones strove to create a strict separation, spatially and legally, between the diplomatic spaces of the capital and the commercial spaces of the two international harbours. They also endeavoured to differentiate between encounters with religious missionaries and those with traders. In China, the Christian mission continued into the 1720s, while in Japan it had been banned almost a hundred years before. By the 1730s, however, Christianity was strictly forbidden in both China and Japan, and missionaries and converts were expelled or even executed.[32]

In China, the focus was not primarily on restricting Christian texts but rather on Christians themselves. In Japan, by contrast, Christian writings were banned, and ships were searched for such material. The Tokugawa officials permitted traders to be Christians 'in their hearts', as long as they did not attempt any outward demonstrations of faith while in the country. Interestingly, in Japan it was Chinese Christian traders rather than the Dutch who were most frequently found carrying bibles. As a result, only the Chinese traders arriving in Japan were made to tread on pictures of the Holy Virgin, a trial called *fumi-e*, in order to prove that they were not Christian; the Dutch were exempt. Suspicion that the Chinese were Christians (and the practice of treading on a Christian idol) directly stemmed from 16th-century Portuguese and Spanish missionary activity in Japan and southern China.[33]

Alongside religion, gender relations were considered a potential threat to the stability of the trading environment, resulting in the policing of the female presence in both Canton and Nagasaki. Foreign men and local women from similar social backgrounds were therefore strictly separated, and foreign women either discouraged or forbidden to enter the trade quarters. Women were considered to be the key to settling and integrating into another society in the early modern world. Neither the Japanese nor Chinese authorities wished to encourage this – their goal was profit, not racial or cultural hybridity.[34] In other early modern international ports, women played a crucial role as linguistic and cultural go-betweens. While some foreign women arrived in Canton and Nagasaki, they were few in number, and their presence illicit. Thus, they did not intermarry or openly play the role of broker to the degree they did elsewhere.[35] Divisions between genders were consciously used to uphold the ethnic borders of the foreign quarters.

This is not to say that there was no interaction at all. In Nagasaki, *yūjo* (a mix of entertainer and prostitute) frequented the foreign quarters. The interaction between Japanese women and foreign men was not only openly known and talked about, it was clearly regulated by the Japanese authorities; they even licensed a brothel quarter, Maruyama. Again, the concept of 'foreigners' included the Dutch as well as the Chinese; a report from the eighteenth century reads '[a]ll of the food and drink as well as female companionship is provided [for us]'. Even those who only spent a few months in port were reported to have Nagasaki 'girlfriends'. The interaction between foreign men and local women can be found mentioned in Japanese administrative records, revealing the regularity of the contact. The *yūjo* could enter both the Chinese and Dutch foreign quarters and acted as cultural go-betweens; they were even called 'Nagasaki wives', and numerous regulations existed regarding their mixed children. In 1715, one such ordinance ruled that while the fathers should be consulted about the upbringing of their children, they were not allowed to bring them home, nor were women allowed to accompany the foreign men when they left Japan.[36] As a contemporary Japanese verse about the *yūjo's* quarter put it:

A farewell at Maruyama
spans a distance
of thirteen thousand miles.[37]

Women were common in pictorial and textual depictions of daily life in Nagasaki. Indeed, in a typical depiction of Dutch traders having dinner, the *yūjo* are not only present, but are central figures embracing the traders.[38] In addition, *yūjo* were commonly shown on Japanese as well as European maps of the foreign areas of Deshima; in fact, the depiction of a *yūjo* on a map symbolized the location of the foreign quarters. The legibility of this female imagery for Japanese and foreigners alike demonstrates how integrated women were to this particular spatial construction.[39]

Canton offers a stark contrast, with an almost complete silence on interaction between men and local women in the Chinese archival records. There were male and female prostitutes working on floating brothels, called flower boats, and some of the boats catered specifically to the foreign traders, in blatant contravention of the rules. During the season 1747–48, Charles Fredrick Noble, an officer on the British Indiaman *Prince Edward*, described such an interaction: 'At last Jack [the pimp] came creeping in at one end of the tilt and asked *Carei grandi hola, pickenini hola?* What size of a wh–e they wanted?' He then saw their pimp going out to some sampans and fetching two girls 'the one younger and more slender than the other'.[40] On land, the interactions were highly regulated and limited, but on the river, it seems, gender separations became more fluid.

The key aspect, however, is ethnic; many of the women were from the so-called Tanka minority, or 'boat people'. The Tanka were not allowed to settle on Chinese land, so even if European men had contact with them, it still would not lead to foreign settlement; a relationship with a Tanka woman was neither a route to integration into Canton's merchant families, nor to owning land, nor to children who would be considered part of Cantonese society. European contact with Tanka women therefore was by and large disregarded by the Chinese officials. Nevertheless, it was still not talked of openly. Indeed, formal rules regarding mixed-race children did not even exist in China. This is not to say that such children did not exist, only that they were so far out on the margins of society that they were not even regulated.[41]

The borders of the trading spaces were additionally upheld by bans on foreign women. European women were almost non-existent in Japan and China, as a result of the regulations of the harbour, and yet they became an exotic and appreciated symbol of that very space. Within China and Japan, images and objects depicting European women were wide-spread and sold in both national markets.[42] For foreigners, the scarcity of foreign women was part of the experience and a common topic in travelogues. In 1751, Pehr Osbeck, a ship's priest, wrote of China that there were 'strange people in this land, who consider foreign women little different from contraband'.[43] In the early-19th century, however, as the respective trading systems began to deteriorate, and an increasing number of groups took part in the trade, differences developed in how this separation was upheld. In 1817, the Dutch trader Jan Cock Blomhoff caused a stir by arriving in Deshima accompanied by his son, his wife, Titia, and two female servants. During their brief stay on Deshima, Japanese artists painted the couple. As an exotic novelty, this motif spread like wildfire on scrolls and prints throughout Japan.[44]

In contrast, European women in Canton were perceived as a more acute threat, possibly because their presence was used to consciously transgress the regulations of border spaces. When a Mrs Bayne arrived with her husband in

Canton in 1830, conflict ensued instantly. Her illegal presence became tangled up with other European infringements and the Chinese viceroy responded by expelling her. Tensions ran high, and in response 26 British merchants wrote a joint protest. The viceroy held fast to the rules, declaring that should a European woman be found and not leave within two or three days, 'soldiers will be sent to the Factory to seize and drive her out'.[45] The absence and presence of foreign women reflected how the borders around foreign quarters were becoming gradually more permeable, as in Japan, and consciously ruptured in the case of Canton.

The clearly stated aim of the Japanese and Chinese authorities was to create an environment that was most suitable for trade, rather than a space for cultural interaction. In both contexts, the separation between men and women was therefore considered important for policing the space. Despite these particular ports being hubs in a global trading network, hybridity and integration were prevented. Nevertheless, the foreign quarters in Canton and Nagasaki became border spaces with their own distinctive cultures that were perceived as peculiar both locally and beyond. The analysis of who was kept out, and for what reason, benefits from the intersection between the histories of art, trade, and international relations, as well as attention to the gender relations they mirror.

Conclusion

In China, foreign traders were restricted to the foreign quarters outside Canton. In Japan, they were limited to the island of Deshima outside Nagasaki. Both places, located far from the capitals, were primarily spaces created for commercial relationships. The resultant spatial regulations are not entirely dissimilar to the way ghettos were set up in European cities at the same time. In fact, both at the time European traders and contemporary scholars have labelled Canton a 'Golden Ghetto'. This highlights the intertwined construction of the ethnic groups and the spaces in which they lived and acted, but it also stresses the highly charged nature of border making. The trading spaces of Canton and Nagasaki were created just as much through the circumvention of rules as through the adherence to and enforcement of them.

Another issue that spatial restrictions brings to light is the identity of who was mobile and who was not, and who was in charge of enforcing regulations. Notably, the foreign officers and the Japanese and Chinese authorities cooperated to keep the sailors close to the ships. Finally, an intersectional analysis of space also highlights challenges to gendered and ethnic divisions. For example, when European or American women moved beyond their allocated zone and entered the foreign quarters of Canton, it was a means for challenging the system and demanding more space and greater mobility.

The fact that these were maritime spaces is significant in several ways. For one thing, as international ports they were connected to global networks of trade and thus circulations of ideas and people. However, the local constitution of the harbours made use of water as a practical border, separating groups on the basis of status, gender, and ethnicity. At the same time, as seen from the case of the Pearl River in Canton, the water also connected different parts of the harbour and provided living and working environments for groups such as the Tanka. Analysis of the making of borders necessitates awareness of these murky in-betweens.

Importantly, the experience of these borders by the various groups inhabiting them and working in them did not only create a particular space but also shaped the construction of the groups themselves. Borders shut people out, but also keep certain people in and help define those excluded and those writing the rules. The sharing of spaces was intertwined with the process of imposing a homogeneity on foreign trading groups, groups that elsewhere, such as in Europe, were perceived as separate. In Japan, the Dutch trade was affected by the presence of the Chinese, as were the English while they were in Nagasaki. In Canton, the foreigners traded with each other, helped each other, and competed with each other.

Canton and Nagasaki were ruled, experienced, imagined, and thus constructed, in relation to and by multiple groups. These included not only diverse groups of foreigners, including slaves and women, but also various local groups, necessary for our understanding of power relations on the ground. The spatial focus helps to draw us away from an image of a meeting between two homogenous national groups. De Certeau's view of borders as a third space, a continuously constructed in-between, draws attention to how any construction of space is also a relationship of power; this becomes particularly clear when transcultural port environments such as Canton and Nagasaki are compared.

Borders constitute contact zones almost by definition, and while the natural focus might lie on the rules and restrictions of that delineation, a constructivist spatial analysis can take this one step further and show how that process is consistently intertwined with other power relations, such as those of gender and ethnicity. This approach, however, demands attention to processes at a local level. It is for this reason that spatial theorists such as de Certeau favour analyses of daily life and social practices. Only in day-to-day experiences and actions is it possible to perceive not only rules and power hierarchies, but also transgressions and subversions. An intersectional and spatial approach transforms borders from a static presupposition to a multi-party negotiation that creates historical space itself.

Notes

1 See, for example, Evert Groenendijk, Cynthia Viallé and Leonard Blussé (eds.), *Canton and Nagasaki Compared, 1730–1830: Dutch, Chinese, Japanese Relations: Transactions*, Leiden: Institute for the History of European Expansion, 2009; Leonard Blussé, *Visible Cities: Canton, Nagasaki, and Batavia and the Coming of the Americans*, Cambridge, Mass.: Harvard University Press, 2008; Masashi Haneda, 'Canton, Nagasaki and the Port Cities of the Indian Ocean: A

Comparison', in Masashi Haneda (ed.), *Asian Port Cities, 1600–1800: Local and Foreign Cultural Interactions*, Singapore and Kyoto: NUS Press in association with Kyoto University Press, 2009; Shigeki Iwai, 'Nagasaki to Kōshū' 長崎と広州, in Yasunori Kōya, Masatoshi Ishii and Shōsuke Murai (eds.), *Kinseiteki sekai no seijuku (Nihon no taigai kankei)* 近世的世界の成熟 (日本の対外関係), Tokyo: Yoshikawa kōbunkan, 2010. For a fuller account of the subject discussed in this chapter see Lisa Hellman, *This House Is Not a Home: European Everyday Life in Canton and Macao, 1730–1830*, Leiden and Boston, Mass.: Brill, 2018.
2 Mary Louise Pratt, 'Arts of the Contact Zone', *Profession*, 1991, 33–40, here 34.
3 Michel de Certeau, *The Practice of Everyday Life*, Berkeley: University of California Press, 1984, French 1980.
4 De Certeau, *The Practice of Everyday Life*, p. 127; see also Se-Yong Jang, 'The Spatial Theory of de Certeau, a Vagabond in Stray Space', *Localities* 5, 2015, 89–102.
5 Charles S. Maier, *Once within Borders: Territories of Power, Wealth, and Belonging Since 1500*, Cambridge, Mass.: Harvard University Press, 2016; Thomas D. Hall, 'Puzzles in the Comparative Study of Frontiers: Problems, Some Solutions, and Methodological Implications', *Journal of World-Systems Research* 15, 2009, 25–47; David Ludden, 'The Process of Empire: Frontiers and Borderlands', in Christopher Bayly and Peter Fibiger Bang (eds.), *Tributary Empires in Global History*, London: Palgrave Macmillan, 2011, pp. 132–50.
6 Doreen Massey, *For Space*, London: Sage, 2005, pp. 120–2.
7 De Certeau, *The Practice of Everyday Life*, pp. 126–7.
8 Doreen Massey, *Space, Place and Gender*, Oxford: Polity, 1994; this is developed further in Sara Ahmed, *Strange Encounters: Embodied Others in Post-Coloniality*, London: Routledge, 2000.
9 Henry D. Smith, 'Five Myths about Early Modern Japan', in Ainslee Embree and Carol Gluck (eds.), *Asia in Western and World History: A Guide for Teaching*, Armonk: M.E. Sharpe, 1997; Gang Zhao, *The Qing Opening to the Ocean*, Honolulu: University of Hawai'i Press, 2013.
10 Peng Hao, *Kinsei Nisshin tsūshō kankei shi* 近世日清通商関係史, Tokyo: Tokyo Daigaku Shuppankai, 2015, pp. 245–6; W.E. Cheong, *Hong Merchants of Canton: Chinese Merchants in Sino-Western Trade, 1684–1798*, Richmond: Curzon, 1997.
11 Matthew W. Mosca, *From Frontier Policy to Foreign Policy: The Question of India and the Transformation of Geopolitics in Qing China*, Stanford: Stanford University Press, 2013; Chunchen Zhao, (ed.), *Guangzhou shisanhang yu qingdai zhongwai guanxi* 广州十三行与清代中外关系, Guangzhou: Guangdong jingji chuban she, 2012; Matsukata Fuyuko (ed.), *Nichiran Kankeishi o Yomitoku: jūnana seiki chūyō, Yōroppa seiryoku no Nihon kenshi to 'kokusho'* 日蘭関係史をよみとく 一七世紀中葉、ヨーロッパ勢力の日本遣使と「国書」, Kyoto: Rinsen Shoten, 2015; Shigeki Iwai, 'Shindai no goshi to "chinmoku gaikō"' 清代の互市と "沈黙外交, in Susumu Fuma (ed.), *Chūgoku higashiajia gaikō kōryū-shi* 中国東アジア外交交流史, Kyoto: Kyōtodaigaku gakujutsu shuppankai, 2007; Dittmar Schorkowitz and Nin Chia, *Managing Frontiers in Qing China: The Lifanyuan and Libu Revisited*, Leiden: Brill, 2017.
12 Shigeki Iwai, '"Kaihentai"go no kokusaishakai' 「華夷変態」後の国際社会, in Yasunori Kōya, Masatoshi Ishii and Shōsuke Murai (eds.), *Kinseiteki sekai no seijuku, Nihon no taigai kankei*, 近世的世界の成熟 (日本の対外関係), Tokyo: Yoshikawa kōbunkan, 2010.

13 Paul A. Van Dyke, *The Canton Trade: Life and Enterprise on the China Coast, 1700–1845*, Hong Kong: Hong Kong University Press, 2005, p. 36.
14 Katō Shigenori, 'Hirado-Oranda Shōkan to VOC No Shima (Yokoshima) No Hakkutsu (VOC to Nichiran Kōryū VOC Iseki No Chōsa to Tabako)' 平戸オランダ商館とVOC の島（横島）の発掘 (VOCと日蘭交流 VOC 遺跡の調査とたばこ), *Tabakotoshio No Hakubutsukan Kenkyū Kiyō* たばこと塩の博物館研究紀要 10, 2012, 160–62; Blussé, *Visible Cities*, pp. 21–2.
15 Mikio Horikawa, 'Hirado Oranda shōkan shisetsu ni kansuru kōsatsu (sono 1)' 平戸オランダ商館施設に関する考察(その1), *Ōbirin tankidaigaku kiyō* 桜美林短期大学紀要, 1998, 65–80.
16 Michael S. Laver, *The Sakoku Edicts and the Politics of Tokugawa Hegemony*, Amherst: Cambria Press, 2011, p. 178.
17 Shigeki Iwai, '"Kaihentai"go no kokusaishakai' 「華夷変態」後の国際社会, 16; Li Wei, 'Wei zouwen chuyang shangchuan qing youshi' 為奏聞出洋商船情由事, Yongzheng 6/8/8, *Gongzhongdang Yongzhengchao Zouzhe* 宮中檔雍正朝奏摺, Vol. 11.
18 Gustav Fredrik Hjortberg, 'Ost-Indisk Resa 1748 och 1749 förrättad och beskrefwen af Gustaf Fr. Hjortberg', M 281a, Manuscript collection, Royal Library of Sweden, 4, 53.
19 Jürgen Osterhammel, *Unfabling the East. The Enlightenment's Encounter with Asia*, Princeton: Princeton University Press, 2018, German 1998.
20 'Plan of the Dutch Factory in the Island of Desima at Nangasaki' and 'The Chinese Factory in the Street of Teng-chan at Nangasaki, founded 1688', from 'Illustrations of Japan' by Isaac Titsingh (c.1740–1812), published London, 1822 (colour litho). The Stapleton Collection, Bridgeman Images.
21 Kristina Kleutghen, 'Chinese Occidenterie: The Diversity of "Western" Objects in Eighteenth-Century China', *Eighteenth-Century Studies* 47, 2014, 117–35; Nagasaki-shi Dejima shiseki seibi shingikai 長崎市出島史跡整備審議会 (curated by), *Dejima-zu – sono keikan to hensan* 出島図―その景観と変遷, Nagasaki: Nagasaki-shi, 1987.
22 Johnathan A. Farris, 'Thirteen Factories of Canton: An Architecture of Sino-Western Collaboration and Confrontation', *Buildings & Landscapes* 14, 2007, 68–83; Paul A. Van Dyke and Maria Kar-wing Mok, *Images of the Canton Factories, 1760–1822: Reading History in Art*, Hong Kong: Hong Kong University Press, 2015; Ichirō Ōtani, *Nagasakihanga to Ikoku No Omokage* 長崎版画と異国の面影, Tokyo: Itabashi kuritsubijutsukan, 2016.
23 Osamu Ōba, *Nagasaki Tōkanzu shūsei* 長崎唐館図集成 Suita: Kansai Daigaku Shuppanbu, 2003; Ōtani, *Nagasakihanga* 長崎版画.
24 Paul A. Van Dyke, *Merchants of Canton and Macao: Politics and Strategies in Eighteenth-Century Chinese Trade*, Hong Kong: Hong Kong University Press, 2011.
25 Marius B. Jansen, *China in the Tokugawa World*, Cambridge, Mass.: Harvard University Press, 1992, pp. 29–30.
26 Patrizia Carioti, '17th-Century Nagasaki: Entrepôt for the Zheng, the VOC and the Tokugawa Bakufu', in François Gipouloux (ed.), *Gateways to Globalisation: Asia's International Trading and Finance Centres*, Cheltenham: Edward Elgar, 2011.
27 Richard Cocks, *Diary of Richard Cocks, Cape-Merchant in the English Factory in Japan, 1615–1622: With Correspondence*, London: Printed for the Hakluyt Society, 1883.

28 Van Dyke, *Merchants of Canton and Macao*; Matsui Yōko, 'Dejima to Kakawaru Hitobito' 出島とかかわる人々, in Matsukata Fuyuko (ed.), *Nichiran Kankeishi O Yomitoku* 日蘭関係史をよみとく, Kyoto: Rinsen Shoten, 2015, pp. 146–80.
29 David Abeel, *Journal of a Residence in China, and the Neighboring Countries, from 1829 to 1833*, London: James Nisbet and Co., 1835, p. 80.
30 Donald Frederick Lach and Edwin J. Van Kley, Asia in the Making of Europe, vol. 3, book 4: *East Asia*, Chicago: University of Chicago Press, 1993, p. 1880.
31 John E. Wills Jr. and John L. Cranmer-Byng, 'Trade and Diplomacy with Maritime Europe, 1664–c. 1800', in John E. Wills Jr. et al. (eds.), *China and Maritime Europe, 1500–1800: Trade, Settlement, Diplomacy, and Missions*, Cambridge: Cambridge University Press, 2010, p. 230.
32 Liam Matthew Brockey, *The Visitor: Andre Palmeiro and the Jesuits in Asia*, Cambridge, Mass.: Harvard University Press, 2014.
33 John W. Witek, 'Catholic Missions and the Expansion of Christianity, 1644–1800', in John E. Wills Jr. et al. (eds.), *China and Maritime Europe, 1500–1800: Trade, Settlement, Diplomacy, and Missions*, Cambridge: Cambridge University Press, 2010; Takayuki Shimada and Yuriko Shimada, *Fumi-e: Gaikokuhito ni yoru fumi-e no kiroku* 踏み絵: 外国人による踏み絵の記録, Tokyo: Yūshōdōshuppan, 1994.
34 For more on the use of relationship to local women, see Carmen Nocentelli, *Empires of Love: Europe, Asia, and the Making of Early Modern Identity*, Philadelphia: University of Pennsylvania Press, 2013.
35 Douglas Catterall and Jodi Campbell, (eds.), *Women in Port: Gendering Communities, Economies, and Social Networks in Atlantic Port Cities, 1500–1800*, Leiden: Brill, 2012.
36 Yoko Matsui, 'Nagasaki to Maruyama yūjo chokkatsu bōeki toshi no yūkaku shakai' 長崎と丸山遊女 直轄貿易都市の遊廓社会, in Ashita Saga and Nobuyuki Yoshida (eds.), *Santo to chihō toshi* 三都と地方都市, Tokyo: Yoshikawakōbunkan, 2013.
37 Frits Vos, 'Forgotten Foibles: Love and the Dutch at Dejima (1641–1854)', *East Asian History* 39, 2014, 139–51, here 144.
38 See 'De groote party in de kamer van het opperhoofd Zyn op het eiland' A.5219(01) in the Het Scheepvaartmuseum. https://www.maritiemdigitaal.nl/index.cfm?event=search.getdetail&id=101008957 (accessed 15 May 2021).
39 Nagasaki-shi Dejima shiseki seibi shingikai 長崎市出島史跡整備審議会 (curated by), *Dejima-zu – sono keikan to hensan* 出島図―その景観と変遷.
40 Charles Frederick Noble, *A Voyage to the East Indies in 1747 and 1748 Containing an Account of the Islands of St. Helena and Java, of the City of Batavia, of the Government and Political Conduct of the Dutch, of the Empire of China, with a Particular Description of Canton, and of the Religious Ceremonies, Manners and Customs of the Inhabitants: Interspersed with Many Useful and Curious Observations and Anecdotes*, London: Printed for T. Becket and P. A. Dehondt and T. Durham, 1762, p. 240.
41 Van Dyke, *The Canton Trade*, pp. 61, 204; Susumu Murao, 'Kenryū kibō — toshi Kōshū to Makao ga tsukuru henkyō' 乾隆己卯 ― 都市広州と澳門がつくる辺疆, *Tōyōshikenkyū* 東洋史研究 65, 2007.
42 Jan Wirgin, *Från Kina till Europa: kinesiska konstföremål från de ostindiska kompaniernas tid*, Stockholm: Östasiatiska museet, 1998; Ty M. Reese, 'Wives,

Brokers, and Laborers: Women at Cape Coast, 1750–1807', in Catterall and Campbell (eds.), *Women in Port*, pp. 291–314.
43 Pehr Osbeck, *Dagbok öfver en ostindisk resa åren 1750, 1751, 1752*, Stockholm: Rediviva, 1969, p. 236.
44 Otani, *Nagasakihanga*.
45 H.B. Morse, *The Chronicles of the East India Company Trading to China, 1635–1834*, vol. IV, Oxford: Clarendon Press, 1926, p. 237.

Part III
Reflecting on concepts, tools, and approaches

11 Lefebvrean landscapes

Dawn Hollis

Many of the challenges facing the historian interested in the concept of space are also encountered by the historian of past landscapes. Both spaces and landscapes are to some extent physical phenomena, with a tangible reality external to the human mind. At the same time, they both have an experiential reality: human beings see, touch, move amidst, and engage with them on both a physical and a mental level. As such, spaces and landscapes are not merely 'out there' in the physical world, but are also evocative of internal mental and emotional responses. Even more than that, these responses – and thus, in a very real sense, the spaces or landscapes themselves – are shaped by human ideas of them. These are all challenging, abstract concepts, which I argue can be given valuable analytical shape by the work of the philosopher and sociologist Henry Lefebvre (1901–91). Lefebvre divided space into three categories:

1 **Spatial practice**, which embraces production and reproduction, and the particular locations and social sets characteristic of each social formation. Spatial practice ensures continuity and some degree of cohesion. In terms of social space, and of each member of a given society's relationship to that space, this cohesion implies a guaranteed level of competence and a specific level of performance.
2 **Representations of space**, which are tied to the relations of production and to the 'order' which those relations impose, and hence to knowledge, to signs, to codes, and to 'frontal' relations.
3 **Representational spaces**, embodying complex symbolisms, sometimes coded, sometimes not, linked to the clandestine or underground side of social life, as also to art (which may come eventually to be defined less as a code of space than as a code of representational space).[1]

In the following chapter I offer a series of simplified definitions of Lefebvre's tripartite division of space and set out demonstrate its potential application to the study of landscapes as cultural phenomena.

A spatial approach to the history of mountain experience

My research focusses on the experience of landscape, and particularly mountains, during the early modern period (roughly 1450–1750). Within landscape studies broadly defined, there exist a variety of paradigms through which to approach the subject of 'landscape'.[2] The one which particularly intrigues me is the concept of the cultural landscape. The term was first coined in 1925 by the geographer Carl Sauer, who defined landscape as 'a land shape, in which the process of shaping is by no means [...] simply physical'. He argued that cultures inscribed themselves upon the 'natural landscape' through choices based upon the population, housing preferences, production, and communication of each given culture, ultimately producing the 'cultural landscape'.[3] This move was later echoed in the field of historical studies by W.G. Hoskins, who in 1955 defined the landscape as a 'palimpsest' upon which 'successive generations had inscribed their way of life, while half-erasing that of their predecessors'.[4]

Both Sauer and Hoskins addressed the cultural construction of landscape insofar as it possessed a physical reality or had a physical impact on the form and contours of the land itself. Denis Cosgrove, in his book *Social Formation and Symbolic Landscape* (1984), glossed landscape as a 'way of seeing'. For Cosgrove, 'landscape' was more than just an area of physical space and the mountains, trees, or buildings which that space contained: instead, the term denoted the external world as 'mediated through subjective human experience'.[5] The cultural landscape could encompass, therefore, not just the physical alterations caused by human society, but the mental ideas, preconceptions, and values with which humans engaged with the landscape. It is this type of cultural landscape with which I am particularly concerned, for it allows me to ask questions both about how people in past contexts experienced mountains (what did they think and feel about mountains?), how they constructed them (how did they depict mountains in writing or art?), and how experience and construction relate to one another (would the literature and imagery which 'constructed mountains' in, say, 17th-century Britain result in a certain type of common mountain experience?).[6]

I believe that Henri Lefebvre's tripartite division of space, as quoted above, offers the ideal heuristic tool for the study of the cultural landscape, and enables historians of landscape to ask more nuanced questions about different aspects of past landscape experience. In the pages that follow, I will read a series of early modern sources relating to mountains through each of the three lenses offered by Lefebvre. I argue that his concept of spatial practice can encourage historians to read sources with an attention to the ways in which non-elite communities both adapted and adapted to the surrounding landscape. His concept of representational space offers, I suggest, an invaluable tool for articulating the complex web of memory and symbolic associations overlaying the experience of both specific landscape sites and landscape in general. Finally, I propose that his concept of spatial representation (also translated as representations of space) can offer a way of appreciating and analyzing the ways in which ostensibly *descriptive* accounts of landscape can in fact be seen as being *prescriptive* of subsequent individual responses and of the cultural landscape more generally.

Space cut three ways

Henri Lefebvre's most famous work, *The Production of Space* (first published as *Production de l'espace* in 1974, and translated into English in 1991) has had considerable influence in fields as disparate as geography and pedagogy, and its potential value within the field of spatial history has already been indicated.[7] Lefebvre wrote from a socio-political perspective, and sought to foreground the ways in which dynamics of power were enacted through and within urban spaces. He distinguished first between 'absolute space' and 'social space'. The former term refers to the physical, external reality of space, whilst the latter term focusses on the human experience of space, which he divided into three different categories: spatial practice, spatial representation, and representational space. I believe that Lefebvre's 'social space' offers a close analogue to the idea of the cultural landscape and that, as such, his tripartite division of social space can, with some adaptation, be applied to the historical study of landscape.

Spatial practice, put simply, encompasses how people use a given space: how they move within it, and how they travel from it to a different space. Spatial practice is how we inhabit our homes: what rooms we use and for what purposes and at what times. On a larger scale, it is the movement from, for example, one city to another: it is the overcrowding of the motorways compared to the relative quiet of back roads. Spatial practice highlights the ways in which humans distinguish and create specific spaces based on their intended usage, and emphasizes the links created between them by human movement.

The application of such a concept to landscape is obvious for, as highlighted above, landscape is not 'natural' but very significantly shaped (physically as well as symbolically) by human usage. The concept of spatial or landscape practice prompts historians to ask how past societies utilized different forms of landscape: in what ways did they gain sustenance or economic benefit from it, how did they travel through it, manage its risks, and take advantage of its benefits? What did people *do* on or around mountains, rivers, coastlines, plains, or deserts in the past?

Representational space is arguably the most powerful of Lefebvre's three divisions, because it captures that which is at once most abstract but which also has the greatest impact upon our experience of the world: the symbolic, metaphysical makeup of space. 'Representational space' is the thing which makes a church a holy space, rather than simply a large building with pillars. Representational space is in the mind of the person experiencing it, but it is also generally a matter of broader consensus, whether at the level of a small community or an entire culture. Whilst it is metaphysical, representational space is very much real, and in turn shapes the ways in which we use spaces, and the ways we represent them or attempt to regulate them.

From the perspective of landscape studies, this concept of representational space brings into full focus the very issue towards which theoretical progress within cultural geography has been tending: that 'the landscape' is not just something which human beings physically shape but one which they also mentally construct. The concept encourages the historian to pay attention to what stories, texts, and ideas are most strongly associated with different landscape features in different contexts – and to consider what these associations *mean* in terms both of the cultural value placed on a specific landscape and the individual experience of it.

The representation of space, or **spatial representation**, is, for Lefebvre, a category of prescription: the documents, visual and textual, produced by architects, bureaucrats, city planners, politicians, and so on. These are the documents which 'create' spatial practice: when an architect designs a house and places the kitchen in a certain room, then that room will inevitably be the space which its eventual inhabitants utilize for the creation of meals, barring their own subsequent acts of domestic re-design (and thus re-presentation). Spatial representation regulates how we use and experience space: whether innocuous (the location of a kitchen) or sinister (the demarcation of ghettos), they are in important ways an exercise of power.[8]

Within the context of modern history, the historian can certainly identify clear examples of 'landscape representation', or documents which prescribe the use and representational significance of certain landscapes. The delineation of national parks, the Countryside Code, and Ordnance Survey maps all encourage people to behave in and think *about* the landscape in particular ways. Is there, however, a way for the historian of the pre-modern landscape to apply Lefebvre's concept of spatial representation to sources such as travel accounts or works of art? I would argue that yes, there is.

The cultural landscape does not, by definition, come out of nowhere: it is landscape mediated through human experience which is in turn mediated by cultural discourses. These discourses – which could just as valuably be termed 'landscape representations' – survive for historians in the form of textual and visual sources: the same sources out of which we can excavate the realities of landscape practice and the symbolisms of representational landscape. Using Lefebvre's category of spatial representation, then, is not so much an analytical tool for attending to particular themes within a source (as in the case of the other two categories) but rather a way of viewing a source – and its expressions of both spatial practice and representational space – in terms of the power it enacted over subsequent experiences of landscape. Writings about or visualizations of landscape had the potential either to reiterate or add to the existing consensus of the cultural landscape, or – in rare cases – to dramatically modify it. Whether drastic or subtle, however, Lefebvre's category of spatial representation allows the historian of landscape to view their sources as not just descriptive but also prescriptive.

The following three sections will take a selection of early modern sources and, with specific reference to the cultural landscape of mountains, will read these sources against the adapted definitions of Lefebvre's tripartite division of space above. Throughout, attention will be paid to the inter-relation of each of the three categories: although their separation is analytically productive, it is also important to acknowledge the extent to which the three categories operate in a constant feedback loop with one another.

Landscape practice

As emphasized earlier, a significant aspect of spatial practice consists of *movement between or through spaces*. In the modern day, mountains act as barriers to movement: roads and train lines either skirt around them or, in some famous examples in the Alps, drill directly through them. Reading early modern sources for evidence of spatial practice, however, reveals that in this period mountains – or at least mountain passes – acted as important routes of communication and transport, despite the challenging nature of the landscape. Thomas Coryate (c. 1577–1617), an English courtier and self-proclaimed 'leg-stretcher' (walker) travelling in Savoy in 1608, crossed the Mont Cenis pass (2,081 m high), and commented upon the twisting yet crowded nature of the path:

> ...the [ways] were wonderfull hard, all stony and full of windings and intricate turnings, whereof I think there were at least two hundred before I came to the foot. Stil I met many people ascending, and mules laden with carriage, and a great company of dunne kine [cattle] driven up the hill with collars about their neckes[.][9]

Accounts such as Coryate's provide evidence for a most basic point of spatial practice – that travellers and herdsmen alike passed through the mountain landscape.

Other sources, in particular Josias Simler's *Description of the Alps* (1574) reveal the more complex web of local spatial practice which maintained such paths and passes. A Swiss theologian and historian, Simler set out to describe the 'passes which are frequently used in the Alps', alongside 'the difficulties and dangers which await travelers who traverse them, and the manner in which they successfully conquer their difficulties'.[10] In so doing, he revealed a significant level of local activity invested in maintaining the mountain paths and passes. This activity included cutting paths in the rock, constructing bridges across small chasms, and even creating 'suspended pathways' out of timber along sheer mountain faces. Regular work was required to ensure that the passes remained open, as Simler reported:

> in order to assure the passability of the route, the neighboring residents are compelled by the local magistrates to maintain the path...almost every day the men of the neighboring villages on each side of the slope explore the path towards the col, and if they see any danger they warn the travellers and repair the path.[11]

Sometimes local farmers would bring their cattle up to the snow-line – with a long pole dragging behind them – to help clear the path, and merchants with urgent deliveries to make might hire additional labourers to clear snow. The spatial practice of utilizing mountain passes for travel and trade was not incidental or easy: it was deliberately and laboriously maintained.

Moreover, people who inhabited or spent time among mountains were clearly both conscious of the risks presented by their environment and how best to ameliorate them. Simler noted that 'a great number of cows and horses are herded over the transalpine regions from Switzerland to Germany to Italy' – a circumstance which sometimes resulted in multiple herdsmen using the same path simultaneously.[12] Fortunately, etiquette dictated how such meetings should be safely managed: herdsmen either paused in prearranged spots, for example on a plateau, to enable safe passage for two herds, or followed 'rules which determine which of the two groups can stay on the path, and which must yield the right of way'.[13]

Simler likewise makes it clear that those who dwelt amidst the mountains took care to guard against avalanches: being careful not to build their homes beneath steep slopes, and moving quickly, quietly, and early in the morning when 'forced to make a journey in avalanche conditions'. In turn, these locals would 'warn…travelers of the precautions they should take' and, when news of an avalanche is raised, 'the mountain people ask themselves immediately if any travelers had set out that day, and calculating the elapsed time, they can guess where they were buried by the snow', and dig for their rescue.[14] Once again, early modern Europeans, in Simler's account, did more than passively live and move amidst mountains: they adapted to them and had formalized practices in place to respond to potentially dangerous situations.

The risks of the mountain environment were not without their reward. The cattle driven to market, to the economic benefit of the communities who raised them, were frequently at the centre of vertical transhumance practices, which saw the movement of cattle to higher grazing lands in summer to allow lower slopes to produce crops.[15] Travelling in the warmer months, Thomas Coryate admired the sight of just such 'goodly corne fields' located in 'wonderfull steepe places', and spent a long time wondering at the ability of the farmers to bring their ploughs to such high ground. He finally concluded that the Alpine farmers must 'set their corner with their hands', just as he had observed in certain places in England during the course of his domestic travels.[16] This offers not only another example of the adaptation of local practice to the mountain environment, but also highlights the fact that the spatial practice of local communities – i.e., cultivation – had a real and visible effect on the landscape. The mountain-dwellers of early modern France and Switzerland both understood their environment and were able to use and even shape it according to their needs.

The spatial practice of elite travellers such as Coryate – to whom the mountains represented a highway between destinations, or occasionally destinations in themselves – also presented the possibility of financial gain. Simler drily reported that whilst some Alpine-dwellers planted poles in the snow to indicate safe routes for travellers, 'most of the time they neglect to do this, to force the travelers who don't know the route to hire their services'.[17]

Coryate himself fell victim, against his best efforts, to the money-earning instincts of Savoyard locals when climbing what he called 'the Mountaine Aiguebelette', a pass leading towards Chambery. Uncomfortable on a horse, Coryate decided to walk the pass, but soon found himself in the company of 'Certaine poor fellowes which get their liuing especially by carrying men in chairs from the toppe of the hill to the foot thereof'. Sighting a potential sale, the chair-bearers began to increase their pace, until the 'Odcombian Leg-Stretcher', afraid of having to choose between a heart attack and losing sight of the people who knew the way up the mountainside, gave in and engaged their services.[18]

These insights – the result of an attention to traces of 'landscape practice' in just two textual sources – are all relatively quotidian and preliminary; people droved, farmed, and knew how to avoid avalanches among the Alps of the 16th and 17th centuries. At the same time, these are precisely the types of details which are easily lost, particularly when dealing with the early modern period. The textual sources which survive are almost exclusively produced by individuals who – by virtue of their privilege in writing, publishing, and having their texts distributed widely enough to be preserved – were de facto members of an elite representing only a very small percentage of the population of their given society. Approaching these sources with Lefebvre's category of spatial practice, and the simple questions of what they reveal about how people – not just the author, but the members of the local communities which they passed through – used and moved among the mountains allows the historian to reveal a more diverse range of landscape experience.

The representational space of mountains

Representational space, as noted above, is an abstract but immensely meaningful category, and offers a valuable tool for analyzing and articulating the complex accretions of cultural meaning, symbolism, and memory which overlay experiences of the 'physical landscape'. It can lead the historian to ask an important set of questions: What stories, ideas, and meanings were associated with particular features of the landscape? Were certain exemplars of that feature (whether a river, a mountain, or a cave) assigned particular cultural prominence? What characteristics would people naturally tend to think of when imagining a certain landscape feature – whether or not all exemplars of that feature display those characteristics at all times? How, in turn, do these aspects of the representational space of mountains impact spatial practice?

To apply these questions to the modern representational space of mountains, one might say that mountains are generally associated both with ideas of the sublime, and with the spatial practice of mountaineering – a practice itself tied up in abstract ideas of heroism and of 'coming first'.[19] The most famous mountain in the world is inarguably Mount Everest, a peak itself layered with representational associations – its name points uncomfortably to the Great Trigonometric Survey of India, and the whole imperial project behind it, whilst countless books, photographs, and films memorialize significant ascents of the peak, and above all Edmund Hillary and Tenzing Norgay's 'first ascent' in 1953. In terms of characteristics – although small, summery, and grassy mountains exist just as much now as they did in the 17th century – the modern representational vision of mountains which looms largest is of remote, austere, snow-topped, rocky peaks, photographed in dazzling blue colour and adorning a glossy magazine or book cover. Modern spatial practice in turn responds to this compelling representational space, with increasing numbers of hopeful climbers arriving at Everest Base Camp each year, and visitors to less vertiginous mountains seeking out a quiet 'wilderness' which is in fact the subject of careful curation.[20]

These observations point towards the questions which need to be asked in order to understand the early modern representational space of mountains: What specific peaks were most often on the pens of writers, and what distinguished them from others – their height, or something else? What names and stories came to mind when looking at the peaks? How did artists choose to depict mountains – in what shades, tones, and in what contexts? What characteristics came most often to the pens of authors seeking to capture the mountain landscape? How, in turn, did this affect spatial practice in terms of the mountains which elite travellers sought to visit?

One particularly helpful source for providing some preliminary answers to these questions, Joshua Poole's 1657 *English Parnassus*, takes the form not of a travel account but rather of a '*Helpe to English Poesie*': a guidebook for aspiring poets as to the best words, phrases, and ideas to draw upon when writing poetry about different subjects. The book itself was named after a mountain: Mount Parnassus, in Greece, the legendary home of the Muses and thus a prominent metaphor for poetic skill and inspiration. Further classically significant mountains loom large in the extracts cited by the *Parnassus* as excellent examples for young poets to follow: they might choose to write about the 'Scythians snowie mountains on whose top/Prometheus growing liver feeds the Crop'; 'the top of snowy Aldigus'; 'Pindus frozen toppes'; 'Atlas pillars'; 'Athos Mount', and many more.[21] A simple list under '*Hill*' suggests that the poet turn their thoughts to 'Athos, Atlas, Haemus, Rhodope, Ismarus, Eryx, Cithaera, Taurus, Caucasus, Alps, Appenine, Oeta, Tmolus, Aetna. Parnassus, Othrys, Cynthus, Mimas, Dyndimus, Mycale, Pelion, Pindus, Offa, Olympus, Helicon, Ida'.[22] It is apparent that the classical mountains of Parnassus, Pindus, and Helicon were, in fame, the Mount Everest, Mont Blanc, and Eiger of the early-17th century.

These names – and the locations which they represented – were potent because they were imbued with memory, whether of mythical or 'real' events. In the introduction to his *Crudities*, Coryate prominently included a translation of an oration on travel written by Hermann Kirchner, a German professor of history, poetry, and rhetoric. The oration highlighted the benefits that foreign travel offered to young gentlemen. One of Kirchner's justifications of travel offers what could be read as an assertion of the rich representational space of the mountain landscape:

> What I pray you is more pleasant, more delectable unto a man then to behold the height [*sic*] of hilles…to admire *Hercules* his pillers? to see the mountaines Taurus and Caucasus? to view the hill Olympus, the seat of *Jupiter*? to pass ouer the Alpes that were broken by *Annibals* Vinegar? […] to visite Parnassus and Helicon, the most celebrated seats of the Muses? Neither indeed is there any hill or hillocke, which doth not containe in it the most sweete memory of worthy matters: there shalt thou see the place where *Noahs* Arke stood after the deluge: there were God himself dwelt, and promulged his eternall law amongst the thunders and lightnings […].[23]

In advocating for what might be termed a form of mountain tourism, Kirchner emphasised the very idea inherent to modern-day scholarly ideas of the cultural landscape: that a feature of the landscape is more than just a physical landform, but is in fact the sum of its imaginative associations. The associations that carried most weight in 17th-century Europe were those drawn from the classical and Scriptural pasts.

Travellers to 'famous' mountains, then, arrived with preconceptions and ideas already in mind. The representational space of a particular peak, however, could also be constructed for visitors at the point of experience when engaging with local traditions or stories. Thomas Coryate dedicated several pages to his discussion of what he called 'Roch Melow' (now known as Rocciamelone or Rochmelon, 3,538 m), 'said to be the highest mountain of all the Alpes, sauing one of those that part Italy and Germany'. He first admired its height, and the visual illusions this lent it: 'it is covered with a very Microcosme of clowdes' and 'seemeth farr off to be three or four little turrets or steeples in the air'.[24] His experience of its visual aspect was in turn overlain by its representational aspect. He and his travelling companions had been under the care of a conductor from Lyons to Turin, and this man related a 'pretty history' of the mountain. The story told of 'a notorious robber' who, experiencing remorse for his sins, acquired a pair of pictures, one of the Virgin Mary and one of Christ, and took these to Rocciamelone, promising 'to spend the remainder of his life in fasting and prayer, for expiation of his offences to God, upon the highest mountaine of the Alpes'. The hapless sinner, however, had chosen the wrong mountain, but luckily 'two pictures more of Christ and our Lady appeared to him', and miraculously gave him to understand that he would need to remove to a different peak in order to fulfil his vow.[25] Representational space was therefore not only revealed and replicated in the elite writings of professors of German rhetoric, but was also the product of local, oral tradition. (Notably, and perhaps hinting at the durability of representational space, an adapted version of Coryate's 'pretty history' is still attached to Rocciamelone to this day).[26]

As noted above, the different categories of Lefebvrian space, whilst helpful to consider in isolation, are closely interwoven. As urged by Kirchner, representational space could inform spatial practice, with travellers making journeys to specific peaks due to the religious or classical associations with them.

Travellers in the Holy Land in the 17th century went to great lengths to physically engage with the 'sweete memory of worthy matters'. In February 1668, the traveller Jean de Thévenot (1633–67) ascended Mount Catherine (2,629 m), a journey which took 'near three Hours' and followed a route 'full of sharp cutting Stones, and many steep and slippery places to be climb'd up, that hinder people from going fast'.[27] After stopping at a basin filled by a spring which had, according to tradition, spouted to slake the thirst of the exhausted monks who had carried the body of St Catherine down the mountain, Thévenot and his companions ascended to the dome under which the body of the saint had allegedly been placed by angels. Although Thévenot was dubious of a tradition which identified depressions in the rock with the miraculous impression of her body (he thought them more likely to 'hath been done by the Hands of Men'), he paid his devotions at the spot before descending 'with a great deal of trouble'.[28] Two months later, he visited Mount Quarantine (350 m), where – according to representational space, at least – Jesus had been tempted by the devil. Thévenot spoke of the stages of ascending the mountain in terms of the Scriptural events associated with specific spots: whilst his companions, more insouciant than he in the face of the slippery rock path, 'went up to the top of all the Hill, to the place where the Devil carried our Lord', he chose to remain at a cave 'where our Lord fasted forty days'.[29] Whether the physical markers of significance were man-made or not was irrelevant to Thévenot: the sites he visited still bore the representational impressions of saints and saviours.

Classical associations were just as potent as religious ones. William Lithgow, a Scottish pilgrim en route to the Holy Land, admired the sight of Mount Parnassus, 'which is of a wondrous height, whose top euen kiss the Clouds'. His account turned immediately to 'the nine *Muses*' and the symbolism of the physical landscape, with its two 'sterile' tops, the one 'dry and sandy, signifying that Poets are always poore, and needy', the other 'barren, and rocky, resembling the ingratitude of wretched, niggardly Patrons', and the rich 'vale between the tops… which painefull Poets, the *Muses* Plow-men, so industriously manure'.[30] The sight of Mount Etna, another mountain famed in classical mythology as (among other things) the home of Vulcan's forge, inspired similar flights of literary fancy from Lithgow's pen. On arriving in Messina, he encountered an old friend, and as a symbol of his affection composed a stirring sonnet on the volcano: '*High bends thy force, through midst of Vulcans ire, / But higher flies my sprit, with wings of loue…*'.[31] Classical and mythological associations provided the focal point for both the traveller's gaze, and for his recounting of what he saw. The literary scholar Cian Duffy terms this experience of the landscape of Italy as 'classic ground', emphasizing 'the extent to which it is all but impossible for an educated traveller to have a disinterested aesthetic response to a landscape or an environment which has long possessed a range of specific historical and cultural associations'.[32] Representational space did more than inform spatial practice related to mountains; it also informed the experience of them, and in turn the way in which they were represented for subsequent readers.

Landscape re-presentation

The hyphen in the subheading above is not accidental. Adapting Lefebvre's concept of 'spatial representation' – a category linked with urban space and the enaction of power through documents of bureaucracy such as town plans – to the study of the premodern landscape requires more divergence from the original definition than in the case of spatial practice and representational space. Where the other two categories promote a new way of reading the content of historical sources of landscape (with attention either to practice or symbolic associations), I argue that the category of spatial re-presentation can best be understood as a way of considering the extent to which those sources themselves either reiterated existing ideas or values of landscape or promoted new or divergent ones. The category therefore takes a step back from close engagement with a piece of source material, and instead considers it in the context of wider discourses regarding landscape. Did a given source either reiterate the existing cultural consensus regarding a particular landscape feature, or did it *re*-present the landscape in a new light?

This theme is particularly relevant in the case of attitudes towards mountains in (published) 17th-century British discourse. Readers familiar with the traditional historiography of Western mountain engagement may well have found themselves surprised by references in the preceding pages to sources which expressed admiration for mountains and revealed travellers climbing peaks. The traditional narrative states that before the 18th century – before the development of ideas of 'the sublime' and before the first ascent of Mont Blanc in 1786 and the subsequent explosion of mountaineering as a sport – people generally feared and avoided mountains and found them to be visually ugly. One source prominently cited as an example of this 'typical' early modern attitude is Thomas Burnet's *Theory of the Earth* (1684).[33] Burnet, a British natural philosopher, set out the theory that mountains were the result of the Flood, or Deluge, which had split open the smooth shell of the original earth and caused the irregular form of continents and mountains to arise and the seas to fill the spaces in between them. As such, mountains should not be viewed as part of God's design, but rather as a memorial of the sinfulness of mankind which prompted the cataclysm.[34] One of the observations at the root of Burnet's theory was his sense that mountains, although undeniably grand in stature, were 'shapeless and ill-figur'd', and characteristic of 'a World lying in its rubbish'.[35]

Such stark critique of the mountain landscape is nowhere to be found in Lithgow's dizzy sonnet on Mount Etna, or in Coryate's high turrets of Rocciamelone. The *English Parnassus*, offering a list of epithets that the aspiring poet might wish to use when invoking the figure of a mountain, suggests that a mountain could be anything from 'mossie', 'hoary', 'aged', and 'pathless' to 'stately', 'lovely', and 'star-brushing', but never shapeless.[36] By contrast, four years after the publication of Burnet's *Theory*, the traveller John Dennis looked at the Alps and saw 'vast, but horrid, hideous, ghastly Ruins'.[37] Daniel Defoe, in his *Tour Thro' the Whole Island of Great Britain*, looked upon the landscape of Derbyshire as 'a Confirmation' of Burnet's theory of the 'great Rupture of the Earth's Crust or Shell'. The Lake District was 'the wildest, most barren and frightful' place he had ever seen, and the peaks of Scotland were indubitably 'hideous'.[38]

Lefebvre's concept of spatial representation offers, I argue, the analytical framework to understand and the vocabulary to explain this apparent *volte face* in British attitudes towards mountains from the 17th through to the early-18th centuries. For most of the 17th century, spatial re-presentation with regards to mountains largely took the form of replication of longstanding cultural ideas and values, which found both representational memory and visual pleasure in mountains. Burnet, in 1684, offered a re-presentation which represented a revolution in attitudes towards mountains. This radical re-presentation, naturally, evoked controversy: numerous scholars and clergymen wrote in protest against his *Theory*, arguing that the utility and beauty of mountains demonstrated that they were the original creation of God.[39] However, Burnet's vision of mountains slowly gained traction, and indeed provided the intellectual basis for new articulations of the mountain landscape as 'sublime': horrifying, but marvellous. John Dennis, predating similar eighteenth-century articulations of sublimity by some decades, experienced 'a delightful Horrour, a terrible joy' at the sight of Burnet's ruins.[40] Each source cited in this chapter offers a re-presentation of mountains; the only difference between them and Burnet's representation is that his changed discourse, rather than simply reproducing it. In both cases, however, the sources still enacted rhetorical power over space. Kirchner's oration contributed to a discourse of the classically and scripturally significant representational space of mountains, and in turn encouraged spatial practices which incorporated significant mountains in travellers' itineraries. Burnet's *Theory* became part of the representational space of mountains which Dennis and Defoe took with them on their travels and informed the nature of their aesthetic experience.

Each of Lefebvre's categories of space encourages the historian to ask specific and valuable questions. A generic reading of his category of spatial representation, divorced from the context of 20th-century urban planning, asks the historian to consider whether and how certain sources have power to control space in all of its aspects. I therefore argue that an interpretation which encourages an awareness of the potential power of all sources – however apparently descriptive they may be in the first instance – to prescribe space is not a misinterpretation or warping of Lefebvre's category of spatial representation but rather an important expansion of it. The concept of landscape re-presentation allows the historian to consider how the sources themselves reshaped the landscape, and in turn reshaped subsequent spatial practice and representational space. In most cases, re-presentations of landscape follow the extant consensus (just as the domestic architect would rarely break with established tradition and place the kitchen on the top floor of a three-storey house), and thus their enaction of power is virtually invisible. In other cases, they break with tradition, and – in even fewer cases – those breaks promote a new consensus. Spatial representation thus enables the historian to trace the evolution and revolution of ideas of landscape through time.

Conclusion

Many of the above discussions may seem, once all is said and done, quite unsurprising: of course mountain passes were used as routes of transport and travel, of course a mountain is experienced through its stories as well as its physical stature, and of course historical sources, in their own time, shaped as well as described attitudes. If the reader is unawed by these insights, then that is quite as it should be, for the true value of Lefebvre's tripartite model of space lies in its simplicity. It divides an inchoate concept into three manageable categories, categories which in turn draw the historian's attention to specific features of the sources they study. The model prompts simple questions with revealing answers: what does this source tell us about the use of space? What ideas, values, or stories did people associate with that space? How did this source replicate or reimagine the extant discourse or cultural consensus relating to that space – or, more simply, what power did this source enact over subsequent spatial practice and experience?

In the case study above, I have demonstrated that an application of Lefebvre's tripartite division of space to the question of the experience of mountains in early modern Europe reveals a multi-faceted story of mountain engagement and appreciation. The spatial practice of local Alpine communities incorporated a keen awareness of the dangers of the mountain landscape and how best to ameliorate them, as well as an ability to benefit economically both from the landscape and from the practices of elite travel which it promoted. The representational space of mountains, both in literary allusion and in the moment of physical encounter, was continuously overlain by memories of the classical and scriptural past. Although dealt with only briefly, the spatial re-presentation of mountains in the early modern period saw a long period in which positive responses to the mountain were replicated for many decades before a new representation, depicting mountains as symbols of mankind's sinfulness, came to the fore and shaped subsequent discourses.

I believe that Lefebvre's three categories of space have a great deal of potential as analytical tools for the study of the history of landscape. The strength of the concept of 'representational space', in particular, is highlighted by its similarity to other attempts to capture the metaphysical nature of landscape in human experience, such as 'cultural landscape' or 'classic ground'. The difference with representational space is that it brings company: the spatial practice which can in turn be shaped by the cultural prominence of certain landscapes, and the understanding of sources as spatial re-presentations which have the power to reiterate or redefine the values and ideas associated with a specific landscape feature. The exploration above has only scratched the surface of the insights into landscape which the application of a Lefebvrian model of space has to offer.

Notes

1 Henri Lefebvre, *The Production of Space*, trans. Donald Nicholson-Smith, Oxford: Blackwell, 1991, p. 33.
2 These include archaeological, anthropological, ecocritical, cognitive, and phenomenological approaches (or a combination thereof).
3 John Leighley (ed.), *Land and Life: A Selection from the Writings of Carl Otwin Sauer*, Berkeley: University of California Press, 1963, pp. 321, 343.
4 W.G. Hoskins, *The Making of the English Landscape*, London: Hodder and Stoughton, 2006, first published 1955, pp. xvii, xxiii, 4.
5 Denis Cosgrove, *Social Formation and Symbolic Landscape*, London: Croom Helm, 1984, pp. 1–4.
6 These questions are explored in depth in my PhD thesis: *Re-thinking Mountains: Ascents, Aesthetics, and Environment in Early Modern Europe*, University of St Andrews, 2016.
7 See Andy Merrifield, *Henri Lefebvre: A Critical Introduction*, London and New York: Routledge, 2006, pp. 99–120; Sue Middleton, *Henri Lefebvre and Education: Space, History, Theory*, London and New York: Routledge, 2014, pp. 10–11; and Richard White, 'What is Spatial History?', *The Spatial History Project*, February 2010, pars. 7–10. Available HTTP: https://web.stanford.edu/group/spatialhistory/cgi-bin/site/pub.php?id=29 (accessed 4 July 2017).
8 See also Despina Stratigakos's chapter on architectural drawings and the floor-plans of Hitler's Berghof.
9 Thomas Coryate, *Coryat's Crudities Hastily Gobbled up in Five Monethy Trauells in France, Sauoy, Italy* [etc.], London: W. S., 1611, p. 80.
10 Josias Simler, 'Vallesiae et Alpinum descriptio', in Alan Weber (ed. and trans.), *Because It's There: A Celebration of Mountaineering from 200 B.C. to Today*, Lanham: Taylor Trade Publishing, 2003, p. 22. Weber's is a translation of a portion of Josias Simler, *De sedunorum thermis et aliis fontibus medicates de Alpibus commentarius Vallesiae description*, Zurich: excudebat Christophh II Froschaur, 1574.
11 Simler, 'Vallesiae', p. 25.
12 Ibid., p. 23.
13 Ibid., p. 23.
14 Ibid., p. 26.
15 For wider transhumance practices, see Albert Bil, *The Shieling, 1600–1840: The Case of the Central Scottish Highlands*, Edinburgh: John Donald Publishing, 1990; Jesper Larsson, 'Labor Division in an Upland Economy: Workforce in a Seventeenth-Century Transhumance System', *The History of the Family* 19, 2014, 393–410.
16 Coryate, *Crudities*, p. 72.
17 Simler, 'Vallesiae', pp. 24–5.
18 Coryate, *Crudities*, pp. 69–70.
19 Peter Hansen, *Summits of Modern Man: Mountaineering after the Enlightenment*, Cambridge: Harvard University Press, 2013, has articulated the concept of 'the summit position' – the myth of the individual alone and first upon the summit – as an outgrowth of the intertwined phenomena of modernity and mountaineering.
20 William Cronon, 'The Trouble with Wilderness, or: Getting Back to the Wrong Nature', in id. (ed.), *Uncommon Ground: Rethinking the Human Place in Nature*, New York and London: W.W. Norton, 1995, pp. 69–90, has emphasized that wilderness is 'quite profoundly a human creation', both metaphorically and physically.

21 Joshua Poole, *The English Parnassus, or: A Helpe to English Poesie*, London: printed for Thomas Johnson, 1657, pp. 343–4 [sigs. Cc5r-v].
22 Poole, *English Parnassus*, p. 345 [sig. Cc6r].
23 Coryate, *Crudities*, C6r.
24 Ibid., pp. 78–9.
25 Ibid., pp. 79–80.
26 A story of which Coryate's one is an evident cousin shapes the modern-day representational space of Rocciamelone: modern-day guidebooks and newspaper articles relate that the first ascent of the mountain was made by a knight, Bonifacius Rotarius of Asti, who carried an image of the Virgin to the summit in gratitude for his safe return from the crusades. Jonathan Trigell, 'High School: Mountaineering in Chamonix', *The Guardian*, 26 July 2008; 'L'alpinismo? È nato sul Rocciamelone', *La Stampa*, 30 July 2008; Brendan Sainsbury, *Hiking in Italy*, Franklin: Lonely Planet, 2010, p. 26.
27 Jean de Thévenot, *The Travels of Monsieur de Thevenot into the Levant*, London: H. Clark, 1687, p. 168.
28 Ibid., p. 168.
29 Ibid., p. 197.
30 William Lithgow, *The Totall Discourse, of the Rare Aduentures, and Painefull Peregrinations of Long Nineteene Yeares Trauayles from Scotland*, London: Nicholas Okes, 1632, pp. 118–9.
31 Lithgow, *Totall Discourse*, p. 397.
32 Cian Duffy, *The Landscapes of the Sublime, 1700–1830: Classic Ground*, Basingstoke: Palgrave Macmillan, 2013, p. 9.
33 See, for example, Marjorie Hope Nicolson, *Mountain Gloom and Mountain Glory: The Development of the Aesthetics of the Infinite*, Ithaca: Cornell University Press, 1959; and Robert Macfarlane, *Mountains of the Mind: A History of a Fascination*, London: Granta Books, 2003, pp. 22–65.
34 Thomas Burnet, *The Theory of the Earth: Containing an Account of the Original of the Earth, and of all the General Changes Which it hath Already Undergone, or is to Undergo, Till the Consummaton of all Things*, 2 vols., London: R. Norton for Walter Kettilby, 1697.
35 Burnet, *Theory*, pp. 145–6 and pp. 110–1.
36 Poole, *English Parnassus*, p. 137 [sig. K5r].
37 John Dennis, *Miscellanies in Verse and Prose*, London: James Knapton, 1683, p. 139; see also Nicolson, *Mountain Gloom*, pp. 276–89.
38 Daniel Defoe, *A Tour Thro' the Whole Island of Great Britain, Divided into Circuits or Journies*, 3 vols., London: G. Strahan et al., 1724–7, 3:58–59, 223–224, and 216–221 (second pagination).
39 The generally positive early modern attitude towards mountains, and the virulent responses to Burnet, are explored in more depth in my PhD thesis, cited above, and summarised in Dawn Hollis, 'Rethinking Mountain Gloom', *Alpinist* 57, 2017, 101–4.
40 Dennis, *Miscellanies*, p. 134.

12 Maritoriality

Michael Talbot

The language that historical societies used to describe their physical environment tells us much about how they viewed such spaces in political terms. Take the following verse by the 18th-century Ottoman poet Koca Ragıb, part of a maritime-themed love poem:

İmtiyāz-ı ṣabıt ü seyyārı müşkildir ḫayāl
Ẓann eder sükkān-ı keştı sāḥil-i deryā yürür

The distinction between fixed and transient is difficult to imagine,
For the ship-dwellers suppose it is the seashore that moves.[1]

Without really meaning to, Koca Ragıb says something quite interesting about ideas of maritoriality and territoriality, the ability of a state to define and control their liquid and land-based claims respectively. As Koca Ragıb's verse shows, making sense of space is often a matter of perspective. This chapter will consider one specific example of the Ottoman Empire in the Eastern Mediterranean in the later 18th century to demonstrate how a careful reading of the archival record can reveal much about how that particular polity claimed both land and sea through carefully crafted language. In providing a particular set of terms to describe maritime space, the Ottoman Empire was able to define a specific maritory and develop systems to enforce their rule over delineated possessions.

Territories and borders

Territory is a space that is not necessarily dependent on natural boundaries or the physical landscape. A territory is a space claimed by a political entity for the purposes of control over both public life and politics.[2] Territoriality, the means by which that space is defined and controlled, is therefore intimately linked to sovereignty and, of course, vice versa. That means that territoriality, being tied to sovereignty, is all about the exclusive exercise of power by a given entity over a particular space.[3] Sometimes the territory of a state evolves gradually over time, shifts due to changes in political or

DOI: 10.4324/9780429291739-16

demographic situations, or expands or contracts to fit the contours of natural features. For most of history, political territories were rather loosely defined, with borders more porous than fixed, and power as much about influence as direct control.

For the historian Charles Maier, one of the defining features of early modernity in Europe was the frontier, or what may be called 'the great confinement', a term also employed by Michel Foucault: the settling and policing of frontiers and the shift of borders from zones to lines.[4] The often arbitrary nature of the extent of a territory can clearly be seen in colonial and imperial divisions of space. Looking at a map, at the borders between Syria, Jordan, and Iraq, between Egypt, Sudan, and Libya, the straight, ruler-drawn lines show the raw performance of the dividing of space for the sake of territory, rather than any other considerations.[5] This is perhaps even more evident in the territorial composition of the United States of America; almost every state west of the Mississippi is a box, drawn on to bring in yet another parcel of Indian or Mexican land into American control. Maier has described territory as the partitioning of space, with borders defining the specific area of control.[6] Almost every inch of the globe is now claimed as the territory of one state or another, with the aim of monopolizing power over populations and resources, and controlling the access to and of both.

One way of understanding territoriality, therefore, is the morphing of physical space into political territory.[7] This is an important distinction between space and territory, as the human geographer Claude Raffestin articulated in his 1980 study, *Pour une géographie du pouvoir* (Towards a Geography of Power):

> Territory is generated from space; it is the result of an action driven by a syntagmatic actor (an actor carrying out an agenda) at whatever level. In appropriating a space, concretely or abstractly (for example, through representation), the actor 'territorialises' the space.[8]

Territoriality, as Chandra Mukerji reminds us, harking back to the influential tome of Henri Lefebvre, is not therefore the product of some innate sense of territory fixed in humanity's DNA, nor did ideas of political ownership over space emerge from the primordial ooze, but it emerges from specific historical processes.[9] Part of the job of historians is to make sense of those processes, and to do so we often have to turn to other disciplines to help frame and explain what we find in the historical record. The studies of human and political geographers are particularly important in providing a framework for territoriality.

One of the most influential texts on the subject is Robert Sack's *Human Territoriality* (1986). Sack provides us with three key terms for defining territoriality[10]: First, there must be a classification by area. This, then, has to be a physical as opposed to imagined space, and includes everything (and everyone) within that particular area. Second, there must be what Sack calls 'a form of communication', that is, something that denotes the limits of the relevant area. This could be a physical marker or a rhetorical gesture. Third, and perhaps crucially, there must be an attempt to enforce control over the area, by regulating access to it, controlling what is inside it, and having the ability to punish those who transgress or trespass. All of this together provides Sack's clear definition of territoriality as 'the attempt by an individual or group to affect, influence, or control people, phenomena, and relationships, by delimiting and asserting control over a geographic area'.[11] I argue that the same can apply to liquid space, and it is around Sack's conditions that the following discussion will be structured.

Seas and maritories

This framework has some merit on dry land, but what about the sea? The trouble is that it is hard to envisage it as an entity of parts; there is only liquid space, distinct from the stuff we can build on or shove a flag into. This, however, misses something quite important about the sea that requires a bit of consideration.[12] As the 17th-century Ottoman poet Nabi explains: 'When we behold the furrowed brow of the ocean, we lose sight of its pearls'.[13] The sea in terms of the human experience is more than an expanse of endless churning waves. Yet the sea has often been employed as a symbol of chaos, of darkness, of nothingness. The creation process in the Bible gives a narrative of the move towards life and light through God's division and taming of the waters; for, in the beginning, 'the earth was without form, and void; and darkness was upon the face of the deep. And the Spirit of God moved upon the face of the waters'.[14] It was only on the third day, where God gathered together the waters into one place that the chaotic waters (*mayim*) became the tamed seas (*yamim*). Indeed, so symbolic of chaos were large bodies of water that towards the end of the Book of Revelation, the arrival of the new heaven and the new earth is heralded by the glad tidings that 'there was no more sea'.[15] Although the Qurʾān's creation narrative speaks of God separating the heavens and the earth, the taming of the waters for humanity's benefit is a key sign of his benevolence, with the sea subjected by God to man.[16] Yet the sense of a mass of waters threatening divinely protected order is something that found itself into the learned geographies of the medieval Islamic world.[17]

Such views of the sea also occur in its historiography. The historian Jan Rüdiger provides an interesting critique of why it can be fundamentally hard to grasp the idea of maritime polities. Because most of us live in relatively sedentary societies, Rüdiger argues that it is difficult for us to comprehend 'this "sense of space" that knows no place'.[18] If territoriality is the modelling of space into territory, how can this be possible if the space cannot be located? This has significant implications for how the history of the sea might be written. In his thoughtful book *The Red Sea: In Search of Lost Space* (2016) Alexis Wick assesses the declaration of Derek Walcott's 1979 poem, 'The Sea Is History'.[19] 'History may be made upon it', Wick argues, 'within it, through it, by means of it, but it will never make History'.[20] He continues:

The sea, that is, can be subjected to History but cannot be subjectified into history. Hegel, in this sense, was ultimately right: only the state can assume the burden of producing history – if by *state* is meant an entity, an apparatus that is able to coagulate and consolidate in such a way as to produce the effect of its own memory. The sea, by contrast, dislocates, disperses, dissolves – all that remains, all that washes up on the shore, are fragments.[21]

Although this works well for Wick's broader historiographical arguments, particularly in terms of the sea as an historical subject, in a number of ways this vision of the sea returns to its role as a space of primordial chaos. As Antonis Hadjikyriacou argues, one of the results of scholarship within the so-called 'spatial turn' has been a move away from terracentric history, 'examin[ing] aquatic spaces as sites of historical processes, overlooked and marginalised by traditional historiography that privileges land as the stage of history'.[22] Nonetheless, Wick challenges several historiographies that centre on liquid space, particularly the 'new thalassology' proposed by Peregrine Horden and Nicholas Purcell, with 'thalass' coming from the Greek word *thálassa*, meaning the sea. Wick took particular exception to their assertion concerning the political neutrality of maritime spaces, where, in the case of the historiography of the Atlantic Ocean, 'a "white", a "black", a "green" (Irish), and even a "red" (Marxist) Atlantic may exist in equilibrium'.[23] Yet for Horden, Purcell, and indeed to some extent for Wick, the sea remains deep blue. It is in the depths that history dissolves, where there is no place, where rolling waves conceal all.

Maier is unequivocal in stating that the 'open sea' is not a territory, but was a space that permitted travel to distant territories and so formed frontiers of multiple empires.[24] One of the problems with the sea, as Hugo Grotius explained in his treatise on the *Free Seas* published in 1609, is that exerting control over it is almost impossible. For Grotius, if a state cannot build a castle or station a permanent garrison somewhere, and cannot physically plant its flag, then it cannot claim occupation and therefore cannot claim sovereignty; 'all property has arisen from occupation', he tells us.[25] All of this depends on what we consider to be the sea and its limits.

In response to a great debate over sovereignty and the sea that raged in Europe across the 17th century, Cornelius van Bynkershoek came up with a solution that is still the basis of maritime jurisdiction. In his 'Essay on the Dominion of the Sea' (1703), van Bynkershoek argued that whatever could be defended could be ruled, and so coastal fortifications could easily be said to project their authority into the waters within the range of their cannon.[26] Coastal waters *could* be territorialized through the projection of force. Yet, this evidently depended on having something there by which force might be permanently projected, such as a fortification. The ability to do this along an extensive coastline was limited, and so other mechanisms had to be employed to assert real maritoriality.

Whilst 'territory' is a pretty common word, 'maritory' is a rather unusual term. A mixture of 'maritime', or 'marine', and 'territory', maritory therefore refers to partitioning and assertion of political control over a space in the sea. If we think of the sea solely in terms of the open, deep sea, then drawing borders across the blue expanse is a largely symbolic activity without the ability to exercise control. However, recent efforts in coastal and island history have provided a clearer sense of the sea as something that is not exclusively the deep, and not so clearly separated from land – something, in other words, that is not just *the* sea. After all, seas are not separated from land, but joined to them through coasts, coastal waters, and islands. For many historical states, therefore, control over these liquid spaces was as crucial as any terrestrial claim. A state reliant on maritime power is known as a 'thalassocracy', a word again with its roots in the Greek word for the sea, and the term is usually employed to refer to the polities of the ancient Mediterranean.[27] The fact that, like thalassocracy, maritory is, to use Jan Rüdiger's phrase, a *'mot insolite'* – an unusual word – says much about how views about state power are focussed on the land.[28]

This is particularly relevant with regard to the Ottoman Empire. A sprawling mass stretching at its height from Algeria to Iraq and from Ukraine to Yemen, the Ottoman Empire encompassed territory over three continents. Yet it also claimed maritory, sovereignty over the seas, something that has often been side-lined in the historiography of the empire. Part of this is due to the so-called 'decline' narrative, a style of history writing that envisaged a golden age of power on land and sea in the 16th century under the might of Selim I (the Stern/the Grim, r. 1512–1520) and Süleyman I (the Lawgiver/the Magnificent r. 1520–1566), followed by a slow and inexorable deterioration in all aspects of the empire's existence, starting with a great naval defeat to the forces of the Holy League of Venice, Spain, Genoa, Malta, and others at the Battle of Lepanto in 1571. This narrative caused subsequent centuries to be viewed as periods of malaise at best and terminal rot at worst, and also forced the view of the historian to dramas taking place as borders shifted on land, ignoring events at sea. I want to use the example of the Ottomans and the Eastern Mediterranean in the 18th century, a century previously associated with 'decline', to explore how the Sublime State, as the Ottoman called their empire, actively attempted to assert a maritory – i.e., sovereignty over a defined area of the sea.[29] This is not a sea separate from the land, however, but a space where land and sea mix and join together through islands, coasts, and coastal waters.

Sultans of the two seas

On 27 March 1766 Sultan Mustafa III (r. 1757–74) issued an imperial command to his grand admiral, Hüseyin Pasha. It is representative of a number of similar commands, memoranda, and notes found in the Ottoman Archives that indicate Ottoman efforts in the 18th century to classify, delineate, and control a series of maritime spaces in the Eastern Mediterranean.[30] The Ottoman navy faced off against a number of foes, a variety of state and non-state actors. The endemic conflict between the corsairs of the Mediterranean powers continued right until the end of the century, with Maltese and occasionally Spanish ships raiding Ottoman shipping and coastal settlements. There were also domestic pirates, particularly in the Southern Balkans, with major activity coming from the ports of Ülgün (modern Ulcinj in Montenegro) and Mania on the Peloponnese. In addition to these, a new threat emerged from the final years of the 17th century with the arrival of British, French, and other European privateers in time of inter-European conflict. Their presence was particularly felt during the War of the Austrian Succession (1740–48), the Seven Years' War (1756–63), and the War of American Independence (1775–83). As well as attacking each other, these privateers had a detrimental effect on Ottoman commerce, as many Ottoman merchants used French and British ships to freight their goods around the Mediterranean. Protecting these trade routes was always a priority for the Ottoman government, but became particularly urgent during times of major famine, a number of which occurred in Anatolia and Syria in the 1740s and 1760s. All of these threats together came at a time that the Ottoman state itself went through a number of administrative and military reforms aimed at consolidating the imperial presence in the provinces and protecting its borders, especially on the back of the first major negotiated peace treaties with its main European enemies, the Habsburgs and the Russians, from the Treaty of Carlowitz (1699) onwards.

Andrew Peacock reminds us in the introduction to a collection of essays on frontiers in Ottoman history that every state has a particular point, a particular space that provides an entry or exit point to the world beyond.[31] For many merchants and travelers, the first point of contact with the Ottoman realms would have been sighting its coastlines and disembarking in its ports. Coasts, therefore, were a crucial contact zone. One of the phrases in Ottoman Turkish to describe a coast or shoreline is the rather poetic *leb-i deryā*, literally meaning 'the lips of the sea'. We find it, for example, in this verse from the 17th-century poet Nabi in praise of a *yalı*, a waterside villa, built on the Bosphorus by a senior official of the Ottoman state:

Kemāl-ı ḥüsnüyle bir dilrübāya beñzer anıñçün
Demādem būs eder dāmānını ḍurmaz leb-i deryā

By the perfection of its grace it is like a heart-stealer;
And so, from moment to moment, the lips of the sea never cease to kiss its robe.[32]

Nabi's amorous metaphor of the waves constantly kissing the hem of the coastal villa may well be applied to the Ottoman realms as a whole. The waters of various seas embraced the shores of the Sublime State, and formed a contiguous part of its imperial possessions. Although the sultan's titles emphasized his claims over lands, provinces, and cities, one of his core identifiers was *sulṭānü'l-berreyn ve ḫāḳānü'l-baḥreyn*, the 'sultan of the two lands and the ruler of the two seas', monarch of Europe and Asia, and of the White (i.e., Mediterranean) and Black Seas. For those in the central government based in the imperial capital of Istanbul that spanned both Europe and Asia at a meeting point of the Black and Marmara seas and dominated by the Bosphorus and Golden Horn, water must have formed a key part of their understanding of their state and its possessions. Indeed, the 17th-century Ottoman geographer and historian Katib Çelebi gives us a sense of these visions in his book *The Bounty of the Great Ones in Sea Campaigns*, at the beginning of a discussion of a strategy for responding to piracy:

> The great pillar in this Sublime State, that which is indispensably tied and bound to its glory, is the state of its affairs at sea. Indeed, the splendour and title of the Most-Radiant State arises from its command over the two lands and the two seas. As much of the Well-Protected Domains are formed of islands and sea coasts, and especially there can be absolutely no doubt that the benefactor of the Abode of the Exalted Sultanate, that is, the city of Constantinople, is situated in the two seas.[33]

Thus, for Katib Çelebi, the Ottoman realms were *primarily* liquid in nature, and it is important for this discussion to note that the islands and coasts, *cezāʾir ve sevāḥil*, form part of the maritime rather than the terrestrial landscape. From a city like Istanbul, itself defined by different kinds of maritime space, it is not hard to see why he saw the sea as something so pivotal to the identity of the Ottoman Empire. Yet this was not an abstract sea, but one firmly tied to littoral (i.e., coastal) spaces.

Understanding Ottoman maritoriality is an important part of understanding how the Sublime State situated itself in a wider context, and how imperial officials viewed their own state. Throughout the 18th century, a programme of fortress construction and improvement was initiated throughout the Mediterranean provinces, particularly in Syria/Palestine and on the Aegean islands, coupled with an increase in routine patrols of the Ottoman navy, which had usually ventured out into the Mediterranean primarily for the purposes of tax collection. But, as the command of 1766 explains, there were more issues at stake in the patrols of imperial squadrons by the second half of the 18th century:

> The command to the grand admiral, the vizier Hüseyin [Hüsnü] Pasha, is as follows. It is required and necessary to protect and guard the groups of merchants and protected foreigners going back-and-forth in the waters of my Well-Protected Domains in the Mediterranean Sea, especially the galleons going to and from Egypt, the three-masted ships, as well as other ships, from being seized and harmed by pirate brigands, and to protect and secure the servants of God living in all the islands and on the coasts from all atrocities and attacks, bringing tranquillity and comfort.[34]

Within this statement we not only see clear aims of treating maritime and littoral subjects and protected foreigners with the same care as anywhere on land, but, through a careful reading, a complex set of spaces begins to emerge that allows an understanding of an Ottoman maritoriality.

Seas, waters, coasts, and islands

When the 1766 imperial command speaks of the sea, it gives a number of interesting formulations that indicate a complexity of spatial delineation and a clear attempt to assert authority over a defined maritory: 'in the waters of my Well-Protected Domains in the Mediterranean Sea'; 'go out into the islands and other waters of my Sublime State up to the Morea'; 'going and sailing in peace into the Mediterranean Sea, going out through the [Dardanelles] Straits, among the islands, along the coasts, and in the open [seas] as far out as the Morea, and moving and stopping in the other waters of my Sublime State'. Here, the Ottoman text provides different levels that denote a sense of layered space. At the top level, there is a discrete sea or ocean, in this case the *Baḥr-ı Sefīd* – in other texts *Aḵ Deñiz* – the White or Mediterranean Sea. Within this wider space is what the command denotes as *ṣular*, the waters. The final quotation indicates the composition of these waters: islands (*aḍalar*); coasts (*sāḥiller*); open bodies of water (*açıḵlar*); and 'other' waters (*sā'ir...ṣular*). These were all a core part of the Ottoman Empire, but evidently separate from the wider sea and the land beyond. The ways in which the Ottoman state classified, delimited, and controlled these 'waters of the Well-Protected Domains in the Mediterranean Sea' permits a deeper examination of maritoriality, what the sea might mean to this particular historical entity and how it attempted to exert control.

The Ottomans classified their territorial *waters* as something distinct within the *sea* and comprising both liquid and solid entities. However, the liquid part of the waters was still liquid. It could not be stood in, built upon, or extensively controlled, with much of the coast and coastal waters beyond the reach of coastal fortifications. What was important to the Sublime State was the role of people, rather than abstract space. The wording of the 1766 command, in three places, shows the variety of spaces this maritory contained through describing the subjects of imperial protection:

> It is required and necessary to protect and guard the groups of merchants and protected foreigners going back-and-forth in the waters of my Well-Protected Domains in the Mediterranean Sea, especially the galleons going to and from Egypt, the three-masted ships, as well as other ships, from being seized and harmed by pirate brigands, and to protect and secure the servants of God living in all the islands and on the coasts from all atrocities and attacks, bringing tranquillity and comfort.
> [...]
> You are also to protect and secure the ships going back-and-forth from the grip of pirates, and to protect and guard the poor peasantry living in the islands and on the coasts from atrocities and attacks.
> [...]
> All the officers of my imperial fleet, with their galleons, galley ships, and frigates employed in this mission are to fulfil the requirements of these instructions and commands by engaging in the task and effort of

guarding and protecting the merchant galleons going to and from Egypt and other major ports, the small and big boats going among the islands, and the ships of protected foreigners in the waters of my Sublime State, from being seized and harmed by pirates. Moreover, the poor peasantry living on the coasts of all the isles and islands, and the population of the realm, shall not suffer harm or injustice in my imperial era of supreme equity; at no time whatsoever shall this receive my imperial consent.[35]

The command describes three distinct spaces – maritime, insular, and littoral – all within the category of the Ottoman waters. The maritime space is the most liquid. The galleons going to and from Egypt from other key Ottoman ports like Izmir and Istanbul would have crossed parts of the sea away from sight of coasts or islands; after all, the part of the Eastern Mediterranean south of Crete and Cyprus is mostly open water. The phrases used to describe this space denote its transience; the ships come and go, *āmed-şüd eden* and *ẕehab ve iyāb eden*, with the waters acting like a great road without stopping places.

This chimes with a framework put forward by David Harvey in his 1986 book *Justice, Nature, and the Geography of Difference*. For Harvey, such spaces form temporary 'permanences', that is, a particular space created by 'processes', usually the circulation of capital, that allowed usually intangible spaces to be bounded, that is, to be given a physical property that allows for sovereign claims.[36] This is a complex idea, and one that makes more sense with an example. The demand for trade between Egypt and other Ottoman ports created shipping routes, and these shipping routes in turn created temporary roads or paths in the sea that could be defined, and therefore made subject to the control of the Ottoman state. In Harvey's thought, there is a distinction between 'bodies of water' that form an 'absolute' space, the big formless mass of the ocean, and water that flows, like energy, people, information, and money.[37] The processes of commerce, hunger, and politics that hauled those ships between Egypt and elsewhere in the Mediterranean, connecting land, people, and water, therefore created a temporary maritory cutting across the open seas, that could be defined, bounded, and controlled.

The second kind of maritory featured in the command, the insular space centred on the Aegean islands, is similarly dependent on flows. This space was often described in terms of a passage, with the key phrase in this and other commands being *aḍalar aralarında*, going amongst or around the islands. The space between the various splodges of land in the sea was clearly important to the Ottoman state because of its ability to function as a connector. Roxani Margariti argues for the study of islands and insularities, which she terms nesiology, in understanding maritime spaces in general; 'islands connect', she contends, '[and] as long as they are inhabited their study will always link water, matter, and people'.[38] In the Ottoman command, the crucial aspect of the islands in a liquid sense is their ability to channel big and small ships through them, and so once again the flow of commerce linking people, land, and sea, provides the rationale for this particular part of the maritory.

The third space, the littoral, is similarly connective, but in a different way. One of the most famous Ottoman writers in the maritime sphere is Piri Reis, whose magnum opus, the *Kitāb-ı Baḥrīye* (The Maritime Book, 1520s), is an important source for early modern navigation in the Ottoman waters, and provides a number of sailing guides and navigation techniques. In the introduction, he described his motivations for compiling this practical geography as seeking to 'describe the features of the Mediterranean Sea, its shores, the thriving and ruined [settlements] of its islands, its harbours, its waters, and the rocks in the sea'.[39] The *Kitāb-ı Baḥrīye* thus provides some key vocabulary for different kinds of land and sea. There is the land (*ḳara* or *berr*), and there is the sea (*baḥr* or *deñiz*). But there is also the coast (*kenār*, *ḳıyı*, or *sāhil*), the island (*cezīre* – which could also be a peninsula – or *ada*), the islet (*adacıḳ*), the cape (*burūn*), the gulf or inlet (*körfez*), the harbour or port (*limān* or *iskele*), the shallows (*sıġlar*) and, that crucial maritorial term, the waters (*ṣular*). But when Piri Reis describes the maritime landscape, he does so in a literal manner that tells us little explicitly about the language of maritoriality. Here, for example, is his description of a part of the coast around Piyade (today's Néa Epídavros) in the Morea:

> This Piyade, as we call it, is part of the district of Mora in the province of Mora. The said castle is a league from the seacoast, situated on a high piece of land. Below the said castle is a harbour. They call the said harbour Pordina. [...] Thirty leagues east by southeast beyond the said port there is an island. They call that island Damala [Poros]. The said island is situated close to the coast of Morea, and there is also a castle next to the shallows. The said castle is subject to Venice. If one comes to the said island by sea, from five or six leagues out that island appears to be joined to the coast of Morea.[40]

Aside from providing a fascinating floating view of the Ottoman coastline, the most interesting aspect here is how Piri Reis almost casually notes a switch in sovereignty, from Ottoman-held Piyade to Venetian Damala. It is the castle that denotes sovereignty and authority, very much along the lines of van Bynkershoek's later theory, with the details of the landscape included insofar as they affect navigation; when waters are mentioned, they only refer to the liquid space near ports. Crucially, this, and most other supposedly 'maritime' books, focus on the coast. Katib Çelebi's description of the Mediterranean, written over a century later, also sees the maritime landscape as formed of ports, islands, and shores.[41] As the earlier quotation from his *Bounty of the Great Ones* told us, 'much of the Well-Protected Domains are formed of islands and sea coasts' (*ekṣer Memālik-i Maḥrūse cezā'ir ve sevāḥil-i deryā olub*).

This should not be entirely surprising. Until well after the technological advances in longitude calculation at the end of the 18th century, maritime navigation was overwhelmingly littoral, as Piri Reis shows us. Unreliable navigation methods meant that it was far safer to stick as close to the shore as possible. For Fernand Braudel, whose great history of the Mediterranean still dominates scholarly discussion of things maritime, the coasts and the islands of that sea formed its heart, due in part to the nature of navigation, and the coasts were the centre of economic activity of the area, comprised of basic units of small villages and major commercial hubs, thriving spaces in contrast to the barren mountains.[42] The coasts connected the land and the sea, and the big wide waters were avoided if at all possible. If connectivity is crucial to the differentiation of maritime spaces from one monotonous blue mass, then this third category of littoral spaces is pivotal.

The significance of coasts to the Ottoman understanding of maritime space is evident in the 1766 command. The whole purpose of that imperial order, and others like it, was to provide protection from maritime violence for Ottoman and foreign subjects peacefully going about their business. The key category of protected subject mentioned time and time again is the *fuḳarā-yı raʿiyet*, the poor subjects. In part, the focus on protecting the poor and vulnerable represents the shifting relationship between ruler and ruled. In previous centuries, the Ottoman fleet came to the islands and the coasts to collect taxes; now, they were there, in addition, as guardians, in part in response to a growing number of petitions sent by provincial subjects in the maritime provinces pleading for protection. These poor subjects are almost always described as living in a particular space denoted by the recurring couplet, *cezāʾır ve sevāḥil*, the islands and the coasts.[43]

It is perhaps not surprising that we should find these together; after all, coastlines are what make islands, islands. This phrase is found throughout imperial commands, and in Katib Çelebi's text above. Islands and coasts, together, form a cohesive littoral space, separate from the liquid vastness of the maritime and the flows of the insular, but still part of the waters claimed by sovereign power, the maritory. Michael Pearson's eloquent 'Case for the Coast' in 1985 positioned a new coastal history as a way to globalize our understanding of the links between sea and land, emphasizing their connectivities.[44] The work of John Gillis, a champion of coastal and island history, is particularly important in framing an understanding of these littoral spaces. Gillis argues that we should view coasts and islands as ecotones.[45] An ecotone is a space where two ecosystems meet, and in doing so concurrently blur the boundaries between each other and create a new space, an 'ecological continuum'.[46] As such, we might begin to situate the Ottoman waters within what Gillis calls 'brown-water history', a history of the near shore waters that have often been over-shadowed by their deep blue neighbours.[47] Another prominent coastal historian, Isaac Land, has eloquently summed up the importance of differentiating maritime spaces in a review article heralding the advent of coastal history:

> Coastal history will never replace oceanic history; nor should it. [...] However, coastal history can anticipate a remarkable future. In the sea, biomass is concentrated close to the shore and thins out in the deeper water. I would suggest that the same principle applies to human beings.

Historians who cast their nets on the coast will catch considerable numbers of people whose lives and experiences would be missed by a scholar who trawls the oceans. In the end, there is simply more history to be written about the coast than about the deep blue sea.[48]

The Ottoman ideas of space present in the 1766 command fit with Land's view. When the Ottoman government sent its fleet on its regular mission of protecting the Mediterranean Sea – *Baḥr-ı Sefīd muḥāfażası* – they did not mean the whole *baḥr*, but specific waters, *ṣular*, defined for the most part by the littoral ecotone, where land and sea met and merged. The coasts and the islands were as much a part of the sea as the shipping routes in the open ocean, because the sea in the Ottoman understanding was not simply a liquid mass. As Gillis cautions us, 'land and water constitute an ecological continuum; we need to be wary of distinguishing the marine too sharply from the terrestrial'.[49] This is a space that Alison Bashford refers to as the terraqueous, where land and sea are transformed into a connected and connecting space.[50] This is a beneficial category, which is also useful when thinking about maritoriality, i.e., the defining, bounding, and controlling of liquid space. The space is still liquid but it has form – a form shaped by the relationship of communities with both land and sea. Maritoriality is about control of water that links islands and coasts, that allows for movement between terrestrial and maritime spaces, and that provides a function for states and their populations beyond rhetorical or cartographical claims.

Conclusions

It is important to note that none of this is present in the imperial command from Sultan Mustafa III to his admiral Hüseyin Pasha. The Ottoman scribes did not write 'ecotonal', 'permanences', 'terracentric', or 'maritoriality' in their flowing calligraphy. There is a danger of being too prescriptive and ahistorical in applying methodologies from other disciplines, times, and places to this very specific historical moment. Yet from a fairly standard archival document, an order to send out a part of the fleet to protect Ottoman subjects and foreign merchants in the Mediterranean from attacks at or by sea, a careful analytical reading of its spatial categories can help us to understand what the sea meant to the authorities in Istanbul. The sea was made up of different spaces, some liquid, some terrestrial, linked by the coastal ecotone – that mishmash area where land and sea met, where the waters facilitated navigation, and where great varieties of economic activities took place.

The command was composed to specifically list the different sorts of spaces into which the admiral should take his ships. This is all part of maritoriality along the lines that Robert Sack laid out for territoriality. In order to enforce the title as 'ruler of the two seas', the sultan's authority had to be exerted in what he claimed as his maritory. This meant protecting his subjects and foreigners from harm. In order to do that, there had to be a 'form of communication' that delineated the claimed maritory, of which we have a rhetorical example here in the claim over the *ṣular*, the waters. The sultan did not claim *all* the sea, but only his waters within it. In order for that claim to be successful, there had to be a clear classification of the relevant space, and this was given as the coastal waters, the coasts, the islands, and the shipping routes. Because maritory in this case was both terrestrial and aquatic, it was also necessarily littoral, i.e., tied to coastal spaces. But littorality does not mean that we should privilege the land; the coast is as much liquid as it is solid. By classifying, defining, and exerting control over this terraqueous space, the Ottoman state was clearly following a particular process of claiming sovereignty over the sea, and defining a maritory: by being explicit about its borders and the nature of its sovereignty in this and other commands. In doing so, to return to Koca Ragıb's poem, the Ottoman Empire attempted to define what was fixed and what was transient, both for its sailors on the water and its subjects on the land, viewing each other within one maritory.

Notes

1 Koca Ragıb Pasha, *Dīvān-ı Rāġıb*, Cairo: Matbaʻa-ı Bulāḳ, 1252 [1836], p. 23.
2 Charles S. Maier, 'Transformations of Territoriality, 1600–2000', in Gunilla Budde, Sebastian Conrad and Oliver Janz (eds.), *Transnationale Geschichte: Themen, Tendenzen und Theorien*, Göttingen: Vandenhoeck & Ruprecht, 2006, pp. 32–55, here p. 34.
3 Ibid.
4 Ibid., pp. 37–40.
5 For a discussion of the relationship between the Middle East and maps, see Michael E. Bonnie, 'Of Maps and Regions: Where Is the Geographer's Middle East?', in Michael E. Bonnie, Abbas Amanat and Michael Ezekiel Gasper (eds.), *Is There a Middle East? The Evolution of a Geopolitical Concept*, Stanford: Stanford University Press, 2012, pp. 56–99.
6 Charles S. Maier, *Once within Borders: Territories of Power, Wealth, and Belonging since 1500*, Cambridge, Mass.: Harvard University Press, 2016, esp. the introduction.
7 Ismael Vaccaro, Allan Charles Dawson and Laura Zanotti, 'Negotiating Territoriality: Spatial Dialogues between State and Tradition', in id. (eds.), *Negotiating Territoriality: Spatial Dialogues between State and Tradition*, New York: Routledge, 2014, pp. 1–20, here p. 1.
8 Claude Raffestin, *Pour une géographie du pouvoir*, Paris: Libraires Technique, 1980, p. 129.
9 Henri Lefebvre, *The Production of Space*, trans. Donald Nicholson-Smith, Oxford: Blackwell, 1991, French 1974, pp. 46–53; Chandra Mukerji, *Territorial Ambitions and the Gardens of Versailles*, Cambridge: Cambridge University Press, 1997, p. 3.
10 Robert D. Sack, *Human Territoriality: Its Theory and History*, Cambridge: Cambridge University Press, 1986, pp. 21–2.
11 Ibid., p. 19.
12 For an interesting critique of this approach, see Sujit Sivasundaram, Alison Bashford and David Armitage, 'Introduction: Writing World Oceanic Histories', in id. (eds.), *Oceanic Histories*, Cambridge: Cambridge University Press, 2018, pp. 1–28.

13 Nabi, *Dīvān-ı Nābī*, rubāʿī, section 113.
14 Genesis 1:2 (KJV).
15 Revelation 21:1 (KJV).
16 Qurʾān 21:30 and 45:12.
17 Karen C. Pinto, *Medieval Islamic Maps: An Exploration*, Chicago: University of Chicago Press, 2016, pp. 172–4.
18 Jan Rüdiger, 'Thalassocraties médiévales: pour une histoire politique des espaces maritimes', in Rania Abdellatif et al. (eds.), *Construire la Méditerranée, penser les transferts culturels: Approches historiographiques et perspectives de recherche*, Munich: Oldenbourg, 2013, pp. 93–103, here p. 99.
19 Alexis Wick, *The Red Sea: In Search of Lost Space*, Berkeley: University of California Press, 2016, pp. 187–8.
20 Ibid., p. 188.
21 Ibid., pp. 188–9.
22 Antonis Hadjikyriacou, 'Envisioning Insularity in the Ottoman World', *Princeton Papers: Interdisciplinary Journal of Middle Eastern Studies* 17, 2017, vii–xix, here ix.
23 Wick, *Red Sea*, p. 9; Peregrine Horden and Nicholas Purcell, 'The Mediterranean and "the New Thalassology"', *American Historical Review* 111, 2006, 722–40, here 723.
24 Maier, *Once within Borders*, pp. 32–8.
25 Hugo Grotius, *Mare Liberum sive De Jure Quod Batavis Competit ad Indicana Commercia Dissertatio*, Leiden: Ludovici Elzevirii, 1609, p. 18.
26 Cornelius van Bynkershoek, 'De Dominio Maris Dissertatio', The Hague, 1703, p. 13.
27 For an interesting set of papers debating the term, see Evi Gorogianni, Peter Pavúk and Luca Girella (eds.), *Beyond Thalassocracies: Understanding Processes of Mioanisation and Myceanisation in the Aegen*, Oxford: Oxbow Books, 2016.
28 Rüdiger, 'Thalassocraties médiévales', p. 94.
29 For a discussion of the decline paradigm, see Dana Sajdi, 'Decline, Its Discontents and Ottoman Cultural History: By Way of Introduction', in id. (ed.), *Ottoman Tulips, Ottoman Coffee: Leisure and Lifestyle in the Eighteenth Century*, London: I. B. Tauris, 2014, pp. 1–40.
30 See also Michael Talbot, 'Separating the Waters from the Sea: The Place of Islands in Ottoman Maritime Territoriality during the Eighteenth Century', *Princeton Papers: Interdisciplinary Journal of Middle Eastern Studies* 17, 2017, 61–86; id., 'Protecting the Mediterranean: Ottoman Responses to Maritime Violence, 1718–1770', *Journal of Early Modern History* 21, 2017, 283–317.
31 Andrew C.S. Peacock, 'Introduction: The Ottoman Empire and Its Frontiers', in id. (ed.), *The Frontiers of the Ottoman World*, Oxford: Oxford University Press, 2009, pp. 1–30, here p. 2.
32 Nabi, *Dīvān-ı Nābī*, Istanbul: Şeyḫ Yaḥyā Efendi Maṭbaʿası, 1292, 1875, dīvān-ı Nābī section 83.
33 Katib Çelebi, *Tuḥfetü'l-kibār fī esfārü'l-biḥār*, Istanbul: Dārü'ṭ-ṭibāʿa el-maʿmūre, 1141, 1729, 72r.
34 Başbakanlık Osmanlı Arşivleri, Prime Ministry's Ottoman Archives, BOA, Cevdet Tasnifi, Bahriye, Cevdet Series, Naval, C.BH, 13/627.
35 BOA, C.BH 13/627.
36 David Harvey, *Justice, Nature, and the Geography of Difference*, Oxford: Blackwell, 1986, pp. 261, 295–6.

37 David Harvey, 'Space as a Keyword', in Noel Castree and Derek Gregory (eds.), *David Harvey: A Critical Reader*, Oxford: Blackwell, 2006, pp. 270–94, here p. 282.
38 Roxani Margariti, 'An Ocean of Islands: Islands, Insularity, and Historiography of the Indian Ocean', in Peter N. Miller (ed.), *The Sea: Thalassography and Historiography*, Ann Arbor: University of Michigan Press, 2013, pp. 198–229, here p. 219.
39 Walters Ms. W.658, Piri Reis, *Kitāb-ı Baḥrīye*, fol. 5a.
40 Ibid. fol. 121b.
41 Katib Çelebi, *Tuḥfetü'l-kibār*, 2v-3r.
42 Fernand Braudel, *The Mediterranean and the Mediterranean World in the Age of Philipp II*, volume 1, trans. Siân Reynolds, Berkeley: University of California Press, 1995, pp. 138–48.
43 For an interesting comparative discussion on piracy and sovereignty, see Emily Sohmer Tai, 'Piracy and Property in the Premodern West', in Jerry H. Bentley, Renate Bridenthal and Kären Wigen (eds.), *Seascapes: Maritime Histories, Littoral Cultures, and Transoceanic Exchanges*, Honolulu: University of Hawai'i Press, 2007, pp. 205–20, especially 210–2 on petition culture.
44 Michael N. Pearson, 'Littoral Society: The Case for the Coast', *The Great Circle* 7, 1985, 1–8, here 1.
45 John R. Gillis, *The Human Shore: Seacoasts in History*, Chicago and London: University of Chicago Press, 2012, pp. 3–9; John R. Gillis, 'Not Continents in Miniature: Islands as Ecotones', *Island Studies Journal* 9, 2014, 155–66, here 155–6 and passim.
46 Gillis, 'Not Continents', 163.
47 Ibid., 163–4.
48 Isaac Land, 'Tidal Waves: The New Coastal History', *Journal of Social History* 40, 2007, 731–43, here 740–1. For a more recent discussion on the nature of coastal history, see Isaac Land, 'The Urban Amphibious', in David Worthington (ed.), *The New Coastal History: Cultural and Environmental Perspectives from Scotland and Beyond*, London: Palgrave Macmillan, 2017, pp. 31–48.
49 Gillis, 'Not Continents', 163.
50 Alison Bashford, 'Terraqueous Histories', *The Historical Journal* 60, 2017, 253–72, here 255.

13 Regional imaginaries

Konrad Lawson

If we imagine the world, not as an undifferentiated globe viewed from the moon, but as the collection of wooden blocks in a child's puzzle, there are many ways to cut the pieces. One way to divide the world today is into countries, each block of wood corresponding to some manifestation of borders and claims of sovereignty etched on the surface of the earth. Competing national claims to territory would require all manner of tricky choices. Many of the other ways to cut the blocks in the mind's eye – and note the commonly used visual metaphor here – quickly creates a range of other difficulties. These include divisions of continents, regions, or concepts like 'the West' or 'the Global South'.

When we attempt to explain what it is that justifies carving the block one way or another, we are constructing *spatial imaginaries*. They are 'imaginaries' because their shape and, more importantly, their perceived content derives from the cognitive categories of human thought and experience. There is, for example, no 'Asia' or 'Europe' out there in the world, without us claiming it to be so. This is certainly not limited to this grand scale. To ask where 'the neighbourhood', 'the city', or the 'wilderness' begin and end raises many of the same questions and challenges.

This chapter uses the example of *regional imaginaries* in Japanese accounts of the lands to its south and southwest – roughly what we would call Southeast Asia today – during the first half of 20th century to illustrate the *dynamic, relational, culturally normative*, and *situated* features of spatial imaginaries more broadly.[1] After defining each of these, I will show them at work in Japanese geography and history textbooks from this period, as well as in a Japanese philosopher's 1943 wartime reflection on the 'Asian' or 'Oriental' (*Tōyōteki*) character of the Philippine people.[2]

Four features of spatial imaginaries

Spatial imaginaries are not fixed

They are constructed, dynamic, and historically contingent. Relational geographers such as Doreen Massey have struggled against an idea of space as inherently static in contrast to its dynamic other, time. Spatial *history*,

DOI: 10.4324/9780429291739-17

including the history of spatial imaginaries, only becomes possible if we release the spatial 'from the realm of the dead' and accommodate the *motion* and *trajectories* of our subjects.³ As Massey puts it, 'there is no stable moment, in the sense of stasis, if we *define* our world, or our localities, *ab initio* [from the start] in terms of change'.⁴ These words are written from our position as observers, no less embedded in the world, with our own trajectories and perspective. Many of our historical sources, however, do not share this assumption and treat spatial extents as natural, fixed, and timeless.

Spatial imaginaries are not singular

They are multiple, relational, and regularly contested. Edward W. Said, who coined the equivalent term 'imaginative geography', was sensitive to the changing nature of Orientalist discourse, especially after Napoleon's failed occupation of Egypt. Despite this, he found it necessary to respond to criticism of a reductive essentialism in an afterword to his 1994 edition of *Orientalism* (1978), where he emphasized that he was drawn to the study of Orientalism because of its 'variability and unpredictability' as well as its 'combination of consistency *and* inconsistency'.⁵ This hints at the challenges of subsuming a diversity of perspectives under the name of a single overarching spatial imaginary even if, ultimately, the patterns identified in Said's work have been immensely productive. Any spatial imaginary co-exists with other mental maps that they may be examined in relation to. This is true for similar regional concepts or spatial imaginaries that work at different scales as in the case, for example, of the relationship between national 'imagined communities' and regional imaginaries.⁶ The example of Japanese discourses around *Nan'yō* (the south seas), *Tōyō* (the Orient), and *Tōnan Ajia* (Southeast Asia) remind us that spatial imaginaries not only change over time and overlap across scales, but each represents a capacious web of discourse which is the product of human encounters. These, in turn, are often mediated by texts, maps, and other images.

Spatial imaginaries are not merely neutral descriptions of the world

While spatial imaginaries must, by definition, imply some *spatial extent* – the 'where' question that defines an inside and an outside – the content of the *imaginary* itself is impossible to untangle from its relationship with other normative elements. This might include claims about the purported impact of climate and environment on a population, or issues of language, ethnicity, and culture. Indeed, the geographical extent of any spatial imaginary is very likely to be negotiated and evolve in response to the changing answers to the questions of 'who' occupies the space and 'why' they matter. The term *spatial imaginary* draws our attention to the spatial aspects of our cognitive categories, but it is inevitably driven in large part by the broader dynamics of the relationship between *self* and *other*.⁷ In drawing contrasts, a source may claim to remain simply observational and descriptive. In fact, however, they are very commonly attached to evaluative judgments, overt or implied. This *culturally normative* element of spatial imaginaries is one of their most studied features. We should not be surprised when the 'where' drops from the foreground of our source materials, and the language moves quickly, for example, from observing that a custom can be found among some, presumably specific, people 'in the Orient' to general claims about 'Orientals'.

Spatial imaginaries are situated

The vantage point of the observer and the broader context of a source shapes and limits what is captured and what is left absent. Histories of the diverse region 'Southeast Asia' have long engaged this problem.[8] Take, for example, the current understanding of Southeast Asia using its most convenient definition as the region corresponding to the ten current countries in the Association of Southeast Asian Nations (ASEAN), modify this slightly to include the non-member Timor-Leste (East Timor), and offer some awkward explanation for the exclusion of the eastern half of the island of New Guinea from the region – despite the fact that both Papua New Guinea and Timor-Leste aspire to ASEAN membership. This offers a historian a presentist territorial frame that they can project backwards and delimit the scope of their analysis. Using this approach, actors and cultures wander into and out of the historical narrative of a region defined in this way from what are today India, Sri Lanka, Bangladesh, but also China, Taiwan, and a host of independent countries and territories in the Pacific, not to mention colonial powers and other migrated communities. All of these mingled with, influenced, and were influenced by the diverse populations within 'Southeast Asia's' conventionally divided mainland and insular lands.

The arbitrary nature of this, and indeed *any regional designation*, is an inherent and inevitable challenge that has justifiably made many historians, anthropologists, and other scholars rather uneasy, perhaps especially so in the case of 'Southeast Asia', as both the term and the perceived unity of the region has such recent origins. Only rare and scattered references to the term can be found prior to the 20th century.[9] Early 20th-century uses of the equivalent Japanese term *Tōnan Ajia* and German *Südostasien* can be found in Japanese textbooks and some German scholarship, especially after World War I. The term is used in a selection of publications by the Institute of Pacific Relations in the 1930s. The 1943 establishment of the 'South East Asia Command' by the Allies in World War II is seen as a key moment in reinforcing the use of the term, though at its establishment its boundaries excluded Java, Borneo, the Philippines, and New Guinea.[10]

Far more importantly, a boom in Cold War publications about the region along with a rise in area studies programmes funded by major grant providers are usually seen as major contributors to its development. This is especially true in the United States, where American entanglements in the Philippines, Indonesia, and especially Vietnam raised its profile.[11] To these mostly external designations, however, must be added the first signs of what Amitav Acharya calls the shift towards the 'internal construction' of a region after World War II, beginning with early postwar proposals for a Southeast Asian group of nations by Burma's Aung San and Ho Chi Minh in Vietnam.[12] In works on Southeast Asia today, authors may point to the concrete institutional manifestations of the region as well as the increasingly broad self-identification of its peoples, while historians may cautiously embrace the concept of the region as a 'contingent device' or a 'heuristic', and as a 'political construct'.[13]

When studying the history of a spatial imaginary, it is unwise to first fix one's geographical gaze. We must be suspicious of statements that assume that *Tōnan Ajia*, *Südostasien* and 'South East Asia' are all signifiers for the same regional imaginary, even if they were to overlap perfectly when circled on a map. The heart of a spatial history of these imaginaries is found in how each writer grapples, at different moments, with the answer to the challenge of arbitrariness; the answer to the question of *what it is* that affords a region its coherence and, in the view of some who embrace it, a substantive reality as part of a meaningful description of the world. Even when scholars of Southeast Asia define the region in functional or cultural terms, and speak of it as a collection of 'Indianized states' (George Cœdès), a 'cultural matrix' of '*maṇḍala*' polities in which 'men of prowess' each claims universal divine rule (Oliver William Wolters), or a region similarly comprised of 'galactic polities' (Stanley Jeyaraja Tambiah), we are observing a process of regional construction.[14]

Elsewhere, we may witness a throwing up of hands into the air. Martin Lewis and Kären Wigen adopt 'Southeast Asia' into their proposal for a revised framework of world regions but admit that it is in many ways a 'residual and artificial category'.[15] A leading historian of Southeast Asia, Anthony Reid, has produced rich studies of the maritime trade and cultural patterns found across the region. In a textbook on the region's history, however, he opens with the statement: 'Southeast Asia was and is a distinct place, but one of infinite variety'. He continues: 'Its coherence has lain in the fact of diversity, and its genius is managing it.'[16] The 'its' grants an animating agency to a region here defined as coherent by virtue of its very incoherence.

It is important to emphasize here that this challenge is neither unique to Southeast Asia, nor does it, in any way, suggest that we should abandon this region, or regions in general as helpful concepts. In accepting their constructed nature, however useful or 'functional' they may be, we must simply acknowledge that our adoption of them may be scrutinized with exactly the same tools and analysis we might apply to any other spatial imaginary taken as objects of historical inquiry. Spatial imaginaries can be explored through a wide range of sources. In the Japanese cases relevant here, historians have made excellent use of textbooks,[17] newspapers and magazines,[18] works of fiction and ethnographies,[19] or broader combinations of sources that also include government documents, pan-Asianist promotional literature, academic works on geography and history, as well as travel accounts.[20] Studies of spatial imaginaries may just as well focus on maps and other visual images, as in the case of Bernard Smith's use of artistic depictions in his work *European Vision and the South Pacific*.[21]

For a historian working with textual sources, a helpful initial step in the analysis is to identify overt spatial categories deployed in the materials. Even if they are familiar terms, it pays to be attentive to their spatial extent, frequency of appearance, changes in their use, and perhaps most importantly, the associative language which gives a spatial imaginary its substantive content. Japanese geography and history textbooks are a good example of a type of source in which these spatial categories are particularly overt and explicit, while their need to cover topics concisely in very short sections makes it easier to identify a writer's priorities and, across editions, shifts in emphasis.

Tōyō, Tōnan Ajiya, and Nan'yō in textbooks

Through the analysis of Japanese educational materials, the historian Shimizu Hajime has traced the history of the Japanese concept of a Southeast Asian region in the first half of the 20th century. He showed that, after World War I, Japanese state geography textbooks adopted for elementary schools embraced a new unified term for a 'Southeast Asia' (*Tōnan Ajiya*). This term conceived of a single unified mainland (French Indochina, Siam, Burma, and the Malay peninsula) and insular (Dutch East Indies, Portuguese Timor, and the Philippines) region.[22] This is similar to the term (*Tōnan Ajia*) that came into general use in Japan in the 1950s and 1960s and is still used today. As Shimizu notes, this term was a 'contradictory mental construct' that, on the one hand, represented a Japanese regional innovation, adopted though it was for convenience by the textbook editors, but was a concept which embedded within it a Western geographical perspective: these lands lay to the southwest and not the southeast from a Japanese viewpoint.[23] The adoption of the term in this 1919 textbook came just as Japan secured recognition as the trustee for a new League of Nations mandate over a collection of German controlled islands in Micronesia that it occupied in 1914 as an ally of Britain.

This new term 'Southeast Asia' in state elementary school textbooks shows the dynamic and evolving nature of Japanese regional imaginaries. Yet it continued to coexist with other terms found in middle school geography textbooks throughout the first half of the 20th century. Most often in geography textbooks, mainland Southeast Asia was depicted as part of South Asia, with the mainland referred to as 'further India' and, separately, the 'Malay archipelago', listed together with the Philippines. This 'South Asia' was part of a greater conception of Asia or the Orient (*Ajiya, Ajia,* or *Tōyō*) which had multiple meanings and uses.[24] In middle school geography textbooks from the Meiji period (1868–1912) onwards, *Ajiya* or *Ajia* was most often used primarily as a continental or regional term. This was an Asia that stretched from the eastern shores of the Mediterranean to Japan. Separate from this in the textbooks, as we might expect, is Oceania, usually with sections for Australia, New Zealand, Melanesia (including New Guinea), Micronesia, and Polynesia (including Hawai'i).

This Asian continent, in turn, generally paralleled racial categories that were applied to the various peoples in each chapter, alongside sections on climate and resources. Occasionally, however, there was an interesting interaction of geographic, racial, and cultural meanings of terms like *Ajiya* and *Tōyō*. For example, we find in a 1935 textbook's concluding reflections on Asia's people the claim that 'the Hindu ethnicity (*Hinzū zoku*) has been Orientalized (*Tōyōka shite iru*), but since they are of the European race, the Himalayas are the borderline between the Asian (*Ajiya jinshu*) and European race'.[25] This is a likely reference to the racial conception of an expansive Aryan race found in lands from India to Europe. A student may notice some of the many assumptions this short sentence carries: Indians are here being discussed as part of a great region of the world known as Asia; that there is, separately, such a thing as an Asian race; this Asian race predominates on one side of the Himalayas, but not the other; that a people which is not of the Asian race may yet be 'Orientalized', whatever that means. This is an example of the kind of interplay that is often found in engagement with spatial imaginaries: a shifting to and from overlapping concepts that may at once be geographic descriptors, civilizations and cultures, or else racial categories. It is in this interplay that the culturally normative aspect of spatial imaginaries are easiest to identify.

Turning to history textbooks, we see that the main regional imaginary shifts and the term for 'the East' or 'the Orient', *Tōyō*, primarily serves as one side of a grand civilizational binary dividing the world. As a Japanese middle school Asian history textbook from 1908 puts it in its opening line: 'Oriental history (*Tōyōshi*), to be contrasted with Occidental history (*Seiyōshi*), comprises one half of world history'.[26] If the two theoretically encompass the whole of human history, the spatial extent of this 'world' is quite limited compared to the exhaustive, if unbalanced, coverage of geography textbooks. History textbooks place a heavy emphasis on Chinese dynasties in *Tōyoshi* texts, and on European history in *Seiyōshi*.[27] The latter have little to say about the Americas beyond the United States and often nothing on Africa beyond ancient Egypt but even with their limited coverage, they may be read with an eye to what the authors want students to know about 'the East' as it stands *in contrast* with 'the West' in relational terms.

Identifying the explicit categories, boundaries, and terminology is a useful initial step. Attention should also extend, however, to what is absent. It is worth noting that Japan *itself* is not covered in *Tōyō* history textbooks or, naturally, in the many geography middle school textbooks on 'foreign countries' (*gaikoku*) examined here. As with many textbooks, school classes, and university departments in Japan today, national history (*kokushi*) exists as a third space. Japan often floats as a 'transcendental subject' separate from the history of *Tōyō* and *Seiyō* but may nevertheless make cameo appearances in either of its two counterparts, either fully integrated as part of the narrative or an implied contrast to its development.[28] The absence of Southeast Asia, on the other hand, might as well suggest an entire region without a history. The few brief mentions of lands to the south of China are merely introduced as objects of European or Chinese dynastic conquest.[29] Also absent in both geography and history textbooks for middle school students prior to the late

1930s, is any significant deployment of one of the most important regional imaginaries found in Japan since the 1880s: the concept of *Nan'yō* or the 'South Seas'.

Nan'yō is an excellent example of a regional imaginary with a deeply ambiguous spatial extent. This varied at different moments, and in the hands of different Japanese writers. Many works promoted the region as a target for Japanese migration, trade, or imperial expansion as part of a 'Southern advance' (*Nanshin*).[30] The term could refer narrowly to smaller island groups in the Pacific (later the so-called 'inner' South Seas), including the Micronesian islands that came under Japanese occupation, or at its greatest extent, it referred to all of today's Southeast Asia. For many early 'Southern advance' supporters, *Nan'yō* was conceived as a geographic space separate from both *Seiyō* in the West or *Tōyō* in the East, and the maritime character of the regional imaginary was particularly emphasized.[31] Japanese literature on this region in the form of travel accounts, adventure novels, children's comics, and speculative political tracts developed its own rich and flexible repository of stock depictions of the Indigenous peoples of the islands that were only occasionally problematized. More important than the representation of its inhabitants, however, the literature on this region depicted it as a setting for Japanese economic exploitation, adventure, and self-discovery.[32] Any full exploration of Japanese regional imaginaries must recognize the parallel but independent development of discourses around the oceanic spaces of *Nan'yō* which were only ever partially folded into other contenders such as the broader regions of *Tōa*, *Ajia*, and *Tōyō*.

One case of attempting to 'fold' in the *Nan'yō* imaginary can be seen in a middle school geography textbook from August 1941. This is one of only a few examples of its use as distinct region with its own section. This textbook, published on the eve of the Pacific War, was by Tanaka Keiji, a geographer who authored many textbooks and other geography pedagogical materials from the 1930s until well into the postwar period. In this edition Tanaka moved both mainland and insular Southeast Asia from its former home as sub-sections of 'South Asia' to join Manchuria and China as a third *Nan'yō* section incorporated into the 'East Asia' (*Tōa*) chapter. In the digitized version of the text hosted by the University of Hiroshima, we may read the corrections a previous owner of the copy made on the text as territories in the region became 'former' Western colonies under Japanese occupation in 1942.[33]

The term *Nan'yō* itself received limited recognition in most geography and history middle school textbooks, and the peoples of Southeast Asia as a whole were mostly denied historical agency as anything but victims of conquest. The patterns of cultural evaluations of its people that matured in the literature on *Nan'yō*, however, may be found in many editions of middle school geography textbooks. These stand out given the fact that, for any given section, on any part of the world, evaluative statements about its people are relatively uncommon. A typical structure introduced a region in terms of its geographic position, then proceeded to discuss its climate, industry, natural resources, transportation networks, and sometimes had sections for its political system, its people, and its connection to Japan. In the section on its people, variously labelled 'people and culture', or 'inhabitants and population density', the reader was usually offered matter-of-fact claims about the racial, linguistic, and religious distribution of the inhabitants.

Despite the terse format, however, we occasionally find that the author could not resist sharing their derogatory evaluation of the peoples. Most prominent of these, in the case of the peoples of Southeast Asia, is their supposed indolence, a motif that has been explored in depth by Syed Hussein Alatas.[34] In a 1925 textbook, in the section on Indochina, we are told: 'The native inhabitants are Indochinese. They are the most numerous but, being lazy and lacking initiative, the Chinese, English and French hold the real power in industry'.[35] In a 1935 textbook, Tanaka writes: 'The natives are Malay Muslims but, because they are incompetent, economic power is in the hands of the immigrant Chinese, similar to the case of Indochina.'[36] In a wartime 1943 elementary school geography textbook, Filipinos are similarly slandered, '… the people, in general, have a docile nature but they will gradually correct their flaw of laziness under Japanese guidance'.[37] Using guidelines provided in elementary school teaching manuals for geography, Shimizu Hajime has shown how instructors were encouraged to ask students to reflect on the causes for the supposed laziness of Southeast Asians. Rather than locating them in inherent and unchangeable racial features, they were to be attributed to environmental causes. Montesquieu, in his 1748 *Spirit of the Laws*, believed that 'great heat enervates the strength and courage of men', while cold climates made them 'capable of long, arduous, great, and daring actions', thus condemning all of Asia to natural subjugation throughout history.[38] In Japanese schools, instructors were likewise told to warn students that the rich natural blessings of tropical lands could be a threat to human character.[39]

The hazy relationship between racial characteristics and environmental impact would leave the student with questions, however. If the tropical climate and natural fertility of region made its races indolent, how did the communities of the ethnic Chinese supposedly remain vigorous and dominant across the generations? And was the hot climate of southern China, where most of these migrants came from originally, so different that they were more naturally capable of 'arduous' actions? These sporadic comments on the overseas Chinese population of Southeast Asia were not merely opportunities to denigrate the majority ethnicities of the region. Rather, as Yano Tōru has argued, they are suggestive of deep Japanese anxieties about the successes of the Chinese, and to a lesser extent, Indian migrants in *Nan'yō*, so much so that they displaced the Europeans as the main perceived rival for expansion and colonial settlement in the region.[40] Moreover, the broad generalizations about the people of Southeast Asia in geography textbooks were complicated by the exception offered to the people of Siam (Thailand). The 1943 textbook spoke highly of the 'compassionate, kind, and friendly nature' of Japan's wartime allies, while many earlier textbooks were forced to admit its considerable 'progress towards becoming a civilised country' and praise its ability to remain independent in a sea of European colonies.[41] What allowed its people to prevail over the supposed debilitating impact of its tropical climate? These were not the kinds of questions addressed in the highly concise format of a textbook. Instead, we may explore an example from the many writings of Japanese visitors to Southeast Asia.

Miki Kiyoshi on the 'Oriental Character'

For our last example we will use a 1943 article by Miki Kiyoshi, a Japanese philosopher and public intellectual who published widely in magazines and newspapers throughout the 1930s. Miki was a creative thinker and not an easy one to categorize. Over the course of his career, he engaged deeply with Neo-Kantian philosophy, Marxist thought, and Japan's so-called Kyoto School of philosophy. In the 1930s he wrote frequently for newspapers and the more intellectually rich 'general interest magazines' (*sōgō zasshi*) on topics ranging from art, technology, to culture, politics, and foreign policy. In 1937 he became the head of the culture research section of the 'Shōwa Research Association', a research group with diverse membership but close ties to the important wartime prime minister Konoe Fumimaro. During this time, Miki played a leading role in producing the cultural section's main manifestos, including the *Principles of Thought for a New Japan* (1939), with sections reflecting on the significance of the 'world-historical' opportunity presented by Japan's invasion of China; a proposal for the unification of East Asia (*Tōa*); and two sections elaborating the 'principles of Asian thought' (*Tōa shisō*, but *Tōyō* culture is frequently referred to within) and its relation to Japanese culture.[42] After Japan's attack on Pearl Harbour in December 1941, and its rapid conquests across Southeast Asia, Miki was drafted and sent to the Philippines to serve in the Japanese army's Propaganda Corps, reporting on occupation conditions there throughout the year 1942.

One of Miki's first essays published in the course of 1942, 'The Oriental Character of the Filipinos', was republished in the leading Japanese general interest journal *Kaizō* in 1943.[43] Compared to the 1939 manifesto's call to unify 'East Asia' in the form of China, Manchuria, and Japanese imperial territories, this essay offers us a snapshot of Miki's efforts to incorporate the recently occupied Philippines into his understanding of Asian culture at the zenith of Japanese ambitions for Asian unity.[44] Miki's essay has two main points to offer its Japanese readers. The first is indicated in the title. On the basis of claims about Filipino culture, Miki argued that Filipinos have an 'Oriental' (*Tōyōteki*) character. However, due to its deficiencies, along with the pernicious impact of Spanish and American influence over time, the Filipino character was in need of 'refinement and development' from a teacher that could represent the 'essence' (*seizui*) of Oriental culture: Japan. 'Naturally, the Japanese must combine an offer of assistance with the authority and affection of a parent, along with the guidance of an elder brother'.[45]

Miki's evidence for the 'Oriental' character of Filipinos is rife with culturally normative evaluations. Interestingly, many of the links and similarities that he identified were labelled as being 'like the Japanese' rather than being depicted as some common feature of a broader Asian culture. Filipinos are, we learn, a polite people; they show an indifference to death due to a long history of rebellions and revolution; they have a love of cleanliness; they avoid using second person pronouns in addressing someone; and they have a love of melancholy songs.[46] All these attributes would have offered a Japanese reader a sense of warm affinity to the Filipino people, but contrasted awkwardly with Japanese colonial discourses on the supposed uncleanliness and cowardly nature of other peoples of *Tōyō* such as its own colonial subjects in Korea or the Chinese.[47] A few character features were explicitly labelled by Miki as a feature common to all Asians. The Filipinos are a 'quiet and calm' people who 'can be said to be without [facial] expression', in contrast with the Westerner's tendency to exhibit 'lively expressions of emotion'.[48] This 'natural reserve' was threatened, however, by other tendencies that emerged 'as the result of Spanish tyranny and American democratic tendencies'. Filipinos were said to have developed a love of politics, a political 'habit' or 'vice' (*kuse*), and Miki complained of their exasperating habit of spontaneous breaking into long speeches.[49]

Three other *Tōyō* features attributed to Filipinos by Miki link to those emphasized in the 1939 manifesto's analysis of 'East Asian thought': familism (*kazokushugi*), a 'special Oriental symbolism' in their literature, and a 'sense of nothingness' (*kyomukan*). Miki notes approvingly that, 'in the Philippines, the family is the foundation of society'.[50] Before long, however, he again addresses deficiencies here too, noting an alarming 'predominance of women over men' (*joson danpi*) in Filipino society.[51] This was not a new custom, and one found elsewhere in the region, but this distinctly un-Oriental feature was no doubt reinforced, he claimed, in the encounter with Christianity. Filipino familism would therefore need some 'refinement' with the help of Japan.[52]

In the literature of the Philippines, Miki identified the presence of a recognizable 'Oriental symbolism' (*Tōyōteki shōchōshugi*), but he regretted that it had a, presumably inferior, 'Indian' flavour to it, and its use was limited to fables and the fantastical. The solution was for them to study Japan's 'spiritual culture'.[53] Again, Miki claimed that Filipinos shared with other Asians a certain 'sense of nothingness', but here too Miki believed that the Filipinos fell short.[54] Their 'sense of nothingness' matched neither the 'spirituality of the Japanese, nor the pragmatism of the Chinese'. Though he failed to elaborate in detail on what he meant, he claimed that the Filipino 'sense of emptiness' was 'flat' or 'two-dimensional', lacking depth. This, he speculated, was probably attributable to the influence of the southern climate.[55]

The supposed causative role of climate was explored by Miki in depth as he grappled with claims about the 'idleness' (*yūda*) of the Filipinos. It was an issue that he returned to no fewer than three times in the course of the essay, and it is clear that he was at pains to offer his Japanese readers some limited defence of Filipinos on this count. However, ultimately, he neither denied the charge, nor did he reject entirely the claim that climate was part of the explanation.

Clearly climate had a role in making the Filipinos idle. The heat of his land makes one prone to laziness. Moreover, here there are rich natural resources and even if one does not work, they can collect sufficient foods year-round.[56]

In this passage he is restating, and confirming, the long-standing Japanese – and Western – assumptions about the peoples of tropical lands. Using a technique found throughout his journalistic writing, however, he began by first conceding a widely believed position, but then started to offer counter evidence.

Miki challenged the monocausal explanation by adding to it a critique of the Filipino diet, on the one hand, and attempted to seed doubts about the impact of climate by turning to history for an explanation. Bountiful though the fruits of nature in the Philippines may be, Filipinos were apparently not, 'according to the results of an investigation', eating enough of those very fruits and vegetables.[57] Contradicting his claim that anyone could simply gather what they need in a tropical land, Miki reported that only those with high enough salaries were able to achieve a balanced diet. To complement this economic explanation, Miki then reaches into the past to a golden, pre-colonial age. 'Can you really say the Filipinos are inherently idle?' Did not the ancestors of the Filipinos, the Malay race, 'make a great contribution to humanity' by 'recognizing the importance of tropical resources' and 'set out on adventures' of discovery and settlement across the region?[58] Miki's historical analysis uses a similar tactic to one adopted in Philippine national hero José Rizal's own writings about claims of Filipino indolence.[59] In the pages that follow, Miki suggested that it was the heavy extractive policies of Spanish colonial rule which robbed Filipinos of their ambition and their hopes for the future, except to dangle before them the promise of eternal salvation in a Christian heaven.[60] In fact, Miki argued, the Filipino 'sense of nothingness', including their melancholy music and literature, rather than merely being a manifestation of their 'Oriental character', could also be explained by hundreds of years of oppressive subjugation.[61]

This article, filled with its many culturally normative generalizations about the Filipinos, amply served the purposes of wartime propaganda, despite a limited attempt to complicate a simplistic narrative of tropical 'idleness'. It also demonstrates some of the common features of regional imaginaries. From the more limited scope of Miki's Asian imaginary in the 1939 manifesto, we may see him pull the Philippines into his conception of an 'Oriental culture'. To do this, however, he had to engage with the powerful cultural stereotypes of the peoples of *Nan'yō*, hardened in the discourses on that distinct regional imaginary. Whenever Miki encountered difference, he reported it as a deviation from the norms of that very category of 'Oriental' culture he assumed was there.

Most of Miki's claims about the supposed blemishes in the 'Oriental character' of Filipino culture were paired with a simple solution: The Filipinos should learn from a Japanese culture that represented the gold standard. And yet, the Oriental imaginary of this article is also *situated* in much more than merely the context of a Japanese wartime occupation. With the important exception of engagement with the work of José Rizal and, briefly, with historian Epifanio de los Santos, the vast majority of the details in Miki's essay describing Philippine culture were taken from the writings of European and American missionaries, travellers, and scholars such as Pedro Chirino, Francisco Colin, Frederic H. Sawyer, William Gifford Palgrave, and John Crawford, among others. Though Miki's essay aimed to highlight the 'Oriental' features of Filipino character, they were often mediated by his 'Western-tinted eyeglasses'.[62] To take the analysis of this element further, Miki's wartime attempt to incorporate the tropical lands to Japan's south into a broader *Tōyō* may be productively explored in relation to other regional imaginaries at work in the texts he drew upon, including those of his Western sources and Philippine perspective of Rizal, each situated in their own historical context.[63] In this way, the historian may better appreciate the crossing trajectories and mutual influences of spatial imaginaries of 'the Orient', the 'South Seas', and the 'Tropics'.

Notes

1 For an introduction to the wide literature related to spatial imaginaries, please see the 'Spatial Imaginaries' section in Konrad Lawson, Riccardo Bavaj and Bernhard Struck, *A Guide to Spatial History: Areas, Aspects, and Avenues of Research* (2021), Available HTTPS: spatialhistory.net/guide.
2 Japanese names in this chapter are listed in the usual family name, first name order in the body, while endnote citations will follow the first name, last name format. My thanks to Someya Nana for her research assistance and to Nicole CuUnjieng Aboitiz for valuable feedback and suggestions.
3 Doreen Massey, *Space, Place and Gender*, Cambridge: Polity, 1994, p. 4.
4 Ibid., p. 136.
5 Edward W. Said, *Orientalism*, London: Vintage Books, 1994, first published 1978, pp. 339–40. Edward Said's emphasis on the historical change in the discourse is on p. 87, and again emphasized in the afterword, pp. 332–3.
6 This chapter focusses on examples of regions larger than the nation, what Lewis and Wigen refer to as 'world region', as opposed to the frequent use of the term in geography for a sub-national unit. See Martin W. Lewis and Kären Wigen, *The Myth of Continents: A Critique of Metageography*, Berkeley: University of California Press, 1997, p. 13; see also Andrew Herod, *Scale*, London and New York: Routledge, 2011, pp. 126–68; on the ambiguity of the term see Martin Jones and Anssi Paasi, 'Regional World(s): Advancing the Geography of Regions', *Regional Studies* 47, 2013, 1–5.
7 For further discussion on this, see Josh Watkins, 'Spatial Imaginaries Research in Geography: Synergies, Tensions, and New Directions', *Geography Compass* 9/9, 2015, 511.
8 The best recent survey of these debates, together with a detailed argument about the formation of socially constructed regions, offers Amitav Acharya, *The Making of Southeast Asia: International Relations of a Region*, Ithaca: Cornell University Press, 2013; see also the important collection of essays by Henk Schulte Nordholt

and Remco Raben (eds.), *Locating Southeast Asia: Geographies of Knowledge and Politics of Space*, Singapore: Singapore University Press, 2005; for a particularly sharp critique of the Cold War contributions to the study of Southeast Asia as a region, see Simon Philpott, *Rethinking Indonesia: Postcolonial Theory, Authoritarianism and Identity*, Basingstoke: Palgrave Macmillan, 2000, pp. 95–143. For similar debates on Asia as a whole, see Prasenjit Duara (ed.), *Asia Redux: Conceptualizing a Region for Our Times*, Singapore: ISEAS Publishing, 2013, Andrea Acri et al. (eds.), *Imagining Asia(s): Networks, Actors, Sites*, Singapore: ISEAS Publishing, 2019, and Marc Frey and Nicola Spakowski (eds.), *Asianisms: Regionalist Interactions and Asian Integration*, Singapore: NUS Press, 2016.
9 See Donald K. Emmerson, '"Southeast Asia": What's in a Name?', *Journal of Southeast Asian Studies* 15, 1984, 5–7.
10 Ibid., 7–9. For arguments against the overstating of the importance of the founding of the wartime command, see Russell H. Fifield, 'Southeast Asia as a Regional Concept', *Southeast Asian Journal of Social Science* 11/2, 1983, 1–14.
11 Henk Schulte Nordholt and Remco Raben, 'Locating Southeast Asia', in id. (eds.), *Locating Southeast Asia*, pp. 2–3; see also Hajime Shimizu, 'Southeast Asia as a Regional Concept in Modern Japan', in ibid., p. 89.
12 Acharya, *The Making of Southeast Asia*, pp. 38, 108–13.
13 Heather Sutherland, 'Contingent Devices', in Nordholt and Raben (eds.), *Locating Southeast Asia*, pp. 20–59; Nicole CuUnjieng Aboitiz, *Asian Place, Filipino Nation: A Global Intellectual History of the Philippine Revolution, 1887–1912*, New York: Columbia University Press, 2020, p. 5. Aboitiz adds an acknowledgement of its 'tangible, natural reality' as a biotic zone.
14 George Cœdès, *Indianized States of South East Asia*, trans. Sue Brown Cowing, reprint ed., Honolulu: University of Hawaii Press, 1968, French 1944; Oliver William Wolters, *History, Culture, and Region in Southeast Asian Perspectives*, Singapore: Institute of Southeast Asian Studies, 1982; Stanley Jeyaraja Tambiah, *Culture, Thought, and Social Action: An Anthropological Perspective*, Cambridge, Mass.: Harvard University Press, 1985, pp. 252–86.
15 Lewis and Wigen, *Myth of Continents*, p. 176.
16 Anthony Reid, *A History of Southeast Asia: Critical Crossroads*, Oxford: John Wiley & Sons, 2015, p. xvii.
17 Shimizu, 'Southeast Asia as a Regional Concept in Modern Japan'.
18 Noriyuki Ishikawa, 'Senzenki Nihon no "nanshin" ninshiki to media gensetu: *Nihon oyobi Nihonjin* no "nan'yō" kanren kiji o daizai toshite' 戦前期日本の「南洋」認識とメディア言説：『日本及日本人』の「南洋」関連記事を題材として *Studies in Political Science and Economics* 政経研究 56, 2019, 443–68.
19 Robert Tierney, *Tropics of Savagery: The Culture of Japanese Empire in Comparative Frame*, Berkeley: University of California Press, 2010.
20 Tōru Yano, *Nihon no Nan'yō shikan* 日本の南洋史観, Tōkyō: Chūō kōronsha, 1979; Yoneo Ishii 'Tōnan Ajia chiiki ninshiki no ayumi' 東南アジア地域認識の歩み *The Journal of Sophia Asian Studies* 上智アジア学 7, 1989, 1-17; Ken'ichi Gotō *Tensions of Empire: Japan and Southeast Asia in the Colonial & Postcolonial World*, Singapore: Singapore University Press, 2003, pp. 1–106. The most comprehensive survey on Japanese conceptions of Asia is Yamamuro Shin'ichi, *Shisō kadai toshite no Ajia: kijiku, rensa, tōki* 思想課題としてのアジア—基軸・連鎖・投企, Tōkyō: Iwanami, 2001; see also Stefan Tanaka, *Japan's Orient: Rendering Pasts into History*, Berkeley: University of California Press, 1995; and

Sven Saaler and J. Victor Koschmann (eds.), *Pan-Asianism in Modern Japanese History: Colonialism, Regionalism and Borders*, London and New York: Routledge, 2007.

21 Bernard Smith, *European Vision and the South Pacific*, 2nd ed., New Haven: Yale University Press, 1985, first published 1960.
22 Shimizu, 'Southeast Asia as a Regional Concept in Modern Japan', p. 89.
23 Ibid., p. 95.
24 See Yamamuro, *Shisō kadai toshite no Ajia*, for his helpful scheme of Japanese civilizational, racial, cultural, national/ethnic, and expansionist conceptions of Asia.
25 Digitized copies of most of the textbooks below can be found at the University of Hiroshima Textbook Database. Available HTTP: dc.lib.hiroshima-u.ac.jp/text/ (accessed 12 May 2021). Keiji Tanaka, *Chūtō shin gaikoku chiri* 中等新外国地理, revised ed., Tokyo: Meguro shoten, 1935, p. 229.
26 Kyūshirō Nakamura, *Shūtei Tōyō rekishi* 修訂東洋歴史, Tokyo: Yoshikawa Kōbunkan, 1908, pp. 1–3.
27 See, for example, Jitsuzō Kuwabara, *Tōyōshi kyōkasho chūtō kyōiku* 東洋史教科書 中等教育, Tokyo: Tōkyo Kaiseikan, 1921, p. 3.
28 Sangjung Kang, 'The Discovery of the "Orient" and Orientalism', in Richard Calichman (ed.), *Contemporary Japanese Thought*, New York: Columbia University Press, 2005, p. 89.
29 Nakamura, *Shūtei Tōyō rekishi*, pp. 125, 163, 166–70, 177–8, 189; Kuwabara, *Tōyōshi kyōkasho chūtō kyōiku*, pp. 154–5.
30 For an English language overview of both the concept and Japanese engagement with Nan'yō, see Mark R. Peattie, *Nan'yo: The Rise and Fall of the Japanese in Micronesia, 1885–1945*, Honolulu: University of Hawaii Press, 1988.
31 Tōru Yano, *'Nanshin' no keifu Nihon no Nan'yō shikan* 「南進」の系譜 日本の南洋史観, Tokyo: Chikura shobō, 2009, pp. 41, 207.
32 See Tierney, *Tropics of Savagery*, especially chapters 3 and 4.
33 Keiji Tanaka, *Chūtō shin gaikoku chiri* 中等新外国地理, 5th ed., Tokyo: Chūtō gakkō kyōkasho, 1941, pp. 83–5, 88–9. Available HTTP: http://dc.lib.hiroshima-u.ac.jp/text/detail/452020170131184921 (accessed 12 May 2021).
34 Syed Hussein Alatas, *The Myth of the Lazy Native: A Study of the Image of the Malays, Filipinos and Javanese from the 16th to the 20th Century and Its Function in the Ideology of Colonial Capitalism*, London: Frank Cass and Company, 1977.
35 Sanseidō editorial office (ed.), *Chūtō kyōiku saikin sekai chiri* 中等教育最近世界地理, vol. 1, Tokyo: Sanseidō, 1925, p. 70.
36 Tanaka, *Chūtō shin gaikoku chiri* (1935), p. 197. Similar passages can be found in other editions, for example, Amio Moriya *Shinsei chiri gaikokuhen jogakkō* 新撰地理 外国編 女学校用, Tokyo: Tōkyo shoten, 1938, p. 184.
37 Ministry of Education (ed.), *Shotōka chiri* 初等科地理, vol. 2, Tokyo: Ministry of Education, 1943, p. 34.
38 Charles de Secondat, Baron de Montesquieu, *The Spirit of the Laws*, trans. Anne M. Cohler, Basia Carolyn Miller and Harold Samuel Stone, Cambridge: Cambridge University Press, 1989, p. 278.
39 Shimizu, 'Southeast Asia as a Regional Concept in Modern Japan', p. 99; for more on the literature on tropicality, see the 'Nature, Environment, and Landscape' section in Lawson, Bavaj and Struck, *Guide to Spatial History*.
40 Yano, *'Nanshin' no keifu*, p. 287.
41 Ministry of Education (ed.), *Shotōka chiri*, p. 94.

42 'Shin Nihon no shisō genri', in Kiyoshi Miki, *Miki Kiyoshi zenshū*, ed. by Ouchi Hyōe et al., vol. 17, Tokyo: Inwanami Shoten, 1966–68, pp. 507–33; see Lewis Harrington, 'Miki Kiyoshi and the Showa Kenkyukai: The Failure of World History', *Positions: East Asia Cultures Critique* 17/1, 2009, 58–72; and John Namjun Kim, 'The Temporality of Empire: The Imperial Cosmopolitanism of Miki Kiyoshi and Tanabe Hajime', in Saaler and Koschmann (eds.), *Pan-Asianism in Modern Japanese History*, pp. 153–9.
43 Kiyoshi Miki, 'Hitōjin no Tōyōteki seikaku', in *Miki Kiyoshi zenshū*, vol. 15, pp. 478–519. This essay is also discussed in the detailed examination of Miki's time in the Philippines by Gonzalo Campoamor, 'Re-Examining Japanese Wartime Intellectuals: Kiyoshi Miki during the Japanese Occupation of the Philippines', *Asian Studies: Journal of Critical Perspectives on Asia* 53/1, 2017, 1–37.
44 On Miki Kiyoshi's time in the Philippines and the activities of the Propaganda Corps, see Susan C. Townsend, *Miki Kiyoshi, 1897–1945: Japan's Itinerant Philosopher*, Leiden: Brill, 2009, p. 242, and Sven Matthiessen, *Japanese Pan-Asianism and the Philippines from the Late Nineteenth Century to the End of World War II: Going to the Philippines Is Like Coming Home?*, Leiden: Brill, 2015, pp. 115–9.
45 Miki, 'Hitōjin no Tōyōteki seikaku', p. 519.
46 Ibid., pp. 493, 481, 498, 496, 479, respectively.
47 See, for example, Todd A. Henry, 'Sanitizing Empire: Japanese Articulations of Korean Otherness and the Construction of Early Colonial Seoul, 1905–1919', *The Journal of Asian Studies* 64, 2005, 639–75; and Louise Young, *Japan's Total Empire: Manchuria and the Culture of Wartime Imperialism*, Berkeley: University of California Press, 1999, pp. 96–7.
48 Miki, 'Hitōjin no Tōyōteki seikaku', pp. 487–8.
49 Ibid., p. 497.
50 Ibid., p. 484.
51 Ibid., p. 486.
52 Ibid., pp. 514, 518.
53 Ibid., p. 517.
54 Ibid., p. 482. For a discussion of what Miki means by 'nothingness', see Kim, 'The Temporality of Empire', p. 157, and Campoamor, 'Re-Examining Japanese Wartime Intellectuals', 26–8.
55 Miki, 'Hitōjin no Tōyōteki seikaku', p. 482.
56 Ibid., p. 507.
57 Ibid., p. 492.
58 Ibid., p. 506.
59 See Alatas, *The Myth of the Lazy Native*, pp. 98–111.
60 Miki, 'Hitōjin no Tōyōteki seikaku', pp. 510–13.
61 Ibid., p. 512.
62 This is Masaki Tsuneo's term 'Seiyō megane', further developed by Tierney, *Tropics of Savagery*, p. 147.
63 For Filipino perspectives, including Rizal, on Asia and a Malay world, see Aboitiz, *Asian Place, Filipino Nation*, pp. 32–73.

14 Economic geographies

Antonis Hadjikyriacou

Writing in the spirit of provocative hyperbole of which the *Annalistes* were so fond, Emmanuel Le Roy Ladurie declared in 1967 that the historian of the 1980s 'will have to be able to programme a computer in order to survive'.[1] Ironically, it was Lawrence Stone's phlegmatic dismissal of this quote which ultimately popularized it: 'The prophecy has not been fulfilled, least of all by the prophet himself'.[2] The timing of this exchange was not insignificant; it came at the moment of 'the huge tectonic shift from social history to cultural history'.[3] The stern, yet optimistic positivism of cliometrics and the 'new economic history' was about to be replaced by the 'cultural turn's' incredulity towards not only grand narratives, but also the very stuff that such narratives were built on: 'hard' data. Put bluntly, the production of ideas about the world was gaining primacy over quantifiable economic data, and agency took primacy over structure in the long-lasting debate among historians.[4]

Four decades later, the celebrated 'spatial turn' is very much the product of this dialectic, but in ways that are not always obvious. Spatial history originally emerged in the late 1980s to argue for the socially constructed nature of space, emphasizing, for instance, that the idea of 'natural borders' is a mirage.[5] In this line of thinking, geographical categories acquire a fluid and malleable nature. The early stages of spatial history therefore followed the legacy of the 'cultural turn'. By emphasizing the social construction of space, ideas side-lined materiality. Two recent developments have altered this balance: one at the historiographical/conceptual level, and another at the level of managing and handling data. Firstly, discussions in global and transnational history have brought space to the fore as a new conceptual unit of analysis, particularly in non-state-centric ways.[6] This mode of thinking has particularly contributed to the efflorescence of global economic history due to its role in the emergent study of the history of capitalism.[7] Closely connected is the growth of environmental and climate history and the availability and employment of new forms of 'scientific data' to measure and evaluate the relationship between nature and humans. While the nature/culture divide remains very much relevant (and unresolved) in these fields,[8] it reflects the materiality/culture divide mentioned above and reopens those age-old historiographical debates that lie at the core of the assumptions historians make. In sum, this shift constitutes what is known as the 'new materialism'.[9]

DOI: 10.4324/9780429291739-18

Secondly, the growth of digital humanities and big data based on Geographic Information Systems (GIS) methodologies has opened up new vistas for spatial history.[10] The new possibilities for analysis and processing of large amounts of data led to a reinvigorated interest in material factors, while Critical GIS has offered a range of conceptual and technical tools that challenge the spectre of positivism.[11] Data-driven spatial history can be seen as less a call for a return to the statistical/social science approaches to historical study of the 1960s and more as an attempt to *better* employ and understand such data. This reconfiguration of the culture/materiality balance means that material factors are making a comeback, not to side-line cultural factors or the intellectual perceptions of the world, but to re-establish the dialectical nature of this relationship.[12] Informed by the critique of the 'cultural turn', digital humanities approaches have no illusion about the objectivity of their data. But to dismiss this data runs the risk of missing out on the insights and effacing the clues they may offer at both an analytical and heuristic level. In many ways, the historiographical shift that was initiated in the 1970s is now coming full-circle.[13]

This chapter argues that spatial history has the potential to offer a more balanced dialectical relationship between ideas/culture and materialism. It does so by dealing with the task of visualizing economic data from pre-modern fiscal sources. The case study used here is Cyprus, and the primary source is a detailed fiscal survey conducted by the Ottomans in 1572, one year after their conquest of the island.[14] This chapter is a preliminary report on processing this set of data by employing Historical GIS (HGIS). One of its goals is to go beyond the standard statistical analysis of the large corpora of data found in such sources. Through a cartographic representation of this economic data, trends and patterns emerge that are otherwise difficult, if not impossible to discern. This exercise is first and foremost heuristic, with the purpose of generating further questions and pushing the research agenda forward.

Opening with the larger problematique behind the research presented here, the chapter then discusses a series of conceptual questions which prompted the methods proposed. It will present the source from which the data is extracted and explain how the data was organized for GIS analysis. Having discussed the GIS tools employed and the reasons why they were chosen, it will then evaluate the insights and shortcomings cartographic representation can offer. Considering the excitement over the 'age of big data' and the possibilities it offers, this chapter employs an overall grounded approach based on the careful monitoring of, and control over, a large corpus of information.

Conceptual points of departure

My concern with visualizing economic space stems from my previous engagement with Ottoman Cyprus.[15] My starting point was the idea of insularity, which I understand as the aggregate of the spatial attributes of islands. I treat insularity not in its literal sense, i.e., as something isolated or secluded, but as a range of possible conditions that may include, but not be limited to, isolation. In a well-known *topos* for the study of insular spaces, the very sea that circumscribes and secludes an island concurrently connects it to the outside world, should maritime transport be available.[16] Insularity is neither a fixed spatial or geographical condition, nor a state that simply oscillates between connectivity and isolation; it is one that includes many other conditions, which fluctuate and shift depending on temporal, spatial, or, in our case, imperial contexts.[17] The value of insularity as a spatial concept lies on the fact that it allows an appreciation of historical processes at different scales of historical experience (local, regional, imperial, Mediterranean, or global) that a sheer consideration of geography cannot reveal.

Having dealt with the conceptual implications of insularity and its cultural production, I recognized that further engagement with the material basis of spatiality could be fruitfully complemented by analysis of quantitative datasets. While I was able to detect the implications of insularity in the lives of different historical agents (officials at various levels of local administration, capital-based bureaucrats, foreign merchants and consuls, local intermediaries, or peasants), and in some of these examples could even discern the spatial imagination of the actors involved, the question of how the economy and environment influenced these historical experiences remained open. This was largely because my previous engagement with these material aspects of the economy and environment was based more on qualitative than quantitative sources. As a result, some conclusions were tentative and exploratory. For example, I was able to identify certain broad patterns in the organization of production and the fluctuating importance of particular commodities (grain, cotton, silk), which were in turn related to the distribution of social, economic, political, and cultural power. These patterns matched some of the Mediterranean-wide trends that Faruk Tabak discussed in *The Waning of the Mediterranean*, but could not be completely verified without quantitative data.[18] Tabak drew attention to processes like the northern migration of grain trade to the Baltic connected to the contraction of cereal cultivation in the Mediterranean from the 17th century onwards before recovering in the late 18th; the westward movement of sugar away from the Eastern Mediterranean and its gradual relocation in the Caribbean; and the proliferation of cotton and sericulture (silk production) in lands hitherto used for the above-mentioned crops. Furthermore, Tabak includes a discussion of Braudel's 'eternal trinity' of wheat, olive trees, and vines. Braudel introduced this categorization in *The Mediterranean and the Mediterranean World*,[19] providing a model on how to think about the relationship between humans, the economy, and the environment at large temporal and spatial scales. Tabak historicizes these crops by arguing that it was after the 17th century that they acquired the hold on the Mediterranean landscape that Braudel essentially describes as atemporal. At the same time, Tabak proposes an overlapping categorization of another Mediterranean triad consisting of cereals, tree crops (including vines, olive trees, or mulberry trees), and animal husbandry.[20]

Such a conceptualization allowed a different understanding of the role of wheat, olives, and viticulture in Cyprus or, more specifically, the different position held by the latter two in the island's economy. Viticulture, while profitable for external trade, was limited to the mountainous regions of Cyprus. Olives, on the other hand, were rarely exported and production was limited for internal consumption. All indications show that neither reached the volume or the overall impact witnessed in two of Cyprus' neighbours: Palestine in the east and Crete to the west.[21] These arguments were based not on large datasets, but on heterogeneous sources of economic history that offered qualitative rather than quantitative conclusions. What was necessary was a large dataset of quantitative data to confirm, revise, or refine the existing picture.

The source

The Ottomans' detailed fiscal register of 1572 for Cyprus provides information on the Cypriot rural landscape at a very specific moment; as the recording took place on the morrow of the conquest, there are very few Ottoman things contained within it. Or, more precisely, whatever was recorded reflected preceding Venetian conditions rather than subsequent Ottoman ones. From place names to district divisions, pre-Ottoman realities were left largely intact, as was standard Ottoman practice.[22] The biggest source of rupture was the war and its consequences for the population, local infrastructure, and the economy at large.

This particular source belongs to the *tahrir* category of Ottoman registers. They have been evaluated and analyzed as sources of Ottoman economic history since the 1930s and 1940s[23] and have been the subject of long and fruitful debates concerning their use, interpretation, and reliability.[24] Like any kind of fiscal data (whether modern or pre-modern), Ottoman state documentation is limited and does not accurately reflect the contemporary economy. Reconstructing trends and patterns in the economy from the assessments of state officials requires serious consideration of the challenges posed by data quality.

These registers were initially compiled upon the conquest of a new province and were then updated periodically, albeit inconsistently. Surveyors assessed each settlement's average production over the previous three years. While the Ottoman register did not precisely report production for 1572, the data it contains provide an approximate indication of the longer term, since any given year could be exceptionally good or bad. In this sense, this 'inaccuracy' can actually benefit present-day economic historians.[25]

While such registers can supply information on demography, taxable production, economy, toponymy, or onomastics, the reliability of the data for examinations of each of these elements can vary significantly. The information contained reflects the concerns and questions of the Ottoman fiscal bureaucracy. For example, male heads of households (*hane*) formed the largest category of taxpayers. As a category, households varied significantly in time and place. Moreover, they should not be assumed to simply refer to a nuclear family, since a household was a fiscal, and by implication, a productive unit. Other taxpayers included widows, bachelors, elders, and people with different kinds of disability. Given the heterogeneous nature of this data, it is extremely difficult to use it for the purposes of demography, despite previous attempts at calculating a multiplier pointing towards the total population.[26] Similarly, the villages and settlements listed are only those which paid taxes to the state; other land categories are not included, leaving significant gaps. Non-taxable products are also not registered. Consequently, it is easy to see how a portion of economic production (potentially a significant one, as we will see) can fall into oblivion.

Historians have argued for different ways of utilizing these documents. The positivism of early contributions was followed by varying degrees of scepticism in subsequent decades.[27] The current consensus is that, despite their limitations, the registers may provide a valid foundation for our knowledge of rural society in the period.[28] The challenge is how to best utilize this rich corpus of data.

Digital humanities tools offer several advantages here. Traditional statistical methods do not sufficiently illustrate the spatial dimensions of economic data. As the following analysis will show, HGIS tools open up possibilities and reveal patterns that would otherwise be impossible to identify or even imagine. On a different level, traditional publishing methods which present masses of data in table after table are both impractical as well as expensive. Making data available in digital rather than printed form makes doublechecking and further analysis much easier than manually copying data series from a book.[29] With these general observations in mind, what may the source at hand tell us about the economy of the region we are examining? The register lists a total of 1,158 taxable settlements organized by subdistrict. Of these, 969 are towns and villages for which a detailed account of the quantity and value of taxed products is provided. The tithe (in-kind taxation of agricultural production) in this case was 20 per cent; multiplying this by five gives us the total amount of the assessment. The remaining 189 entries comprise settlements that were taxed by lump sum in cash (city quarters, monasteries, farms, and other forms of land). For this reason, they are excluded from the present analysis given their lack of information on agricultural products (Table 14.1).

What do in-kind taxed settlement entries look like, and what kind of information do they contain? Table 14.1 shows the example of the village of Sotira in the subdistrict of Mesaoria (Tk. Mesarya). Like all such entries, it starts by listing the names of male heads of households, followed by other categories of taxpayers. The entry then lists the monetary taxes according to taxpayer category and records the amount and value of taxation per product or productive activity (such as watermills, taverns, tanneries, dye houses), followed by other fines and dues for which every settlement was responsible. In total, there were 51 categories of taxable products, fines, and dues in the Cyprus register. Finally, these entries record the total value of the revenue collected from each village.

Table 14.1 Sample entry: The village of Sotira in the Mesaoria (Tk. Mesarya) subdistrict

Category	Type of data	Unit of measurement	Entry value
Subdistrict	Toponym		Mesarya
Settlement type	Toponym category		Village
Settlement name	Toponym		Sotira
Households	Quantity		57
Bachelors	Quantity		10
Widows	Quantity		3
Blinds	Quantity		1
İspence tax	Monetary value	Akçe (Ottoman silver coin)	2,010
Widow tax	Monetary value	Akçe	18
Wheat	Quantity	Keyl (Ottoman weight measure)	75
Wheat	Monetary value	Akçe	900
Barley	Quantity	Keyl	22.5
Barley	Monetary value	Akçe	1,350
Vetch	Quantity	Keyl	15
Vetch	Monetary value	Akçe	90
Lentils	Quantity	Keyl	2
Lentils	Monetary value	Akçe	26
Broad beans	Quantity	Keyl	2.5
Broad beans	Monetary value	Akçe	25
Bee-hives tax	Monetary value	Akçe	50
Sheep tax	Monetary value	Akçe	1,000
Pigs fine	Monetary value	Akçe	85
Fruit orchards tax	Monetary value	Akçe	180
Tavern Tax	Monetary value	Akçe	300
Stray animals duty	Monetary value	Akçe	43
State taxes	Monetary value	Akçe	67
Charges and fines	Monetary value	Akçe	75
Total monetary value	Monetary value	Akçe	6,219

Source: *Kıbrıs Tahrir Defterleri*, p. 394.

Conspicuous absences

The construction of space is very much about exclusions. The representation of a spatial order offered by these sources determines which individual and collective agents are visible, invisible, or in the margins. In this case, the source at hand reproduces certain hierarchies in the composition of Ottoman society inasmuch as they are reflected in fiscal liability towards the state. What can we learn about how the Ottoman state constructed its taxpayers as economic actors? Where is the register silent and what are we to make of this?

As far as the source is concerned, women only existed under one condition: that they were non-Muslim widows. This is because a particular monetary tax (*ispence*) applied to non-Muslims, and therefore non-Muslim widows paid a widow tax that presumably came at a reduced rate. Widowers, however, were not recorded as a separate fiscal category – something that implies that they still counted as heads of a household if they had their own family. This fiscal/gender bias by the Ottoman state meant that not only were women stripped of any autonomous economic or productive capacity from a fiscal-administrative point of view, but that their recording in this historical source was contingent upon the life (or death) of their husband. Spinsterhood, unlike bachelorhood, was not recorded. In other words, women were not independent economic actors if these sources are taken at face value. Fortunately, other forms of Ottoman state documentation testify to the economic agency of women.[30]

By uncritically applying these sources, the early pioneers of Ottoman social and economic history failed to account for the absence of women as fiscal or economic actors.[31] Since widows, by definition, had suffered the loss of a husband and were not counted as heads of economically active households, they have been ignored in attempts to calculate population multipliers. Historians have therefore consciously, subconsciously, or unconsciously limited female economic and demographic agency and existence. It was not until the 1980s that this was remedied and that widows were finally counted as heads of an economically active household.[32]

Social hierarchies are also absent. Ottomanist historiography has been unable to document pre-modern class in the Marxist sense. This is not, however, to exclude social stratification, diverging interests, or socio-economic asymmetries. To some extent, this is because of the nature of the sources. The overwhelming majority were state-produced, and fiscal registers are quintessential examples of sources that reproduce official discourse. As such, the image they project is one of a social monolith of taxpayers. Far from the European socio-historical paradigm of institutionalized hierarchically organized groups, Ottoman political theory and practice divided subjects into the tax-paying *reaya* (literally, flock) and the tax-exempt *askeri* class of bureaucratic and military personnel. Taken at face value, the register presents a picture of heads of households who paid the same amount of monetary taxes and the same portion of in-kind taxes regardless of how much they produced or owned.

Beyond the mass of undifferentiated male heads of household, only bachelors, widows, and various categories of disabled citizens were recorded. Their lower fiscal burden reflected Sultanic magnanimity and mercifulness in line with state ideology and rhetoric. Devoid of evidence of social tensions and diffusing the fragmentation of economic power, the survey is unsurprisingly more lucid about Ottoman ideals and self-perceptions of social peace and justice than any other aspect of social organization. Interestingly, this image suits not only the Ottoman state but also post- (and usually anti-) Ottoman nationalist discourse; the body of the nation under Ottoman rule had a unified and homogenous character. In other words, even nationalist observers of the post-Ottoman lands failed to critically approach this aspect of the quantitative data, for it suited their national-historical vision of a cohesive community.

HGIS and the 1572 fiscal survey of Cyprus

The first step towards processing and analyzing the source was to create a database. Using a spreadsheet, settlements were entered in rows and columns that contained data on their fiscal burden: monetary taxes for different categories of taxpayers, followed by various taxable productive activities (agriculture, manufacturing, services), fines, and dues.

Two more columns were added adjacent to that listing village names according to the 1572 register. The first provided the present toponym, and the second the name as it appeared in the Venetian survey of 1565,[33] the last conducted under the pre-Ottoman regime. Such an exercise is useful for deciphering the rendering of place names in Ottoman Turkish, and for permitting certain (limited) comparisons between the two surveys.

The most vital piece of information for any GIS analysis is the geographic coordinates of any given place name (geo-locating).[34] Coordinates were entered on the basis of present-day geo-data. Of the 969 villages recorded in the register, 534 are extant. This was the first major hurdle encountered; how does one deal with the fact that nearly half of the villages cannot be geo-located? There are several ways of dealing with this problem.

The first and most obvious solution is to use historical maps to identify currently abandoned villages. One extremely detailed historical map of use here is Kitchener's 1885 trigonometrical survey of Cyprus.[35] Its geo-referencing is not only a major asset for researchers of the late-19th century, but also for those of much earlier periods.[36] For example, the map often records ruins, a valuable piece of information which, in combination with other clues, can help identify the location of abandoned villages. It is also possible to find some villages on early cadastral maps of Cyprus by looking for micro-toponyms in the vicinity of the abandoned settlements.

Another way to mitigate the challenge of a large portion of non-geo-referenced villages is to choose an appropriate method of spatial analysis to deal with the problem of missing points on the map. Inverted Distance Weighted (IDW) Interpolation allows this possibility.[37] The basic assumption behind this algorithmic tool is that points on the map close to each other behave similarly.[38] This method of calculation clusters points on the map (12 in this instance) and averages a numerical attribute (for example, wheat production) in relation to distance within a given neighbourhood. In the context of this case study, this assumption means that the agricultural production, economic activity, or geographical characteristics of villages in close proximity do not significantly diverge. In effect, it allows an estimate for our missing points by interpolating nearby points on the assumption that the missing value is close to the average of the region concerned. Put differently, since the visual representation aims to generalize the features of a region rather than specific points on the map, the characteristics of the regions are not expected to radically change whether the non-geo-referenced villages are included or not. It must be emphasized that IDW Interpolation is not the only method available for dealing with the challenges of the data, but it is an appropriate one given the challenges of the data. Figure 14.1 illustrates the geo-located villages, indicating a good overall distribution across the island, with the exceptions of the tip of the Karpasia (Tk. Karpas) peninsula in the northeast, some regions in the west, and the less accessible areas of the two mountain ranges.

Going back to the database, the more than 200 columns that recorded the amounts for different products, services, dues, or fines needed to be consolidated. This is because goods were sometimes charged at different rates (probably based on their quality) or because different divisions of measures and weights were used. This consolidation resulted in 51 compartmentalized categories, which were converted to the metric system, where possible.

The next problem was evaluating data quality and seeing whether statistical analysis can help interpret the data. How can we determine the accuracy and reliability of the source? As mentioned above, such surveys provide estimates of three-year averages of taxable production. John H. Alexander has suggested a method for evaluating whether the data was recorded in a prescriptive way by seeing if there is a strong statistical relationship between similar products such as wheat and barley. A strong relationship would suggest that what the surveyors recorded was based on a largely stable ratio between the two commodities. This in turn implies that the figures are not accurate on-the-spot production assessments, but rather approximations based on the surveyor's assumption of a fixed relationship between the two crops.[39]

The relationship between wheat and barley is established by the coefficient of determination (R^2) from a regression analysis, whereby the closer the coefficient

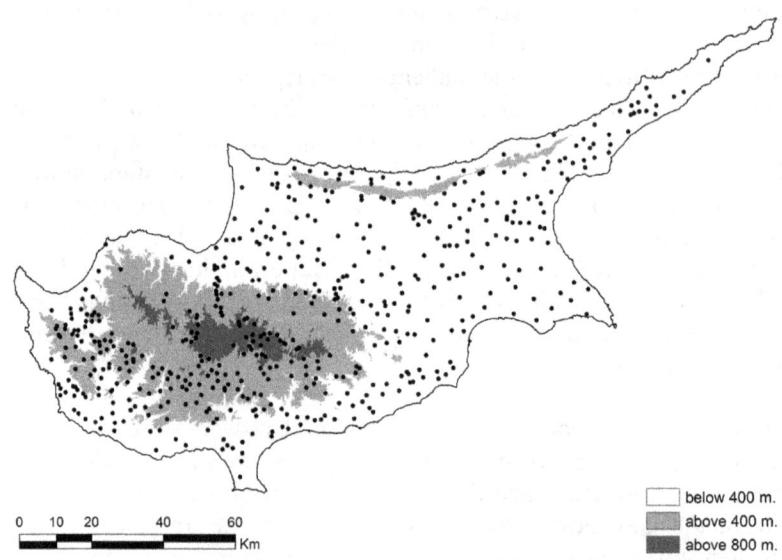

Figure 14.1 Digital Elevation Model (DEM) and distribution of extant villages.

Geodata provided by the Department of Geological Survey of Cyprus.

is to a value of one, the stronger the quantitative relationship.[40] In other words, the value of wheat recorded in any village will rise or fall according to the value of barley (or vice versa). In the case of the two registers examined by Alexander, these were 0.99 and 0.98. This essentially means that the ratio between wheat and barley was almost identical in all the villages – virtually a statistical impossibility unless it was deliberate. This does not necessarily mean, however, that the data is fictional or useless; Alexander interprets the result as an indication that this was a lump-sum-like method of tax calculation.

In the case of the Cyprus register, R^2 is 0.55. Such a result indicates a noticeably strong relationship, but one nowhere near the almost perfect match found by Alexander. The opposite would be more worrying; a value of R^2 close to zero would indicate no relationship, which would not make sense for these two crops. Wheat and barley are closely connected in terms of production, soil type, irrigation, labour, and suitability for human consumption. Randomness thus indicates the presence of arbitrary figures and unreliable data. The result for the Cyprus register (0.55) is encouraging in the sense that it falls between these two scenarios; the data are neither arbitrarily random nor prescribed.

One final remark on the process of data entry is in order. A great number of mathematical errors arose when the scribe was adding or multiplying the figures, a fairly common phenomenon in such sources.[41] In the cases where I detected errors, they were corrected before entry into the database as source fidelity was a lower priority than the re-creation of aspects of the economy.

Reading between the pixels

Every map is misleading to a lesser or a greater extent.[42] The spherical shape of the earth has to be projected in order to be converted into a flat surface. There are different projection methods, each of which carries certain advantages over others. Nonetheless, any projection will depict something accurately at the expense of distorting something else. Beyond the technical side of things, there are also important political implications (and motivations) in the cartographic choices one makes. When the historian employs the tools of cartography, it is helpful to think of the cartographic representation of data along similar lines. How is one to read between the pixels of the digital map?

I have decided to use interpolation (see above) as it permits the visualization of trends, while at the same mitigating the loss of data incurred from non-geo-referenced villages. Employing this method when dealing with an island province carries a major advantage: as pre-modern provincial,[43] district, or even state borders were malleable and fluid, and thus difficult to define, using HGIS for an island entails a neatly defined geographical space, within which the data can be easily depicted. It is worth noting, however, that the cartographic coastline does not exist in reality. This is for two reasons: (a) because of the constant ebbs and flows of the sea and the changing tides; and (b) because of the coastline paradox, i.e. the cartographic condition which changes the length of a coastline depending on the map's scale and/or step size.[44] As a result, there are not one but many measurements of a coastline.

Be that as it may, the cartography of islands creates a neat and defined border that greatly facilitates both analysis and representation.

The main limitation of IDW Interpolation is that it does not account for topography, since it assumes that the land it represents is flat. What this method depicts is the concentration and distribution of data (in our case, specific agricultural products or taxes) in clusters of villages. Given the limitations of the data and the density of the points on the map, our aim is to establish broad trends and patterns.

In what follows, I shall discuss a small selection of maps that show the spatial distribution of certain goods and commodities calculated by means of IDW Interpolation. Figure 14.2 shows the distribution of wheat, which appears ubiquitous. 86 per cent of all villages (including those not geo-located) were wheat producers. Another striking feature is the extremely high concentration in the north-eastern tip of the island. Going back to Figure 14.1, which shows all geo-located villages, we notice that the last village on that peninsula is situated some distance from the easternmost tip. As there is no other point in that whole region, IDW Interpolation assumes that the area is homogenous and covers it with the values of that last village. As explained above, however, this does not mean that wheat was cultivated across that whole area. Unless one is aware of this detail, this can be a misleading representation. While there are ways of mitigating this problem, these are beyond the scope of this chapter.

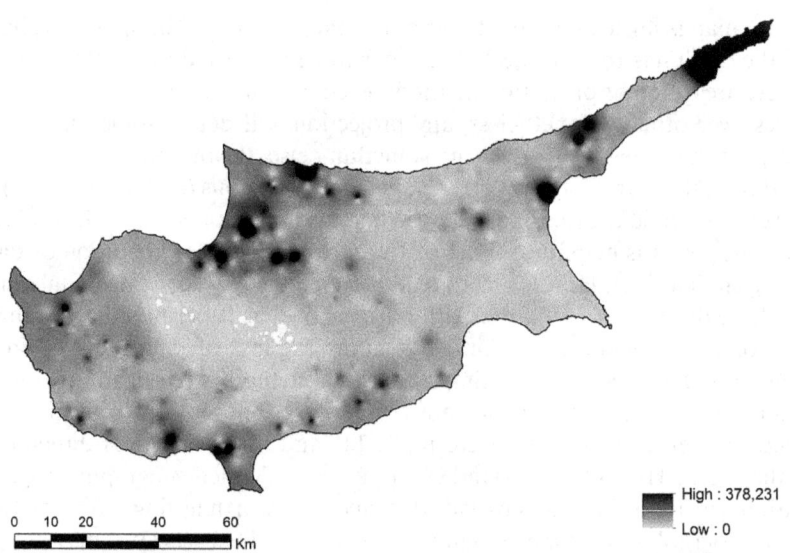

Figure 14.2 Heatmap showing spatial distribution of wheat production.

Geodata provided by the Department of Geological Survey of Cyprus. Source: *Kıbrıs Tahrir Defterleri*, pp. 36–390.

Also noticeable is the surprisingly low concentration of wheat, and indeed any other product, in the Mesaoria (Tk. Mesarya) plain. Stretching across the western and eastern coasts of the island between the two mountain ranges, Mesaoria has historically been a very fertile region and the breadbasket of the island. One possible explanation for this discrepancy is the fact that most of the fighting of the War of Cyprus (1570–71) took place during the sieges of Nicosia, at the centre of the Mesaoria plain, and Famagusta, at its eastern end. After two years of warfare, this region was particularly hit in comparison with parts of the island that did not witness fighting, severe casualties, requisitioned provisions, plundering, or destruction.

Finally, it is important to note that such observations become apparent *because* of this visualisation. Glancing at the rows and columns of a table may provide hints, but it is only through cartographically representing the relevant data that a much fuller spatial understanding of the relationship between agricultural production and geography may be acquired.

What are the implications of these observations for larger debates about the Mediterranean? It is worth returning to the Braudelian trinity of wheat, olive trees, and vines. While the 1572 register indicates the omnipresence of wheat, thus confirming Braudel, the status of the other two crops is more complex. The quantitative data available here shows the limited presence of vines in Cyprus. Figure 14.3 provides further evidence of the high concentration of vines in the south and southwestern parts of the Troodos mountains.

Figure 14.4 depicts the distribution of the Tavern Tax, indicating the difference between production and consumption in Cyprus. The former seems like an inverted image of the latter (see Figure 14.3). The distribution of olive production also presents an interesting picture; as Figure 14.5 demonstrates, olive trees were fairly widespread. However, other sources on the rural history of Cyprus indicate that olive exports were very limited, and that production was largely internally consumed. Indeed, the heat map reveals that the distribution is rather uniform at low levels of concentrations, but for certain notable exceptions at the central axis of the island. Be that as it may, the scale of olive production can only be understood when compared with quantitative data from major olive producers such as Crete or Palestine, a task that remains to be completed. Figure 14.6, showing the amount of tax paid for olive oil, further complicates the picture. While Figure 14.5 shows a widespread presence of olive production, its conversion into oil (Figure 14.6) appears to be limited to a single, small region. What is more, the comparison of the two maps reveals another inconsistency: the one region where olive oil production is recorded lies west of the one with the highest concentration of olive production. One consideration is the fact that wild olive trees, the crops of which were poor for oil production, were much more prolific in Cyprus.[45] Yet, this factor alone cannot explain this conundrum, and one may speculate that olive oil was not recorded by surveyors in other regions, perhaps on account of its limited volume.

Revising Braudel, Faruk Tabak proposed an additional categorization of crops that acquired important positions in the mediaeval and early modern Mediterranean. Classifying them as oriental crops, he discussed the role of

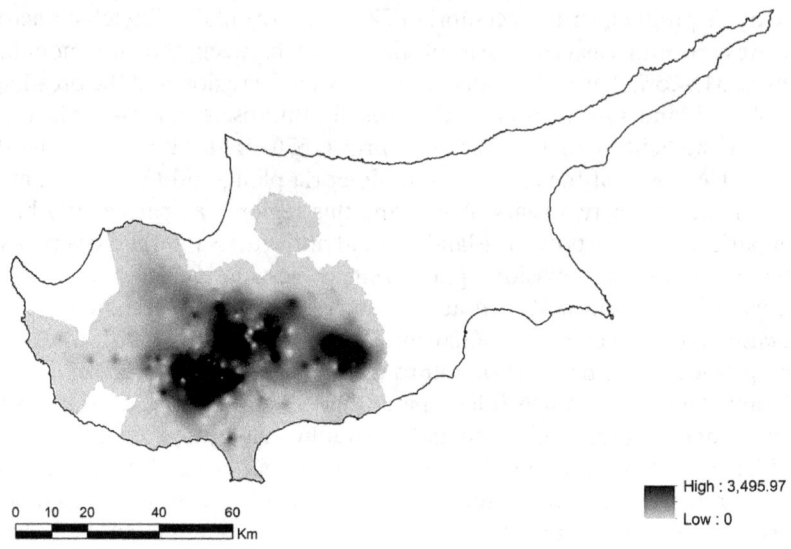

Figure 14.3 Heatmap showing spatial distribution of wine production on the basis of the grape juice tax.

Geodata provided by the Department of Geological Survey of Cyprus. Source: *ibid.*

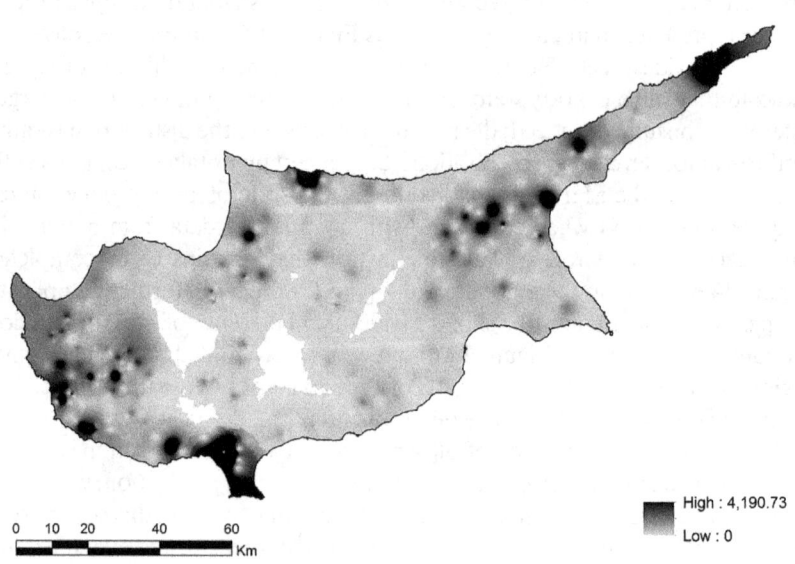

Figure 14.4 Heatmap showing spatial distribution of wine consumption of the basis of the tavern tax.

Geodata provided by the Department of Geological Survey of Cyprus. Source: *ibid.*

Economic geographies 265

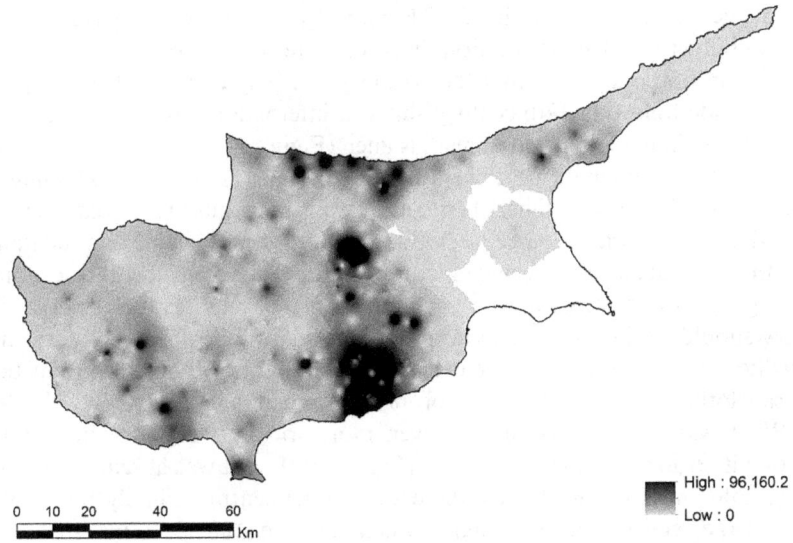

Figure 14.5 Heatmap showing spatial distribution of olive production.
Geodata provided by the Department of Geological Survey of Cyprus. Source: *ibid*.

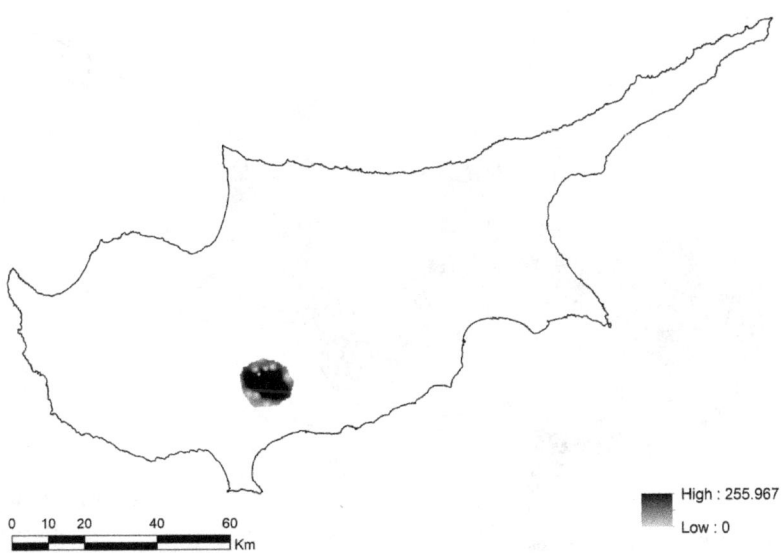

Figure 14.6 Heatmap showing spatial distribution of olive oil production.
Geodata provided by the Department of Geological Survey of Cyprus. Source: *ibid*.

cotton plants, sugarcane, and mulberry trees (for sericulture). What was the status of these crops in Cyprus in 1572? Figure 14.7 reveals a wide distribution for cotton but with high concentrations in specific areas. The more uniform overall low distribution confirms that the cotton plant only conquered the Cypriot countryside following 17th-century shifts in international trade. Closely connected, and in fact more widespread, is linen (Figure 14.8), with a large area of high concentration in the northwest. Linen is perhaps an unexpected finding of the project. This was a time when cotton was still acquiring a hold over the island, and linen was a source of clothing fibre, also providing the foundations for the subsequent cotton manufacturing infrastructure in the 18th century. This generates an important research question that requires further research: How should we integrate linen into the larger narrative of both Cypriot and Mediterranean history? To what extent was it used for clothing, the production of sailcloth, or to satisfy the needs of internal or external demand?

What was true for cotton was even more so for silk. Sericulture had a strong if regionally limited presence (Figure 14.9), but was absent in most of the island and not consolidated there for another century. Finally, the biggest gap in the register concerns sugar, which is entirely absent. This is because sugar became a state monopoly upon the Ottoman conquest and was therefore not taxed. Sugar, however, was a commodity produced exclusively for the external market. The Cypriot peasants and workers who toiled to produce it never had access to it. Honey and carob syrup were the preferred sweetener, or what may be called 'the poor person's sugar' (Figures 14.10 and 14.11).

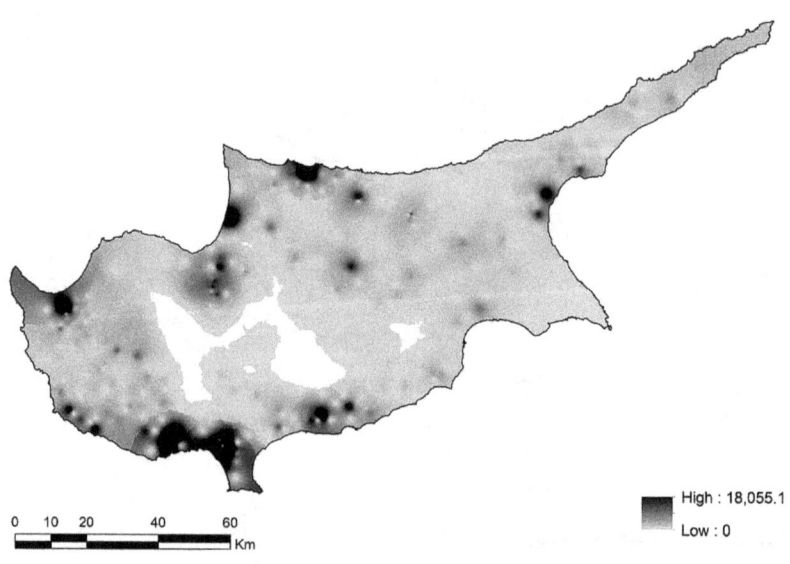

Figure 14.7 Heatmap showing spatial distribution of cotton production.
Geodata provided by the Department of Geological Survey of Cyprus. Source: *ibid*.

Economic geographies 267

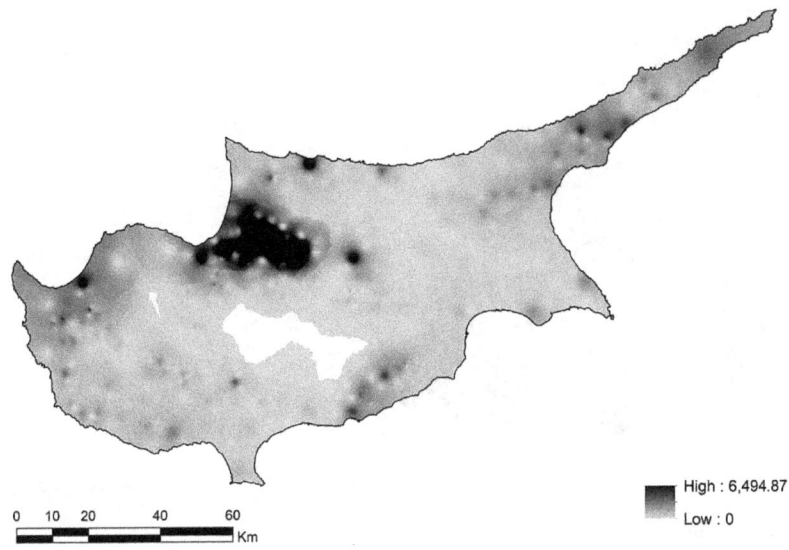

Figure 14.8 Heatmap showing spatial distribution of linen production.
Geodata provided by the Department of Geological Survey of Cyprus. Source: *ibid.*

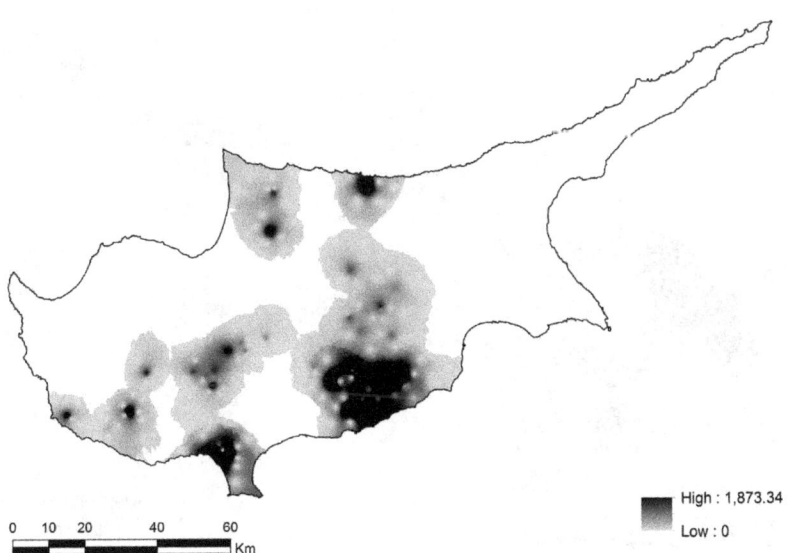

Figure 14.9 Heatmap showing spatial distribution of silk production.
Geodata provided by the Department of Geological Survey of Cyprus. Source: *ibid.*

268 *Antonis Hadjikyriacou*

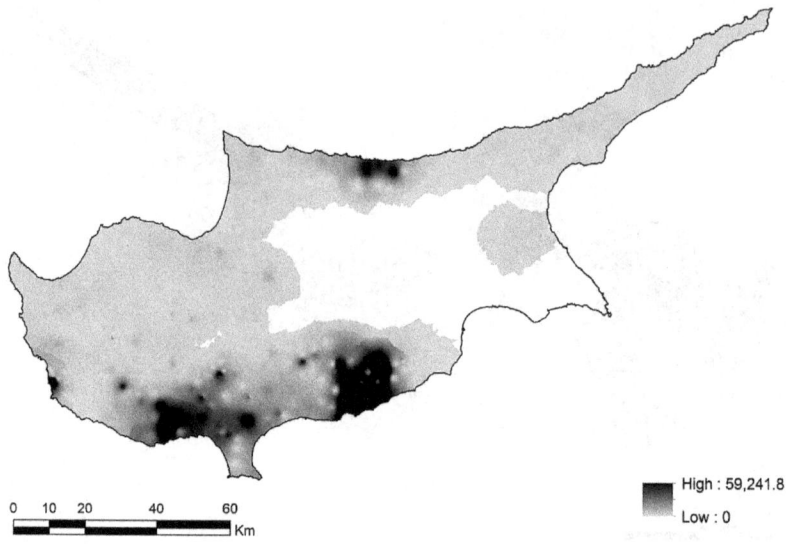

Figure 14.10 Heatmap showing spatial distribution of honey production.
Geodata provided by the Department of Geological Survey of Cyprus. Source: *ibid*.

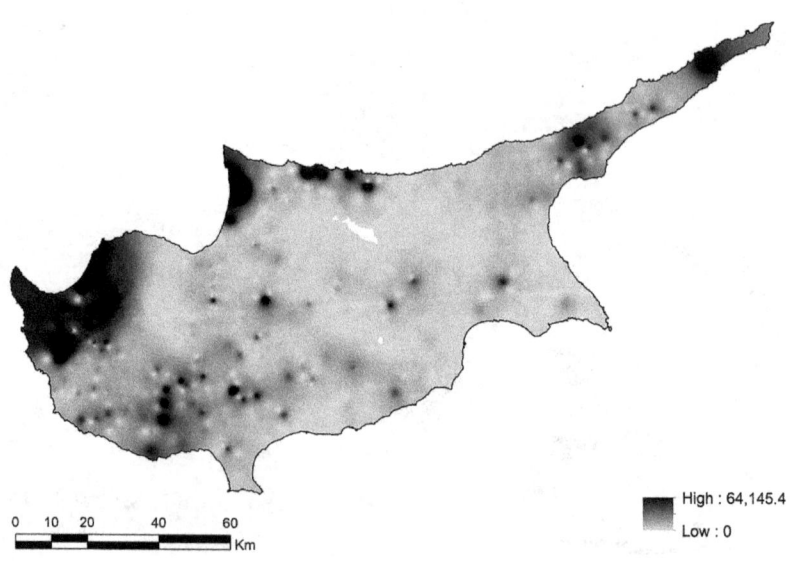

Figure 14.11 Heatmap showing spatial distribution of carob production.
Geodata provided by the Department of Geological Survey of Cyprus. Source: *ibid*.

Further questions: from economic and environmental to spatial history via HGIS

The present chapter is inspired by current historiographical and methodological debates in economic and environmental history and seeks to connect them to the recent development of spatial history. The renewed interest in material factors in historical processes fits well with the data-driven methodologies of GIS. In this context, this chapter has demonstrated ways of using HGIS to visualize patterns of economic production based on pre-modern fiscal data. Such an endeavour is fraught with problems, pitfalls, and limitations. The challenge is to mitigate the limited accuracy and reliability of data extracted from Ottoman fiscal surveys. The visualization attempted here is cartographic and a major caveat concerns the canonical and authoritative status that maps have as a visual source; they are often perceived as accurate depictions of reality. Just like any other visual representation, however, they are laden with choices and omissions that carry significant biases. Any analytical approach addressing existing questions must be aware of these limitations, but the pay-out lies in the heuristic value of this exercise, generating new questions.

In this case, the project considers Mediterranean-wide historical processes at the material level of the economy and society, namely, the movement of crops and shifts in patterns of economic activity. A number of well-established arguments were confirmed or refined. For example, this investigation confirms the statement that cotton already had an established presence in the 16th century, and by consulting other available data we can deduce that cotton cultivation expanded in subsequent centuries.[46] The quantitative data analyzed here lends weight to the argument presented by Tabak on the Mediterranean-wide trends in cotton cultivation. In other cases, we have a more refined idea of the importance of viticulture. It shows that Braudel's argument for the omnipresence of vines across the Mediterranean must be regionally qualified. HGIS analysis reveals the high concentration of viticulture only in a particular region of Cyprus. Though remuneratively important thanks to international trade, vines had a limited presence in the local landscape. New questions also emerged on the prolific nature of linen, which demand further research, particularly concerning its role in the agriculture, manufacturing, and trade of the island and the wider Mediterranean. These conclusions, nevertheless, must be qualified by the fact that the picture presented here is only a snapshot of a very specific and extraordinary moment in the history of the Cypriot rural landscape.

The starting point of this inquiry was a conceptual engagement with the notion of insularity, and more specifically the cultural construction of Cypriot insular space by different individual and collective actors. Cultural aspects of insularity based on qualitative impressions of a diverse set of sporadic data are insufficient if one wishes to draw more solid conclusions or produce further questions that go beyond the shores of Cyprus and extend to larger spatial contexts such as the Mediterranean. Such a broadening and deepening of the research agenda needs to pay closer attention to economic and environmental dimensions and the materiality of spatial categories. HGIS and digital mapping go far beyond the quantitative processing of large datasets while adding the spatial dimension to the economic or environmental transforms analytical and heuristic possibilities. Situated at the nexus between economic, environmental, and spatial history, HGIS offers unprecedented levels of complex analysis and calculations, stretching the horizon for historical research through the medium of spatial history. Such a mode of inquiry can further elucidate what Lucien Febvre called 'the vicissitudes of possibility' when he discussed the relationship between geography, human agency, and historical processes several decades before the advent of spatial history.[47]

I opened this essay with an episode from transition from the quantitative-driven methods of economic history to the 'cultural turn'. Writing at that precise juncture in 1980, and after the 'new economic history' had reached its peak, Eric Hobsbawm published a rebuff to Lawrence Stone's dismissal of quantitative data.[48] In the same year, Hobsbawm made a case for the heuristic value of economic theory.[49] Many of those arguments, questions, and concerns remain highly relevant today. Four decades later, historiography has come full circle, and spatial history offers improved data-driven methods that can be synthetically and harmoniously combined with cultural-critique informed approaches. The task before us is to devise and develop new tools that aid in the reconsideration of old questions and the generation of new ones. Spatial history has the potential to do just that.

Acknowledgement

Research for this essay was funded by the Sylvia Ioannou Foundation ('Economy, environment and landscape in the Cypriot *longue durée*' project, hosted at Harokopio University, Athens, 2018–21), and the European Commission's 7th Framework Programme Marie Skłodowska-Curie Actions ('MedIns: Mediterranean Insularities' project, reference ID: 630030, hosted at the Institute for Mediterranean Studies, Foundation for Research and Technology-Hellas in Rethymno, Greece, 2014-2016). I am grateful to Amy Singer, Christos Chalkias, Ece Turnator, Athos Agapiou, Harry Paraskevas, and Tuna Kalayıcı for their helpful suggestions and support during the process of writing this essay.

Notes

1 Emmanuel Le Roy Ladurie, *The Territory of the Historian*, trans. Ben and Sian Reynolds, Hassocks, Sussex: The Harvester Press, 1979, p. 6.
2 Lawrence Stone, 'The Revival of Narrative: Reflections on a New Old History', *Past & Present* 85, 1979, 13.
3 Geoff Eley, *A Crooked Line: From Cultural History to the History of Society*, Ann Arbor: The University of Michigan Press, 2005, p. xii.
4 William H. Sewell Jr, *Logics of History: Social Theory and Social Transformation*, Chicago and London: University of Chicago Press, 2005, pp. 22–80.
5 See, for example, Martin W. Lewis and Kären Wigen, *The Myth of Continents: A Critique of Metageography*, Berkeley: University of California Press, 1997; Kären

Wigen, *A Malleable Map: Geographies of Restoration in Central Japan, 1600–1912*, Berkeley: University of California Press, 2010.
6 Matthias Middell and Katja Naumann, 'Global History and the Spatial Turn: From the Impact of Area Studies to the Study of Critical Junctures of Globalization', *Journal of Global History* 5, 2010, 149–70; see also the special issue 'Spatializing Transnational History: European Spaces and Territories', *European Review of History* 25, 2018, no. 3/4.
7 Christopher A. Bayly et al., 'AHR Conversation: On Transnational History', *American Historical Review* 111, 2006, 1441–64.
8 Andrew C. Isenberg, 'Introduction: A New Environmental History', in id. (ed.), *The Oxford Handbook of Environmental History*, Oxford: Oxford University Press, 2014, pp. 1–20; Christian Pfister, Sam White and Franz Mauelshagen, 'General Introduction: Weather, Climate, and Human History', in id. (eds), *The Palgrave Handbook of Climate History*, London: Palgrave Macmillan, 2018, pp. 1–16.
9 Julia Adeney Thomas, 'AHR Forum Comment: Not Yet Far Enough', *American Historical Review* 117, 2012, 794–803, here 802.
10 Richard White, 'What Is Spatial History?', *The Spatial History Project*, February 2010. Available HTTP: https://web.stanford.edu/group/spatialhistory/cgi-bin/site/pub.php?id=29 (21 August 2018); Ian Gregory, Don DeBats and Don Lafreniere (eds.), *The Routledge Companion to Spatial History*, New York: Routledge, 2018; William Rankin, 'How the Visual Is Spatial: Contemporary Spatial History, Neo-Marxism, and the Ghost of Braudel', *History and Theory* 59, 2020, 311–42.
11 Matthew W. Wilson, *New Lines: Critical GIS and the Trouble of the Map*, Minneapolis: University of Minnesota Press, 2017; William Rankin, *After the Map: Cartography, Navigation, and the Transformation of Territory in the Twentieth Century*, Chicago and London: University of Chicago Press, 2016.
12 Angelo Torre, 'Un "tournant spatial" en histoire? Paysages, regards, ressources', *Annales: Histoire, Sciences Sociales* 63, 2008, 1127–44; see also the two *American Historical Review* forums 'Historiographic "Turns" in Critical Perspective', 117, 2012, 698–813, and 'Geoff Eley's *A Crooked Line*', 113, 2008, 391–437.
13 Thomas, 'AHR Forum', 803.
14 The register is kept at the Land Register and Cadastre Archive (Tapu ve Kadastro Arşivi) in Ankara, Turkey, T.T. 64, Kıbrıs Mufassal Defteri. It was published in facsimile as *Kıbrıs Tahrir Defterleri: Mufassal, İcmal ve Derdest. Tıpkıbasım*, Ankara: T.C. Çevre ve Şehirlik Bakanlığı, Tapu ve Kadastro Genel Müdürlüğü, Arşiv Dairesi Başkanlığı, 2013, pp. 36–490. It has been partially statistically analyzed by Ronald C. Jennings who selected a sample of the highest-taxed villages in his *Village Life in Cyprus at the Time of the Ottoman Conquest* (ed. Akif Erdoğru and Ali Efdal Özkul), Istanbul: The Isis Press, 2009. Vera Costantini statistically analyzed the register in her *Il sultano e l'isola contesa. Cipro tra eredità veneziana e potere ottomano*, Torino: UTET, 2009. I am unable to consult this work due to language limitations.
15 Antonis Hadjikyriacou, 'Society and Economy on an Ottoman Island: Cyprus in the Eighteenth Century', PhD thesis, School of Oriental and African Studies, University of London, 2011.
16 Fernand Braudel, *The Mediterranean and the Mediterranean World in the Age of Philip II*, trans. Siân Reynolds, vol. 1, London and New York: Fontana, 1972, p. 150.
17 Antonis Hadjikyriacou, 'Envisioning Insularity in the Ottoman World', in id. (ed.), *Insularity in the Ottoman World*, special issue of *Princeton Papers* 17, 2017, vii–xx;

id., 'The Respatialization of Cypriot Insularity during the Age of Revolutions', in Megan Maruschke and Matthias Middell (eds.), *The French Revolution as a Moment of Respatialization*, Berlin: De Gruyter, 2019, pp. 149–65.
18 Faruk Tabak, *The Waning of the Mediterranean, 1550–1870: A Geohistorical Approach*, Baltimore: Johns Hopkins University Press, 2008.
19 Braudel, *Mediterranean*, vol. 1.
20 Tabak, *The Waning of the Mediterranean*, pp. 89–133.
21 Hadjikyriacou, 'Society and Economy', pp. 32–49.
22 Halil Inalcik, 'Ottoman Methods of Conquest', *Studia Islamica* 2, 1954, 103–29.
23 For a broad overview of the development of the study of these sources see Suraiya Faroqhi, *Approaching Ottoman History: An Introduction to the Sources*, Cambridge: Cambridge University Press, 1999, pp. 97–101; Nikolay Antov, *The Ottoman 'Wild West': The Balkan Frontier in the Fifteenth and Sixteenth Centuries*, Cambridge: Cambridge University Press, 2017, pp. 10–11.
24 The literature is quite extensive. See indicatively Amy Singer, *Palestinian Peasants and Ottoman Officials: Rural Administration around Sixteenth-Century Jerusalem*, Cambridge: Cambridge University Press, 1994, 17–19. For the latest summary of approaches see Nora Lafi, 'Organizing Coexistence in Early Ottoman Aleppo: An Interpretation of the 1518, 1526 and 1536 Tahrîr Defteris and the 1536 Qanunname', in Hidemitsu Kuroki (ed.), *Human Mobility and Multiethnic Coexistence in Middle Eastern Urban Societies*, vol. 2, Tokyo: ILCAA, 2018, pp. 103–20.
25 Metin M. Cosgel, 'Ottoman Tax Registers (Tahrir Defterleri)', *Historical Methods* 37/2, 2004, 87–100, here 89.
26 Bekir Kemal Ataman, 'Ottoman Demographic History (14th–17th Centuries): Some Considerations', *Journal of the Economic and Social History of the Orient* 35, 1992, 187–98.
27 Lafi, 'Organizing Coexistence', p. 104.
28 Singer, *Palestinian Peasants*, pp. 17–19.
29 Antonis Hadjikyriacou and Elias Kolovos, 'Rural Economies and Digital Humanities: Prospects and Challenges', in Elias Kolovos (ed.), *Ottoman Rural Societies and Economies: Halcyon Days in Crete VIII*, Rethymno: Crete University Press, 2015, pp. 415–21, here p. 420.
30 Madeline C. Zilfi, 'Muslim Women in the Early Modern Era', in Suraiya N. Faroqhi (ed.), *The Cambridge History of Turkey*, vol. 3: *The Later Ottoman Empire, 1603–1839*, Cambridge: Cambridge University Press, 2006, pp. 226–55, here pp. 238–42; Ronald C. Jennings, 'Women in Early 17th Century Ottoman Judicial Records: The Sharia Court of Anatolian Kayseri', *Journal of the Economic and Social History of the Orient* 18, 1975, 53–114.
31 A notable exception is Hedda Reindl-Kiel, 'A Woman *timar* Holder in Ankara Province during the Second Half of the 16th Century', *Journal of the Economic and Social History of the Orient* 40, 1997, 207–38.
32 Heath W. Lowry, *Trabzon şehrinin İslamlaşma ve Türkleşmesi 1461–1583*, Istanbul: Boğaziçi Üniversitesi Yayınevi, 1981, pp. 186–7.
33 Gilles Grivaud, *Villages désertés à Chypre (Fin XIIe-Fin XIXe siècle)*, Nicosia: Idryma Archiepiskopou Makariou G', 1998.
34 See Phokion Kotzageorgis, 'Ottoman Tax Registers and Geo-History', *Historicogeographica* 16–17, 2018, 1–14.
35 Christos Chalkias, Christoforos Bradis and Athanasios Psaroyiannis, *Κατασκευή βάσης χαρτογραφικών δεδομένων και διαδικτυακής εφαρμογής παρουσίασης των*

γεωγραφικών πληροφοριών του Χάρτη του Kitchener για τη νήσο Κύπρο (έκδοσης 1885)/Creation of the Cartographic Database of a Web Application Presenting the Geograpfical Information of Kitchener's Map of the Island of Cyprus (Edition of 1885), Athens: Harokopio Panepistimio and Sylvia Ioannou Foundation, 2017.
36 The map was digitized by the Sylvia Ioannou Foundation in collaboration with Harokopio University, Athens. The web application is available at https://geoprojects.hua.gr/kitchener/ (accessed 8 September 2018).
37 For an important critique of this method and the measures proposed to mitigate its problems see William Rankin, 'The Accuracy Trap: The Values and Meaning of Algorithmic Mapping, from Mineral Extraction to Climate Change', *Environment and History*, 2020. Available HTTP: https://doi.org/10.3197/0967340 20X15900760737275 (accessed 9 May 2021).
38 Jochen Albrecht, *Key Concepts and Techniques in GIS*, Los Angeles: Sage, 2007, pp. 65–6.
39 John H. Alexander, 'Counting the Grains: Conceptual and Methodological Issues in Reading the Ottoman *Mufassal Tahrir Defters*', *Arab Historical Review for Ottoman Studies* 19–20, 1999, 55–70.
40 Pat Hudson and Mina Ishizu, *History by Numbers: An Introduction to Quantitative Approaches*, 2nd ed., London: Bloomsbury, 2017, pp. 184–9.
41 Heath W. Lowry, 'The Ottoman *Tahrîr Defterleri* as a Source for Social and Economic History: Pitfalls and Limitations', in id. (ed.), *Studies in Defterology: Ottoman Society in the Fifteenth and Sixteenth Centuries*, Istanbul: The Isis Press, 1992, pp. 3–18, here p. 10.
42 J.B. Harley, 'Deconstructing the Map', *Cartographica* 26/2, 1989, 1–20.
43 See, however, an interesting approach in Luca Scholz, 'Deceptive Contiguity: The Polygon in Spatial History', *Cartographica* 54, 2019, 206–16.
44 L.F. Richardson, 'The Problem of Contiguity: An Appendix to Statistics of Deadly Quarrels', *General System Yearbook* 6, 1961, 139–87.
45 See the 1833 land and property survey in the Ottoman Archive (Osmanlı Arşivi, OA), Istanbul, Turkey, ML.VRD.TMT.d. 16152–16155.
46 Hadjikyriacou, 'Society and Economy', pp. 33–72.
47 Lucien Febvre, in collaboration with Lionel Bataillon, *A Geographical Introduction to History*, trans. E. G. Mountford and J. H. Paxton, London: Kegan Paul, 1925, p. 172.
48 Eric J. Hobsbawm, 'The Revival of Narrative: Some Comments', *Past & Present* 86, 1980, 3–8.
49 Eric J. Hobsbawm, 'Historians and Economists: II', in id. (ed.), *On History*, London: Weidenfeld & Nicolson, 1997, pp. 109–23, here p. 118.

15 Digital mapping

Tim Cole and Alberto Giordano

Writing spatial histories does not necessarily involve either using, or making, maps. Spatial history can mean working with spatial concepts to frame analysis, which is then disseminated through the dominant medium used by historians – the written text.[1] However, thinking spatially about the past does suggest the value of maps and mapping as source, method, or output. In this chapter we do not focus on using historical maps as sources – something we have done elsewhere – but rather on how maps created from a variety of historical sources can play an integral part in the research process as tools of analysis.[2] In order to illustrate the potential specifically of digital mapping for spatial history, we draw on the experience of being members of an interdisciplinary team that made use of Geographical Information Science (GIS) methods to examine the spatialities of the Holocaust.[3] Alongside assessing the value of digital mapping as a way to visualize spatial patterns in historical experiences – focussing here in particular on the example of mapping Jewish ghettoization in wartime Budapest – we also aim in this chapter to critically reflect on the limits of this approach.

Digital mapping and the geovisualization and analysis of data have become possible with advances in computer technologies, the rise and development of Geographical Information Science, new spatial analytical methods and developments in spatial statistics, and a reconceptualization of what a map is.[4] In brief, GIS is a system for capturing (including *from* a map), storing, displaying, analyzing, and representing (including *with* a map) geographic information. Different tasks are accomplished in a GIS using different toolsets. Thus, for example, data are organized, stored, and retrieved using database technologies, principles, and designs, while spatio-temporal patterns are unearthed using quantitative methods and techniques, including spatial analysis and spatial statistics. It is the integration and almost seamless passage between one set of tools and another, from one method and technique to another, that makes GIS a powerful tool within a broader range of digital humanities approaches that draw on the processing power of computer technologies to read the archive differently.[5]

As this brief description suggests, one important distinction between a traditional paper map and a digital map is the separation of data from their representation. While in a paper (analogue) map, data and visual

DOI: 10.4324/9780429291739-19

representation are one and the same, in digital mapping the two are distinct, with data organized in a database and recalled, visualized, and analyzed as desired by the map creator. This means that a map on a screen can be easily manipulated at the level of a single data piece, while a printed map needs to be redrawn in its entirety every time a single change (for example a boundary change) is necessary. The separation of data and representation – and the capability for analyzing geographic data looking for spatial patterns – are the main features of a Geographic Information System, or GIS. As with a host of other digital humanities methods, the database is critical to digital mapping.

Building a database, and specifically the type of database we are dealing with here that includes spatial and temporal attributes, requires a deep understanding of how the data were created, by whom and for what purpose. It needs to consider to what degree the database is complete (what pieces of information, if any, are missing or lost, and what is the implication of this incompleteness), accurate in its spatial, thematic, and temporal attributes, and consistent and comparable in the context of what we are aiming to do with it. This is because a map, whether it is created with a computer or not, is only as 'truthful' as the information used to create it, which means that the knowledge we acquire from it is only as reliable as its building blocks – that is the individual pieces of data used to make it. This is especially critical as we apply increasingly complex spatial analytical methods and tools to explore the mapped phenomenon's spatial patterns. The search for spatial patterns with formal quantitative methods requires a deep understanding of the characteristics of the data in the database. For example, there are many methods to measure spatial concentration or dispersion and they often give different results. It is up to the researcher to choose the method more appropriate for the data in the database and for the kind of questions they are interested in answering.[6]

As is clear, digital mapping does not involve simply visualizing the archive, but rather visualizing the database that is constructed and filled with geographical information gleaned from the archive. Although there is nothing new to the researcher playing this mediating role, what is striking about digital mapping projects is that this interpretive labour is oftentimes laid bare. Each step of the process is revealed because the claims of the map are so clearly tied to the structure and nature of the database. This could be critiqued as generating a tireless focus on methodology within work – drawing on scientific notions of replicability – but there is also a methodological transparency here which reveals each step of the process rather than indulging in the sleight of hand of moving seamlessly from source to argument in a carefully footnoted sentence.

But the act of database design and filling does more than insert a sense of methodological self-reflection on the historian or geographer. The act of constructing the database is part and parcel of the research process, and not merely a means to an end (the digital map).[7] Perhaps we can give an example from the work of some of our colleagues within the wider *Geographies of the Holocaust* project team. Here one group, led by historical geographer Anne Kelly Knowles and architectural historian Paul Jaskot, created a database of the opening and closing of SS concentration camps that would enable them to map the shifting SS camp system from 1933 to 1945. As with any historical research, what began as a simple question – where (and when) was the SS camp system? – became more complex through the process of research that assumed a particularly pointed form through database design and completion. As they reflected on this initial part of their research process, 'for the members of our team the process of establishing rules for database construction became an engrossing and intimate engagement with our source material and the dynamic history of the SS camps. We came to see the database not as a passive receptacle for data or a neutral tool for geovisualization and analysis, but as a work of interpretative scholarship', in particular when it came to identifying intended and actual labour performed by prisoners.[8]

Even with a question as seemingly straightforward as what Auschwitz was in definitional terms, the team were left with the problem of how to separate out the various meanings and functions of this place. As they reflected,

> while there is no question that it was a concentration camp, at various and overlapping times Auschwitz was also a death camp, a labour camp, a work-education camp, a penal camp, a prison, and a transit camp. The place called 'Auschwitz' included a historical town, a pre-war facility that became the basis for the first concentration camp at the site, and Birkenau and Monowitz, two other vast facilities with main-camp status.[9]

The complexity of this one site over both time and space proved a challenge in constructing something as seemingly simple as a database of where SS concentration camps were located, what their function was, and when they opened and closed. The complexity of this site over both time and space demands far more than simply a few columns in even the most complete and sophisticated database. Here, and this is something that we discuss in more detail in the close of this chapter, the complexity of historical reality clashed with the needs of digital mapping to work not simply with a single point on the map expressed in longitude and latitude but also with a set of values and variables identified in the database.[10]

Given this complexity, there is a need for sophisticated probing of what geographical data a set of sources contain that might be mapped (and what this might enable analytically) as well as the recognition of what cannot be mapped and how this can be worked with in a range of other ways that may well – as we suggest at the end – include digital and spatial methods beyond GIS. Moreover, the act of constructing and filling the database requires an attentiveness to the (partial, messy, and complex) sources that is more akin to close rather than distant reading which oftentimes stimulates asking new questions because the evidence does not neatly fit with the initial categories.[11] There is an irony to much digital humanities work in that it involves the slow and painstaking process of database construction and completion with the rapidity of computer analysis. But the slow and painstaking labour of database design and data entry can be a critical iterative and productive part of the research process.[12]

Although the construction and completion of the underlying database is both slow and challenging, it is critical to unlocking the potential afforded by digital mapping. Two important aspects of this potential can be illustrated with examples drawn from work done by the Holocaust Geographies Collaborative. Firstly, the separation of data and representation means that maps can be integrated and paired with other types of representations with the objectives of testing hypotheses, exploring datasets from multiple perspectives, identifying spatio-temporal patterns, and thereby raising and answering research questions. The interpretive value of maps can be greatly enhanced by this pairing, as illustrated by the map of the arrests of Italian Jews available on the website of the Spatial History Project at Stanford University, which allows the viewer to simultaneously map, chart, and explore patterns of arrests both spatially and temporally.[13] The potential for comparing data is a core possibility opened up by digital mapping which allows for the filtering of any number of layers overlaid on to the base map. As we will suggest more fully later, this layering process is less a final research product – the map that will end up in the finished book, article, or chapter – and more an iterative and active research process that allows for a hypothesis to be tested, or seemingly disparate sources and data to be brought together on the common base map to interrogate whether there are spatial patterns and congruences or differences worth further exploration. In this way, digital mapping is very much an act of experimentation.

Secondly, the union of computers and maps has opened up new possibilities for exploring narratives through dynamic mapping. A limitation of paper maps is that a single map is capable of representing simultaneously variations in space and theme, but not also in time (although that is what historical atlases do as serialized maps). Digital methods enable more dynamic, or animated, maps that permit the production, visualization, and exploration of as many intermediate states as needed and the insertion of chronological change into the map. This, of course, can only occur on a computer monitor. One example of a dynamic map is the mapping of Jews walking from ghetto houses to market halls in Budapest in 1944, something we return to below.[14] Here the map shows spatial changes as they could occur over a short period of time – in this case only a few hours when Jews were permitted to leave their homes for example to go shopping – over a network of city streets.

In order to explore the potential of digital mapping in greater detail we turn to our own work as part of the wider *Geographies of the Holocaust* project which involved constructing a historical GIS of the Budapest ghetto, using the most commonly adopted proprietorial software, Arc GIS. The process, common to all digital mapping projects, involved creating a base map of Budapest, and then adding map layers drawn from a database that featured geographical data extracted from a range of historical sources. Of particular interest to us were the various iterations of the multiple ghettos imagined and implemented in Budapest across 1944, as well as the places where Jews were permitted to go when they could leave the ghetto houses for a few hours each day. Alongside creating layers of these various sites, we created a layer of the historic road network of the city which enabled us to work with network analysis in what turned out to be highly productive ways. Rather than going through each element of analysis that we were able to run through the historical GIS, we will focus in on two different examples that illustrate the potential of digital mapping to generate new questions and avenues of analysis. They also, as we reflect in closing, point to the value of mixed methods that combine distant and close reading, and signal the limits – as well as possibilities – of digital mapping in driving forward the agenda of spatial history.

Digital mapping and spatial analysis: mapping the Budapest ghettos

It is perhaps worth very briefly sketching out the historical and historiographical context to the places – a series of Holocaust ghettos – that we analyzed through digital mapping. Although not enacted everywhere, in hundreds of towns and cities in occupied central and eastern Europe Jews were separated out from non-Jewish populations in ghettos. While ghettos varied from place to place in form, they broadly shared two key characteristics. Firstly, they involved the concentration of Jews within demarcated sites. Secondly, they entailed the physical segregation of Jews from non-Jews, generally through some kind of border.[15] Early historiographical debate focused on whether ghettos were intentionally set up in occupied Poland in 1940 as collecting points prior to deportation, but the important work of Christopher Browning has tended to lead to acknowledgement that far from being part of a Nazi master plan, ghettos were initially set up for different reasons in different places.[16]

The importance of local distinctiveness when thinking about ghettos carries over to the context of Hungary. In what has been dubbed the 'last chapter' of the Holocaust, Hungarian Jews were placed into ghettos in the aftermath of the German occupation of their somewhat reluctant ally in the spring of 1944.[17] National legislation authorized local officials in towns and cities across Hungary with a population of more than 10,000 to identify in their locale 'parts, or rather specified streets, or perhaps designated houses' where ghettos were to be established and Jews relocated.[18] This explicit creating of room for manoeuvre on the part of local authorities meant that ghettos in Hungary looked different from place to place. Nowhere was this more strikingly visible than in the capital city, Budapest, where a highly dispersed form of ghetto was ultimately adopted after weeks of planning – and wrangling – in June 1944. Here, on 22 June 1944, the city's close to 200,000 Jews were ordered to move into just under 2,000 separate apartment buildings dispersed across every district of the city, each one marked with a large yellow star.[19]

In earlier work, Tim Cole had made much of the changing shape of the ghetto in Budapest as it was first sketched out by local officials and then implemented through first one ghetto list issued on 16 June, and then another less than a week later. Ultimately the ghetto continued to change shape in minor ways across the summer and fall of 1944, before a radically different model of two ghettos – one a looser group of houses, the other a closed area around a few streets – was adopted in the winter of 1944. Emphasizing a history of change, Cole pointed to shifting concerns underlying the shifting shape of ghettoization, signalling in particular the different stress laid upon whether creating a ghetto was primarily about removing Jews from certain parts of the city (outside the ghetto), or placing Jews in certain parts of the city (within the ghetto): aspects that he dubbed ghettoization as creating spaces of Jewish absence and/or Jewish presence.[20]

Ideas of continuity and change are core to much historical work. Historians above any other discipline are attuned to pay attention to chronology and the ways that things change – or stay the same – across time. Spatial history, rather than eschewing chronology, adds another dimension, with a concern with thinking about continuity and change over time and space. Focussing on space does not mean that chronology disappears, but assumes heightened importance within the context of focussed attention being paid to one place and the ways that it changed – or did not change – over time (facilitated by the possibilities of animating digital maps to show change over time at a range of temporal resolutions). This was certainly true of our work on ghettoization in Budapest, where digital mapping of the ghetto in the city drew us into rethinking about how – and if – ghettoization as a process evolved across the course of 1944.

The changing shape of ghettoization across 1944 can be mapped on paper or digitally. In earlier work Cole had used paper maps to narrate a story of ghettoization in Budapest as one primarily of change. For example, he had mapped out the number and placing of ghetto houses on 16 June and then 22 June for one district, to show both the reduction in number and nature of houses making up the ghetto. Both did happen. There was a dramatic reduction in the total number of ghetto houses – particularly visible in outlying districts of the city where suburban villas were excluded from the ghetto.[21] However, digital mapping – drawing on multiple data – enabled more nuanced and sophisticated representation of the changing nature of ghettoization that pointed to a narrative of continuity.

Firstly, digital mapping meant that we could integrate a greater sense of the varied nature of the housing stock of the city into our mapping of the shifting ghetto. Budapest, like other central European cities, was characterized by varied housing stock that ranged from large multi-storey apartment buildings in the centre through to smaller single-family villa homes in the suburbs. However, in Cole's earlier analogue mapping of the ghetto, and in our initial digital mapping of the city, this variety was flattened by each individual address from our database being represented by a single, equally sized point on a map.[22] In order to rectify this, we worked with census data from 1941 to weigh the size of buildings by district and thus show the spatial variability of the phenomenon – something that would have been impossible without digital mapping. This changed the picture of the changing nature of ghettoization markedly. Although the number of buildings designated for Jewish use fell dramatically – from over 2,600 to just under 2,000 buildings – between 16 and 22 June, the weighted maps suggested that the changes were not as dramatic as unweighted visualizations suggested, with the reductions largely being confined to smaller buildings in the outlying suburbs rather than the larger apartment buildings in the central districts of the city (which saw additions, rather than cancellations).[23]

Rather than our simply eyeballing these weighted maps, GIS enabled the adoption of a range of spatial analytical tools to analyze the maps. Three such tools to measure the distributional characteristics of spatial data were mean centre analysis (that identifies the geographic centre of a distribution), directional distribution (that shows the overall directional pattern of the data), and standard distance (that measures the degree of concentration). We ran these techniques for a mapping of the ghetto weighted by building size on 16 June and 22 June respectively.[24] This showed a moderate shift in the concentration of the ghetto (not surprising given the reduction in the number of ghetto houses). However, what was more striking and surprising were the continuities between these two dates. The mean centre – or the centre of gravity – of ghettoization more or less stayed in exactly the same place, and the directional orientation of the ghetto also changed little. In order to further test this picture of the ghetto's centre of gravity hardly changing at all despite the seeming radical cutting of the number of ghetto houses from over 2,600 to just over 1,900 we used kernel density analysis. Rather than returning a single value – as for example mean centre does which identifies the one point which is the geographic centre of all the data – kernel density analysis returns a set of continuous values, and this returns a more detailed and nuanced 'hotspot map' of any phenomenon.[25] Running this more nuanced method returned very similar results. It seemed that, a small cluster of houses in the outlying suburban 14th district of the city aside, the centre of gravity of the ghetto across the city's 5th, 6th, 7th, and 8th districts did not change much from 16 June to 22 June.

Seeing these maps generated by a range of analytical methods came as something of a eureka moment. Seeing how maps generated by a range of analytical methods and a plurality of datasets all pointed to continuity rather than change, and reinforced the validity of findings obtained with a single analytical method. This contrasted with Cole's own sense from extensive work with the archival records from Budapest which led him to see this week in the middle of June 1944 as a critical moment when the ghetto changed shape dramatically on the basis not only of the different ghetto lists produced by city officials, but also by working with the hundreds of petitions sent by non-Jews and Jews in the aftermath of the initial ghetto list being issued, and the remaining evidence of a door-by-door rechecking of individual buildings.[26] There was, in short, plenty of activity in the middle of June by a lot of different people over the question of ghettoization in Budapest, and the archives contained the paper traces of all this activity which engaged with, and involved, city officials. Cole wondered if he had been seduced by the archive and its sheer quantity of paperwork in equating volume of activity over ghettoization with a story of radical change in the shape and nature of the ghetto. Traditional archival work suggested one thing, but digital mapping caused a need for rethinking.

Revisiting this week with a new set of methods and a different way of seeing that digital mapping affords, Cole was struck by spatial patterns that spoke of continuity rather than dramatic change. As a result, we needed to reframe our initial question. Rather than asking 'Why did ghettoization change in mid-June 1944?', we now asked 'Why did ghettoization stay the same in mid-June 1944?'. The answer, it seemed to us, lay in a persisting tendency across 1944 to 'bring the ghetto to the Jew', rather than 'the Jew to the ghetto', by locating the ghetto in those parts of the city where Jews traditionally lived in the largest numbers.[27] This appears to have been driven by pragmatic concerns with the costs to non-Jewish residents of ghettoization (who petitioned the authorities in their hundreds), with a desire to limit the number of forced relocations of non-Jews from those apartment buildings included in the ghetto.

As can be seen from this example, digital mapping, and the variety of spatial techniques that allow for interrogating the data, can be a critical part of research process in reframing questions. But digital mapping can also generate entirely new questions and thus provoke new analysis and understanding. An example comes from our decision to create the street network as one layer in the Historical GIS of the Budapest ghetto. This enabled us, for example, to quantify the proportion of the streets in the city that were 'ghetto streets' at any point during 1944.[28] But what was more striking as our analysis developed was how incorporating the street network into analysis was suggestive in reframing our understanding of ghettoization. Given the unusual nature of this highly dispersed ghetto that Jews could – and did – leave for a limited number of hours each day in order, for example, to go shopping, we became interested in mapping out distance within this ghetto that stretched across the entire city. Working with the layer of the street network and an average walking speed of 1.3 metres per second, we were able to calculate walking distances to and from a number of sites we knew to be critical in 1944.

Using network analysis we calculated the 'least cost' route between these sites and individual ghetto houses and then mapped out 30-minute and 60-minute walking distances for market halls, the Swedish legations, and functioning Jewish hospitals.[29] This allowed us to model the potential for interaction between those dubbed victims, bystanders, and perpetrators, as well as proximity to and distance from three critical resources – food, paperwork, and medical treatment/visiting family members in hospital. What was striking was how it became clear that within this dispersed ghetto individuals and families living in different places had greater or lesser access depending on where they lived. The growing sense of the importance of distance and proximity in understanding day-to-day experiences in this dispersed ghetto that this mapping revealed to us also came through a mapping out of the distances between ghetto houses. Taken together, these maps suggested a reconceptualization of the dispersed ghetto in Budapest as one that did not simply divide Jews from non-Jews (as is generally imagined by the concept of segregation) but also divided Jews from other Jews, as well as access to critical resources.[30] It ultimately led us to imagine 'invisible walls' within this dispersed ghetto that resulted from the temporal limits placed by a curfew that meant that Jews could only leave their homes for a few hours each day. Mapping out those 'invisible walls' of distance – and thinking about what the implications of these were on daily life for Jews in the summer and fall of 1944 – became a new focus in our work.

Before saying a little more about that, it is worth noting more about this experimental mapping which differs in important ways from the earlier example we gave. In the first example – examining continuity and change in the shape of the ghetto across 1944 – we had, in a sense, more to go with. In the database we had a complete list of the buildings identified for inclusion in the ghetto at the various moments of planning and implementing ghettoization across 1944. It was a relatively simple process to attach these places to the base map and then explore both the spatial patterns that emerge within any chronological period as well as – most critically – across chronological time periods. Throughout this process we had a fairly strong sense of certainty about the data we were working with. We were visualizing a database rather than the archive, but that database was populated from the series of plans and pronouncements emanating from city officials. We could say with confidence that this particular address was there on the 16 June ghetto list, but not there on the revised and definitive 22 June ghetto list. However, we were far less certain when we began mapping out the movement of Jews through the city. We did know where Jews came from (the ghetto houses) and where they were going (for example the Swedish legation buildings). The latter became particularly important in the summer and fall of 1944 as Jews looked for protective paperwork – but we did not know how they got from a to b. Here the digital map became a much more speculative model to experiment with ideas. Working out the 'least cost' route between a and b was something we could do across the model, as well as working with average walking speeds, but this left many unknowns. For example, did Jews use the most direct route to get to their destination to avoid the long lines that built up, or did they use side streets to avoid officials and non-Jews? In important ways, then, these later maps are not representations of historical realities but more experimental models within an interdisciplinary research process that explored whether ideas such as potential for interaction, as well as distance and proximity, might be relevant in thinking about Jewish experience in the dispersed ghetto. Exploring these ideas was not entirely random. It worked with the reality of a ghetto spread across the city as well as with the importance of time given the limits of the curfew, but it was still more a way to visualize ideas of movement rather than to seek to faithfully represent movement.

Pointing out these distinctions is not simply delving into methodological minutiae, but critical in the wider act of laying bare the construction of digital maps from the initial mediation that can be seen in database design or choosing map layers, through to explaining a set of assumptions – such as 'least cost' route or walking speed – that underlie the model. Doing so is a reminder of the geographer's perspective on maps as experimental and exploratory elements of research process rather than final research products. They are not so much the dissemination of research as a critical part and parcel of framing and shaping that research. It was through these experimental maps that we started to think about the dispersed ghetto as being divided by 'invisible walls' and so began to rethink the lived experience of segregation in the dispersed Budapest ghetto.

Again, as with the mapping of continuity and change, this process forced us to return to the archives with an attentiveness to a new set of questions. In particular, we turned to oral history interviews and memoirs with an ear and eye to how individuals and families got hold of food or Swedish paperwork in the ghetto or managed to visit family members in hospital. Here we became aware of the importance of staying put with a set of pre-existing social relationships and in particular the significance of non-Jewish neighbours in negotiating the city outside of the curfew.[31] What became noticeable were the variety of ways in which Jews (and non-Jews) could – and did – breach the 'invisible walls' within the dispersed ghetto that we were able to visualize through digital mapping, particularly through mobilizing pre-war social networks. The lived reality of daily life in the ghetto was, as we soon came to realize, far more complex than our model (as is the case with any model) suggested.

Spatial visualization

This sense of the limits of digital mapping framed around the tool of GIS was one that emerged across the project team, in particular in the strand that dealt most closely with oral history sources. In their study of the evacuations on foot from Auschwitz in January 1945, Simone Gigliotti, Marc Masurovsky, and Erik Steiner quickly came to the limits of GIS mapping. They were able to use this to trace the evacuation routes on the map, but this did little in terms of revealing what the experience of evacuation – or 'death marches' as survivors have tended to call these – was like.[32] In order to visualize these spatial experiences, they left more traditional digital mapping methods behind and adopted a range of more experimental maps and visualizations.[33] One of these was particularly striking. It placed the textual content of six post-war testimonies onto a simplified map of the evacuation route from the Rajsko camp to Wodzisław. This enabled a visualization of which place (and events) along the route the women recalled in their testimony. Mapping this out along the journey reveals the blank spaces along the march route that are separated out by tight clusters of memory focused on particular sites along the road.[34]

This practice of working with more non-representational visualizations of spatial experience is something that project members have experimented with further in more recent work that has, in particular, shifted focus from developing databases from the paperwork of the perpetrators (with the tendency thereby to represent – and arguably replicate – Nazi space) to the post-war testimony of the victims (and an interest in uncovering Jewish place). In part, as we were well aware, digital mapping raises ethical questions related to any representation of the Holocaust. As Todd Presner has suggested, 'To turn Holocaust victims into quantifiable entries in a database and to visualize their lives as data points using coloured pixels on a bitmap is, on the face of it, problematic' given its uncanny parallels with the dehumanizing nature of the Holocaust.[35] But alongside such ethical self-reflection, the broadening of the sources we sought to map to post-war testimony, led us to seek a broader range of digital mapping tools.

Anne Kelly Knowles, drawing on Henri Lefebvre's identification of three kinds of 'space' in his seminal work *The Production of Space*, suggests that 'GIS is ideal for studying representations of space but has limited utility for studying spatial practice' and representational space.[36] It led her to adopt a broad range of what she terms 'inductive visualization' methods which she suggests 'revealed qualities of place, time, and scale in survivor narratives that were very different from the coordinate space, regular sequential time, and consistent geographical scale of GIS representation'.[37] Some of these experimental visualizations were sketched out on paper, while others made use of a range of visualization software packages.

What Gigliotti, Masurovsky, Steiner, and Knowles point to is a need for spatial historians to reckon with the complexity of what we understand by the words 'space' and 'place' and to draw on the range of digital mapping methods and approaches available accordingly. There is not a single 'spatial history' that demands a single digital mapping tool – be it ArcGIS or some other GIS software. Rather, there are multiple spatial questions to be asked of a wide variety of sources which contain differing kinds of geographical information. Some of those spatial questions can, we suggest, be digitally mapped and analyzed using a variety of tools. As we have pointed out, digital mapping can be a powerful tool in answering and generating research questions. However, as with historical methodology more generally, choosing when to map and how to map depends on what the questions are and what the archive contains. Rather than defaulting to a technological solution, there is need for critical reflection on when digital mapping is appropriate and when other methods are more suited to both the questions being asked and the sources being examined.

Notes

1 See, for example, Tim Cole, *Holocaust Landscapes*, London: Bloomsbury, 2016.
2 On the use of maps as sources see Tim Cole, *Traces of the Holocaust: Journeying In and Out of the Ghettos*, London: Continuum, 2011, especially prologue and ch. 4.
3 For the results of this work see, in particular, Anne Kelly Knowles, Tim Cole, and Alberto Giordano (eds.), *Geographies of the Holocaust*, Bloomington: Indiana University Press, 2014; more recent surveys and relevant volumes include Tim Cole, 'Geographies of the Holocaust', in Simone Gigliotti and Hilary Earl (eds.), *A Companion to the Holocaust*, Hoboken: John Wiley & Sons, 2020, pp. 333–47; Tim Cole and Simone Gigliotti (eds.), *The Holocaust in the Twenty-First Century: Relevance and Challenges in the Digital Age*, Evanston: Northwestern University Press, 2021; Claudio Fogu, 'A "Spatial Turn" in Holocaust Studies?', in id., Wulf

Kansteiner and Todd Presner (eds.), *Probing the Ethics of Holocaust Culture*, Cambridge, Mass.: Harvard University Press, 2016, pp. 218–39; and 'Interview with Anne Knowles, Tim Cole, Alberto Giordano, and Paul B. Jaskot, Contributing Authors of *Geographies of the Holocaust*', in ibid., pp. 240–56.

4 For an introduction to GIS and Geographic Information Science, see, for example, Michael Goodchild, 'GIScience, Geography, Form, and Process', *Annals of the Association of American Geographers* 94, 2004, 709–14; Nadine Schuurman, *GIS: A Short Introduction*, Malden, Mass.: Blackwell, 2004; Paul Bolstad, *GIS Fundamentals: A First Text on Geographic Information Systems*, Ann Arbor: XanEdu, 2016; Paul A. Longley et al., *Geographic Information Science and Systems*, 4th ed., London: Wiley, 2015.

5 For an introduction to the use of GIS in historical scholarship, see, for example, Anne Kelly Knowles (ed.), *Past Time, Past Place: GIS for History*, Redlands: ESRI Press, 2002; Anne Kelly Knowles (ed.), *Placing History: How Maps, Spatial Data, and GIS Are Changing Historical Scholarship*, Redlands: ESRI Press, 2008; Ian Gregory and Alistair Geddes (eds.), *Toward Spatial Humanities: Historical GIS and Spatial History*, Bloomington: Indiana University Press, 2014; Ian Gregory and Paul S. Ell, *Historical GIS: Technologies, Methodologies, and Scholarship*, Cambridge: Cambridge University Press, 2007. For examples of applications, see Nicholas Terpstra and Colin Rose, *Mapping Space, Sense, and Movement in Florence: Historical GIS and the Early Modern City*, London and New York: Routledge, 2016; Michele Tucci, Rocco Walter Ronza and Alberto Giordano, 'Fragments from Many Pasts: Layering the Toponymic Tapestry of Milan', *Journal of Historical Geography* 37, 2011, 370–84; Alberto Giordano and Tim Cole, 'On Place and Space: Calculating Social and Spatial Networks in the Budapest Ghetto', *Transactions in GIS* 15, 2011, 143–70; Michele Tucci, Alberto Giordano, and Rocco Walter Ronza, 'Spatial Analysis and Geovisualization to Reveal Urban Changes: Milan, Italy, 1737–2005', *Cartographica* 45, 2010, 47–63.

6 For an introduction to spatial analytical methods and an application, see, for example, A. Stewart Fotheringham and Peter A. Rogerson (eds.), *The SAGE Handbook of Spatial Analysis*, Thousands Oaks: SAGE, 2009; Michele Tucci and Alberto Giordano, 'Positional Accuracy, Positional Uncertainty, and Feature Change Detection in Historical Maps: Results of an Experiment', *Computers, Environment, and Urban Systems* 35, 2011, 452–63.

7 See, for example, Sonja Cameron and Sarah Richardson, *Using Computers in History*, Basingstoke: Palgrave Macmillan, 2015, esp. ch. 4; Charles Harvey and Jon Press, *Databases in Historical Research: Theory, Methods and Applications*, Basingstoke: Palgrave Macmillan, 1995; Patrick Manning, *Big Data in History*, Basingstoke: Palgrave Macmillan, 2013.

8 Anne Kelly Knowles and Paul B. Jaskot, with Benjamin Perry Blackshear, Michael De Groot, and Alexander Yule, 'Mapping the SS Concentration Camps', in Knowles, Cole and Giordano (eds.), *Geographies of the Holocaust*, p. 24.

9 Knowles and Jaskot, 'Mapping the SS Concentration Camps', p. 23.

10 Although as Knowles and Jaskot, 'Mapping the SS Concentration Camps', p. 24 point out, the concept of 'multi-instantiation' enabled them to include something of that complexity within their analysis.

11 On distant reading, see Franco Moretti, *Distant Reading*, London: Verso, 2013.

12 On the geohumanities, see David Bodenhamer, John Corrigan and Trevor M. Harris (eds.), *The Spatial Humanities: GIS and the Future of Humanities Scholarship*, Bloomington: Indiana University Press, 2010; Meghan Cope and

Sarah Elwood, *Qualitative GIS: A Mixed Methods Approach*, Thousand Oaks: SAGE, 2009; Michael Dear et al. (eds.), *GeoHumanities: Art, History, Text at the Edge of Place*, London: Routledge, 2011.

13 Eli Katz, Melissa Wiggins, Erik Steiner, Anna Holian and Alberto Giordano, 'Arrests of Italian Jews, 1943–1945'. Available HTTP: http://web.stanford.edu/group/spatialhistory/cgi-bin/site/viz.php?id=383&project_id=. See also Alberto Giordano and Anna Holian, 'Retracing the "Hunt for the Jews": A Spatio-Temporal Analysis of Arrests during the Holocaust in Italy', in Knowles, Cole and Giordano (eds.), *Geographies of the Holocaust*, pp. 52–87.

14 Tim Cole, Alberto Giordano and Erik Steiner, 'Mapping Mobility in the Budapest Ghetto'. Available HTTP: http://web.stanford.edu/group/spatialhistory/cgi-bin/site/viz.php?id=411&project_id=.

15 Tim Cole, 'When (and Why) Is a Ghetto not a "Ghetto"? Concentrating and Segregating Jews in Budapest, 1944', in Wendy Z. Goldman and Joe William Trotter Jr. (eds.), *The Ghetto in Global History, 1500 to the Present*, New York: Routledge, 2018, pp. 169–86.

16 Christopher R. Browning, 'Nazi Ghettoization Policy in Poland, 1939–1941', *Central European History* 19, 1986, 343–63. For a different perspective, although one that also stresses difference, see Dan Michman, *The Emergence of Jewish Ghettos during the Holocaust*, New York: Cambridge University Press, 2011. For more on the broad brushstrokes of ghetto historiography see Tim Cole, 'Ghettoization', in Dan Stone (ed.), *The Historiography of the Holocaust*, Basingstoke: Palgrave Macmillan, 2004, pp. 65–87.

17 Christian Gerlach and Götz Aly, *Das letzte Kapitel: Der Mord an den ungarischen Juden, 1944-1945*, Stuttgart: Deutsche Verlags-Anstalt, 2002.

18 Cole, *Traces of the Holocaust*, pp. 57–9.

19 Tim Cole, *Holocaust City: The Making of a Jewish Ghetto*, New York: Routledge, 2003.

20 Id., *Holocaust City*; id., 'Ghettoization'.

21 Id., *Holocaust City*, p. 106.

22 Ibid., e.g. pp. 106, 160; Tim Cole and Alberto Giordano, 'Bringing the Ghetto to the Jew: Spatialities of Ghettoization in Budapest', in Knowles, Cole and Giordano (eds.), *Geographies of the Holocaust*, p. 124.

23 Cole and Giordano, 'Bringing the Ghetto to the Jew', esp. pp. 124, 133–4.

24 Ibid., p. 135.

25 Ibid., pp. 136–7.

26 Cole, *Holocaust City*, pp. 131–67.

27 Cole and Giordano, 'Bringing the Ghetto to the Jew', pp. 137–9.

28 Ibid., pp. 139–41.

29 Ibid., pp. 146–8.

30 Tim Cole and Alberto Giordano, 'Rethinking Segregation in the Ghetto: Invisible Walls and Social Networks in the Dispersed Ghetto in Budapest, 1944', in Hilary Earl and Karl Schleunes (eds.), *Lessons and Legacies XI: Expanding Perspectives on the Holocaust in a Changing World*, Evanston: Northwestern University Press, 2014, pp. 265–91.

31 Cole and Giordano, 'Rethinking Segregation in the Ghetto'; Tim Cole and Alberto Giordano, 'Microhistories, Microgeographies: Budapest, 1944, and Scales of Analysis', in Claire Zalc and Tal Bruttmann (eds.), *Microhistories of the Holocaust*, New York: Berghahn, 2017, pp. 113–27.

32 Simone Gigliotti, Marc J. Masurovsky and Erik B. Steiner, 'From the Camp to the Road: Representing the Evacuations from Auschwitz, January 1945', in Knowles, Cole and Giordano (eds.), *Geographies of the Holocaust*, p. 206.

33 Ibid., pp. 212–14, 216, 220–1.
34 Ibid., p. 216.
35 Todd Presner, 'The Ethics of the Algorithm: Close and Distant Listening to the Shoah Foundation Visual History Archive', in Fogu, Kansteiner and Presner (eds.), *Probing the Ethics of Holocaust Culture*, p. 179.
36 Anne Kelly Knowles, Levi Westerveld and Laura Strom, 'Inductive Visualization: A Humanistic Alternative to GIS', *GeoHumanities* 1, 2015, 233–65; Henri Lefebvre, *The Production of Space*, Oxford: Blackwell, 1991.
37 Knowles, Westerveld and Strom, 'Inductive Visualization', pp. 14–15.

Selected bibliography

This bibliography lists a small selection of titles that are discussed in this open-access publication: Konrad Lawson, Riccardo Bavaj and Bernhard Struck, *A Guide to Spatial History: Areas, Aspects, and Avenues of Research* (2021), URL: spatialhistory.net/guide.

Territoriality, infrastructure, and borders

Anzaldúa, Gloria, *Borderlands/La Frontera: The New Mestiza*, San Francisco: Spinsters/Aunt Lute, 1987.

Bassi, Ernesto, *An Aqueous Territory: Sailor Geography and New Granada's Transimperial Greater Caribbean World*, Durham: Duke University Press, 2016.

Burnett, D. Graham, *Masters of All They Surveyed: Exploration, Geography, and a British El Dorado*, Chicago and London: University of Chicago Press, 2000.

Carse, Ashley, *Beyond the Big Ditch: Politics, Ecology, and Infrastructure at the Panama Canal*, Cambridge, Mass.: MIT Press, 2014.

Gandy, Matthew, *The Fabric of Space: Water, Modernity, and Urban Imagination*, Cambridge, Mass.: Harvard University Press, 2014.

Herzog, Tamar, *Frontiers of Possession: Spain and Portugal in Europe and the Americas*, Cambridge, Mass.: Harvard University Press, 2015.

Howard, Allen M. and Richard M. Shain (eds.), *The Spatial Factor in African History*, Leiden: Brill, 2005.

Laak, Dirk van, 'Infrastructures' (Version: 1.0), *Docupedia-Zeitgeschichte*, 20 May 2021, Available HTTP: http://docupedia.de/zg/Laak_infrastructures_v1_en_2021 (accessed 24 June 2021).

Maier, Charles S., *Once within Borders: Territories of Power, Wealth, and Belonging since 1500*, Cambridge, Mass.: Harvard University Press, 2016.

Marung, Steffi and Matthias Middell (eds.), *Spatial Formats under the Global Condition*, Berlin: De Gruyter, 2019.

Mukerji, Chandra, *Territorial Ambitions and the Gardens of Versailles*, Cambridge: Cambridge University Press, 1997.

Mukhopadhyay, Aparajita, *Imperial Technology and 'Native' Agency: A Social History of Railways in Colonial India, 1850–1920*, London: *Routledge*, 2018.

Nugent, Paul, *Boundaries, Communities and State-Making in West Africa: The Centrality of the Margins*, Cambridge: Cambridge University Press, 2019.

Paasi, Anssi, *Territories, Boundaries and Consciousness: The Changing Geographies of the Finnish-Russian Border*, New York: John Wiley & Sons, 1996.

Park, Alyssa M., *Sovereignty Experiments: Korean Migrants and the Building of Borders in Northeast Asia, 1860–1945*, Ithaca: Cornell University Press, 2019.
Sack, Robert David, *Human Territoriality: Its Theory and History*, Cambridge: Cambridge University Press, 1986.
Sahlins, Peter, *Boundaries: The Making of France and Spain in the Pyrenees*, Berkeley: University of California Press, 1989.
Scott, James C., *Seeing Like a State: How Certain Schemes to Improve the Human Condition Have Failed*, New Haven: Yale University Press, 1998.
Simpson, Thomas, *The Frontier in British India: Space, Science, and Power in the Nineteenth Century*, Cambridge: Cambridge University Press, 2021.
Song, Nianshen, *Making Borders in Modern East Asia: The Tumen River Demarcation, 1881–1919*, Cambridge: Cambridge University Press, 2018.
Tischler, Julia, *Light and Power for a Multiracial Nation: The Kariba Dam Scheme in the Central African Federation*, Basingstoke: Palgrave Macmillan, 2013.
Wenzlhuemer, Roland, *Connecting the Nineteenth-Century World: The Telegraph and Globalization*, Cambridge: Cambridge University Press, 2013.

Nature, environment, and landscape

Applegate, Celia, *A Nation of Provincials: The German Idea of Heimat*, Berkeley: University of California Press, 1990.
Arnold, David, *The Tropics and the Traveling Gaze: India, Landscape, and Science, 1800–1856*, Seattle and London: University of Washington Press, 2006.
Blackbourn, David, *The Conquest of Nature: Water, Landscape and the Making of Modern Germany*, London: Jonathan Cape, 2006.
Cohen, David William and E.S. Atieno Odhiambo, *Siaya: The Historical Anthropology of an African Landscape*, London: Currey, 1989.
Cosgrove, Denis E., *Social Formation and Symbolic Landscape*, with a new introduction, Madison: University of Wisconsin Press, 1998, first published 1984.
Courtney, Chris, *The Nature of Disaster in China: The 1931 Yangzi River Flood*, Cambridge: Cambridge University Press, 2018.
Cronon, William, *Nature's Metropolis: Chicago and the Great West*, New York: W.W. Norton, 1991.
Cronon, William, 'The Trouble with Wilderness; or, Getting Back to the Wrong Nature', in id. (ed.), *Uncommon Ground: Rethinking the Human Place in Nature*, New York: W.W. Norton, 1995, pp. 69–90.
Duncan, James S., *The City as Text: The Politics of Landscape Interpretation in the Kandyan Kingdom*, Cambridge: Cambridge University Press, 1990.
Huang, Fei, *Reshaping the Frontier Landscape: Dongchuan in Eighteenth-Century Southwest China*, Leiden: Brill, 2018.
Hughes, David McDermott, *Whiteness in Zimbabwe: Race, Landscape, and the Problem of Belonging*, New York: Palgrave Macmillan, 2010.
Matless, David, *Landscape and Englishness*, London: Reaktion, 1998.
Merchant, Carolyn, *The Death of Nature: Women, Ecology, and the Scientific Revolution*, New York: Harper & Row, 1980.
Mitchell, W.J.T. (ed.), *Landscape and Power*, 2nd ed., Chicago and London: University of Chicago Press, 2002, first published 1994.
Pratt, Mary Louise, *Imperial Eyes: Travel Writing and Transculturation*, London and New York: Routledge, 1992.

Shapiro, Judith, *Mao's War against Nature: Politics and the Environment in Revolutionary China*, Cambridge: Cambridge University Press, 2001.
Shetler, Jan Bender, *Imagining Serengeti: A History of Landscape Memory in Tanzania from Earliest Times to the Present*, Athens: Ohio University Press, 2007.
Spence, Mark David, *Dispossessing the Wilderness: Indian Removal and the Making of the National Parks*, Oxford: Oxford University Press, 2000.
Thomas, Julia Adeney, *Reconfiguring Modernity: Concepts of Nature in Japanese Political Ideology*, Berkeley: University of California Press, 2001.

City and home

Chattopadhyay, Swati, *Representing Calcutta: Modernity, Nationalism, and the Colonial Uncanny*, London: Routledge, 2005.
Chauncey, George, *Gay New York: Gender, Urban Culture, and the Making of the Gay Male World, 1890–1940*, New York: Basic Books, 1994.
Flather, Amanda (ed.), *A Cultural History of the Home*, 6 vols., London: Bloomsbury, 2021.
Giles, Judy, *The Parlour and the Suburb: Domestic Identities, Class, Femininity and Modernity*, Oxford: Berg, 2004.
Hall, Peter, *Cities of Tomorrow: An Intellectual History of Urban Planning and Design since 1880*, 4th ed., Malden, Mass.: John Wiley & Sons, 2014, first published 1988.
Hayden, Dolores, *The Grand Domestic Revolution: A History of Feminist Designs for American Homes, Neighborhoods, and Cities*, Cambridge, Mass. and London: MIT Press, 1981.
Helly, Dorothy O. and Susan M. Reverby, *Gendered Domains: Rethinking Public and Private in Women's History*, Ithaca: Cornell University Press, 1992.
Henry, Todd A., *Assimilating Seoul: Japanese Rule and the Politics of Public Space in Colonial Korea, 1910–1945*, Berkeley: University of California Press, 2014.
Jacobs, Jane, *The Death and Life of Great American Cities*, New York: Random House, 1961.
King, Anthony D., *The Bungalow: The Production of a Global Culture*, London: Routledge, 1984.
Lewis, Su Lin, *Cities in Motion: Urban Life and Cosmopolitanism in Southeast Asia, 1920–1940*, Cambridge: Cambridge University Press, 2016.
Lynch, Kevin, *The Image of the City*, Cambridge, Mass. and London: MIT Press, 1960.
Myers, Garth Andrew, *Verandahs of Power: Colonialism and Space in Urban Africa*, Syracuse: Syracuse University Press, 2003.
Robinson, Jennifer, *Ordinary Cities: Between Modernity and Development*, London: Routledge, 2006.
Sand, Jordan, *House and Home in Modern Japan: Reforming Everyday Life 1880–1930*, Cambridge, Mass.: Harvard University Press, 2003.
Schorske, Carl E., *Fin-de-Siècle Vienna: Politics and Culture*, New York: Knopf, 1979.
Shah, Nayan, *Contagious Divides: Epidemics and Race in San Francisco's Chinatown*, Berkeley: University of California Press, 2001.
Simmel, Georg, 'The Metropolis and Mental Life' (1903), in Sharon M. Meagher (ed.), *Philosophy and the City: Classic to Contemporary Writings*, Albany: SUNY Press, 2008, pp. 96–101.
Skinner, G. William, *The City in Late Imperial China*, Stanford: Stanford University Press, 1977.

Spain, Daphne, *Gendered Spaces*, Chapel Hill: University of North Carolina Press, 1992.
Stanek, Łukasz, *Architecture in Global Socialism: Eastern Europe, West Africa, and the Middle East in the Cold War*, Princeton: Princeton University Press, 2020.
Wiese, Andrew, *Places of Their Own: African American Suburbanization in the Twentieth Century*, Chicago and London: University of Chicago Press, 2004.
Williams, Raymond, *The Country and the City*, London: Chatto and Windus, 1973.
Yeoh, Brenda S. A., *Contesting Space in Colonial Singapore: Power Relations and the Urban Built Environment*, Singapore: NUS Press, 1996.

Social space and political protest

Elden, Stuart, 'There Is a Politics of Space Because Space Is Political: Henri Lefebvre and the Production of Space', *Radical Philosophy Review* 10, 2007, 101–116.
Epstein, James, 'Spatial Practices/Democratic Vistas', *Social History* 24, 1999, 294–310.
Featherstone, David, *Resistance, Space and Political Identities: The Making of Counter-Global Networks*, Oxford: Wiley-Blackwell, 2008.
Forster, Laura C., 'The Paris Commune in London and the Spatial History of Ideas, 1871–1900', *The Historical Journal* 62, 2019, 1021–1044.
Lefebvre, Henri, 'Reflections on the Politics of Space' (1970), in id., *State, Space, World: Selected Essays*, eds. Neil Brenner and Stuart Elden, Minneapolis: University of Minnesota Press, 2009, pp. 167–184.
Lefebvre, Henri, *The Urban Revolution*, Minneapolis: University of Minnesota Press, 2003, French 1970.
Lefebvre, Henri, *The Production of Space*, trans. Donald Nicholson-Smith, Oxford: Blackwell, 1991, French 1974.
Lefebvre, Henri, *Writings on Cities*, trans. Eleonore Kofman and Elizabeth Lebas, Oxford: Blackwell, 1996.
Navickas, Katrina, *Protest and the Politics of Space and Place, 1789–1848*, Manchester: Manchester University Press, 2016.
Parolin, Christina, *Radical Spaces: Venues of Popular Politics in London, 1790-c.1845*, Canberra: ANU Press, 2010.
Ross, Kristin, *The Emergence of Social Space: Rimbaud and the Paris Commune*, London: Verso, 1988.
Sewell, Jr., William H., 'Space in Contentious Politics', in Ronald R. Aminzade et al. (eds.), *Silence and Voice in the Study of Contentious Politics*, Cambridge: Cambridge University Press, 2001, pp. 51–88.

Spaces of knowledge

Dorn, Harold, *The Geography of Science*, Baltimore: Johns Hopkins University Press, 1991.
Easterby-Smith, Sarah, *Cultivating Commerce: Cultures of Botany in Britain and France, 1760–1815*, Cambridge: Cambridge University Press, 2017.
Finnegan, Diarmid A., 'The Spatial Turn: Geographical Approaches in the History of Science', *Journal of the History of Biology* 41, 2008, 369–388.
Finnegan, Diarmid A. and Jonathan Jeffrey Wright (eds.), *Spaces of Global Knowledge: Exhibition, Encounter and Exchange in an Age of Empire*, Farnham: Ashgate, 2015.

292 Selected bibliography

Geertz, Clifford, *Local Knowledge: Further Essays in Interpretative Anthropology*, New York: Basic Books, 1983.

Gieryn, Thomas F., *Truth Spots: How Places Make People Believe*, Chicago and London: University of Chicago Press, 2018.

Golinski, Jan, *Science as Public Culture: Chemistry and Enlightenment in Britain, 1760–1820*, Cambridge: Cambridge University Press, 1992.

Hochadel, Oliver and Agustí Nieto-Galan (eds.), *Urban Histories of Science: Making Knowledge in the City, 1820–1940*, London: Routledge, 2019.

Knorr Cetina, Karin, *Epistemic Cultures: How the Sciences Make Knowledge*, Cambridge, Mass.: Harvard University Press, 1999.

Latour, Bruno, *Science in Action: How to Follow Scientists and Engineers through Society*, Cambridge, Mass.: Harvard University Press, 1987.

Latour, Bruno, *The Pasteurization of France*, Cambridge, Mass.: Harvard University Press, 1988.

Latour, Bruno and Steve Woolgar, *Laboratory Life: The Social Construction of Scientific Facts*, London: Sage, 1979.

Lightman, Bernard (ed.), *A Companion to the History of Science*, Oxford: Wiley-Blackwell, 2016.

Livingstone, David N., *Putting Science in Its Place: Geographies of Scientific Knowledge*, Chicago and London: University of Chicago Press, 2003.

Livingstone, David N. and Charles W. J. Withers (eds.), *Geographies of Nineteenth-Century Science*, Chicago and London: University of Chicago Press, 2011.

Ophir, Adi and Steven Shapin, 'The Place of Knowledge: A Methodological Survey', *Science in Context* 4, 1991, 3–21.

Outram, Dorinda, *The Enlightenment*, Cambridge: Cambridge University Press, 2012.

Raj, Kapil, *Relocating Modern Science: Circulation and the Construction of Knowledge in South Asia and Europe, 1650–1900*, Basingstoke: Palgrave Macmillan, 2007.

Shapin, Steven and Simon Schaffer, *Leviathan and the Air-Pump: Hobbes, Boyle, and the Experimental Life*, Princeton: Princeton University Press, 1985.

Smith, Crosbie and Jon Agar (eds), *Making Space for Science: Territorial Themes in the Shaping of Knowledge*, Basingstoke: Palgrave Macmillan, 1998.

Withers, Charles W.J., *Placing the Enlightenment: Thinking Geographically about the Age of Reason*, Chicago and London: University of Chicago Press, 2007.

Spatial imaginaries

Bassin, Mark, Sergey Glebov and Marlene Laruelle (eds.), *Between Europe and Asia*, Pittsburgh: University of Pittsburgh Press, 2015.

Bavaj, Riccardo and Martina Steber (eds.), *Germany and 'The West': The History of a Modern Concept*, New York: Berghahn, 2015.

Cosgrove, Denis, *Apollo's Eye: A Cartographic Genealogy of the Earth in the Western Imagination*, Baltimore: Johns Hopkins University Press, 2003.

Gingeras, Ryan, 'Between the Cracks: Macedonia and the "Mental Map" of Europe', *Canadian Slavonic Papers* 50/3–4, 2008, 341–358.

Gregory, Derek, 'Imaginative Geographies', *Progress in Human Geography* 19, 1995, 447–485.

Lewis, Martin W. and Kären Wigen, *The Myth of Continents: A Critique of Metageography*, Berkeley: University of California Press, 1997.

Mishkova, Diana, *Beyond Balkanism: The Scholarly Politics of Region Making*, London and New York: Routledge, 2018.
McDonald, Kate, *Placing Empire: Travel and the Social Imagination in Imperial Japan*, Oakland: University of California Press, 2017.
Moe, Nelson, *The View from Vesuvius: Italian Culture and the Southern Question*, Berkeley: University of California Press, 2002.
Osterhammel, Jürgen, *Unfabling the East: The Enlightenment's Encounter with Asia*, Princeton: Princeton University Press, 2018, German 1998.
Owens, Patricia and Katharina Rietzler (eds.), *Women's International Thought: A New History*, Cambridge: Cambridge University Press, 2021.
Said, Edward W., *Orientalism*, with a new preface, London: Penguin, 2003, with a new afterword 1995, first published 1978.
Schenk, Frithjof Benjamin, 'Mental Maps', *European History Online*, 2013. Available http://www.ieg-ego.eu/schenkf-2013-en (accessed 10 March 2021).
Todorova, Maria, *Imagining the Balkans*, New York: Oxford University Press, 1997, updated ed. 2009.
Wolff, Larry, *Inventing Eastern Europe: The Map of Civilization on the Mind of the Enlightenment*, Stanford: Stanford University Press, 1994.

Cartographic representations

Akerman, James R. (ed.), *Decolonizing the Map: Cartography from Colony to Nation*, Chicago and London: University of Chicago Press, 2017.
Anderson, Benedict, *Imagined Communities: Reflections on the Origin and Spread of Nationalism*, 2nd ed., London: Verso, 1991, first published 1983.
Craib, Raymond B., *Cartographic Mexico: A History of State Fixations and Fugitive Landscapes*, Durham: Duke University Press, 2004.
Dunlop, Catherine, *Cartophilia: Maps and the Search for Identity in the French-German Borderland*, Chicago and London: University of Chicago Press, 2015.
Edney, Matthew H., *Mapping an Empire: The Geographical Construction of British India, 1765–1843*, Chicago and London: University of Chicago Press, 1997.
Foliard, Daniel, *Dislocating the Orient: British Maps and the Making of the Middle East, 1854–1921*, Chicago and London: University of Chicago Press, 2017.
Harley, J.B., *The New Nature of Maps: Essays in the History of Cartography*, ed. Paul Laxton, Baltimore: Johns Hopkins University Press, 2001.
Harley, J.B. and David Woodward (eds.), *The History of Cartography*, Chicago and London: Chicago University Press, 1987-present.
Herb, Guntram Henrik, *Under the Map of Germany: Nationalism and Propaganda, 1918–1945*, London: Routledge, 1997.
Konvitz, Josef W., *Cartography in France, 1660–1848: Science, Engineering, and Statecraft*, Chicago and London: University of Chicago Press, 1987.
Kopp, Kristin, *Germany's Wild East: Constructing Poland as Colonial Space*, Ann Arbor: University of Michigan Press, 2012.
Murphy, David Thomas, *The Heroic Earth: Geopolitical Thought in Weimar Germany, 1918–1933*, Ohio: Kent State University Press, 1997.
Ramaswamy, Sumathi, *The Goddess and the Nation: Mapping Mother India*, Durham: Duke University Press, 2010.
Rankin, William, *After the Map: Cartography, Navigation and the Transformation of Territory in the Twentieth Century*, Chicago and London: University of Chicago Press, 2016.

Schulten, Susan, *The Geographical Imagination in America 1880–1950*, Chicago and London: University of Chicago Press, 2001.
Schulten, Susan, *Mapping the Nation: History and Cartography in Nineteenth-Century America*, Chicago and London: University of Chicago Press, 2012.
Seegel, Steven, *Mapping Europe's Borderlands: Russian Cartography in the Age of Empire*, Chicago and London: University of Chicago Press, 2012.
Speich, Daniel, 'Mountains Made in Switzerland: Facts and Concerns in Nineteenth-Century Cartography', *Science in Context* 22, 2009, 387–408.
Wigen, Kären, *A Malleable Map: Geographies of Restoration in Central Japan, 1600–1912*, Berkeley: University of California Press, 2010.
Wigen, Kären and Caroline Winterer (eds.), *Time in Maps: From the Age of Discovery to Our Digital Era*, Chicago and London: University of Chicago Press, 2020.
Winichakul, Thongchai, *Siam Mapped: A History of the Geo-Body of a Nation*, Honolulu: University of Hawaii Press, 1994.

Historical GIS

Berman, Merrick Lex, Ruth Mostern and Humphrey Southall (eds.), *Placing Names: Enriching and Integrating Gazetteers*, Bloomington: Indiana University Press, 2016.
Bodenhamer, David J., John Corrigan and Trevor M. Harris (eds.), *The Spatial Humanities: GIS and the Future of Humanities Scholarship*, Bloomington: Indiana University Press, 2010.
Bodenhamer, David J., John Corrigan and Trevor M. Harris (eds.), *Deep Maps and Spatial Narratives*, Bloomington: Indiana University Press, 2015.
Cunfer, Geoff, *On the Great Plains: Agriculture and Environment*, College Station: Texas A&M University Press, 2005.
Curry, Michael R., *Digital Places: Living with Geographic Information Technologies*, London: Routledge, 1998.
Gregory, Ian and Paul S. Ell, *Historical GIS: Technologies, Methodologies and Scholarship*, Cambridge: Cambridge University Press, 2007.
Gregory, Ian and Alistair Geddes (eds.), *Toward Spatial Humanities: Historical GIS and Spatial History*, Bloomington: Indiana University Press, 2014.
Gregory, Ian, Don DeBats and Don Lafreniere (eds.), *The Routledge Companion to Spatial History*, London: Routledge, 2018.
Knowles, Anne Kelly (ed.), *Past Time, Past Place: GIS for History*, Redlands: Esri Press, 2002.
Knowles, Anne Kelly and Amy Hillier (eds.), *Placing History: How Maps, Spatial Data, and GIS are Changing Historical Scholarship*, Redlands: Esri Press, 2008.
Lünen, Alexander von and Charles Travis (eds.), *History and GIS: Epistemologies, Considerations and Reflections*, New York: Springer, 2013.
Pickles, John (ed.), *Ground Truth: The Social Implications of Geographic Information Systems*, New York: Guilford Press, 1995.

Index

Page numbers in *Italics* refer to figures

Aberdeen 134
Académie des Sciences 43
Acharya, Amitav 239, 249
Addams, Jane 77
Aegean Islands 228, 230
Agnew, John 16
agriculture 90, 155–157, 212, 214, 254, 256, 259–263, 266, 269
aircraft/airports 127, 171
Alatas, Syed Hussein 244
Alcott, Louisa May 76
Alexander, John H. 260–261
Alexander I, Tsar 45
Alexander II, Tsar 181
Algeria 226
Alps *see* mountains
Alsace 39
Alt, Rudolf von 108
Amazon, the *see* rivers
American Association of Geographers 6, 8
Amerindians 158, 161–168
Anatolia 227
Ancient Egyptians 155, 160, 242
Andes, the *see* mountains
Anglo-America 2, 13, 15
Anglo-German Review 110–111
Anglosphere 5, 7, 13
Annales/Annalistes 15, 252
Antwerp 177
aquatic environments 158, 162, 225, 234
Arc GIS 277
archaeology/archaeologists 6, 18, 102–103, 155–158, 160, 164, 166–167
architects/architecture 5–7, 11–14, 17, 62, 67, 105–106, 112, 115, 173, 177, 210, 218; drawings 19, 102–116, 173, 178; plans 103, 151

architectural history/historians 7, 19, 103, 275
archives 40–41, 87–98, 105, 113, 127, 130, 135–136, 141–142, 147, 155, 165, 198, 222, 226, 233, 274–275, 280, 282–284
Arctic 121, 125, 128–129, 133, 135–136
area studies 239
Arezzo 149
Arias, Santa 18
Armenia 195
Art Nouveau 177
Aryan race 242
Asia 19, 175–177, 179–180, 183, 188, 192–193, 195–196, 227–228, 237–247
asylums 13
Atlantic 8, 124, 128, 130, 160, 168, 176, 225
atlases 40–51, 214, 277
Auschwitz 276, 283
Australia 3, 87–88, 90, 241
Austria 40, 43, 45, 60, 112, 144, 227
autobiographies 75–76, 78, 81, 83

Baar, Monika 50
Bachmann-Medick, Doris 17
Back Bay 82
Baedeker 56, 59
Baffin Bay/Island 122, 124, 133
Baker, Alan 5
Balkans 1, 57, 227
Baltic Sea 44, 46–47, 254
Banat 59–60, 63–64
Bangladesh 239
barrier 166, 211
bars 19, 139–151
Barthes, Roland 12

Index

Barth, Hans 143–144
Bashford, Alison 233
Bavaria 103, 105, *107*, 113, 115
Beaumont, Matthew 173
Beijing 191
Belarus 44, 49
Belém 164
Bengal 192
Bentham, Jeremy 103
Bergamo 148–149
Berger, John 9
Berghaus, Heinrich 42
Berghof 103–105, 108–115
Berlin 40–41, 58, 103, 109, 113
Berry, Stephen 126
Black Sea 62, 227
boarding houses 79–81
Board of Trade 123, 125, 131–132, 135
Bode, Katherine 88
body 11, 43, 102, 110, 177, 188, 216, 258
Bolivia 163
Book of Revelation 224
border/borders 1, 4, 10–11, 16, 18–19, 41, 44–47, 49, 65, 96, 144, 188–200, 222–223, 226–227, 234, 237, 252, 261–232, 278; 'border-making' 188, 199; zones *see* zones
borderlands 6, 50, 175, 179, 182–183, 189–190
Bordone, Paris 113
Borneo 239
Bosporus 227–228
Boston 77–82
Bosworth, Louise 79
Bottai, Giuseppe 144
boundary/boundaries 5, 7–9, 14, 47, 50, 57, 76, 82, 154, 160, 166, 168, 172, 174, 222, 232, 239, 242, 275; 'spanning'/'spanners' 7–9
Bourdieu, Pierre 16
Brașov 62–63
Braudel, Fernand 15, 232, 254, 263, 269
Braun, Eva 105, 112
Brazil 155, 163–165
bridges 6, 171–172, 181, 211
Britain 2–3, 6–7, 14, 18, 89–90, 95–96, 105, 108–111, 121, 123–124, 130–131, 133, 173, 195, 198–199, 208, 217–218, 227, 241
British East India Company 195
British Empire 124
British Union of Fascists 108
Browning, Christopher 278
Brzhosovskii, Stanislav 177
Bucharest 60, 62

Budapest 274, 277–283
buildings 16, 18, 77–78, 82, 94, 102–103, 105–106, 113, 115, 157, 167, 177–178, 183, 193, 208–209, 278–282
built environment 6, 73, 75, 77
Burma 240–241
Burnet, Thomas 217–218
Burzenland 63–65
bush/bushland 87–88, 91–96
Butler, Christopher 12
Buttimer, Anne 8
Bynkershoek, Cornelius van 225, 231

cadastral maps 259
Cahokia 158
Canton 188, 190–200; 'Canton System' 191
capital 168, 176, 196, 230, 254
Caporetto 144
Caribbean 157, 254
Carpathians 62–63, 66, 69
Carte Générale de la France 43
Carter, Ian 173
Carter, Paul 3–5
Cartesian coordinate system 14
cartography 4, 6–7, 12, 14, 39–43, 46, 49, 51, 160, 162, 253, 261–263, 269
Carvajal, Gaspar de 157–160
Cassini (family) 43
Castello 145
Catherine II 49
Catholicism 47–49, 51, 81, 144, 182
Caucasus 181–182, 214–215
Ceaușescu, Nicolae 59, 61, 65, 68
Ceaușescu-Schmidt agreement 65
Çelebi, Katib 228, 231–232
census/census records 74–75, 77–78, 279
Certeau, Michel de 140, 189–190, 200
Chamberlain, Neville 109–111
Chambery 213
Charlemagne, Emperor 108, 114
childhood/children/childscapes 11, 13, 61, 73–75, 87–97, 109, 133–134, 139, 197–198, 243
China 155, 177, 188–200, 239, 242–246
Chirino, Pedro 248
Chita 183
Christianity 192, 196–197, 246–247
chronology 1, 4, 12, 57, 139, 147, 277, 279, 282
churches 18, 49, 62–63, 65–66, 73, 81–82, 140, 209
Cioc, Mark 154

cities 1, 11, 13, 18, 45, 47, 58, 67, 73–83, 88, 103, 124, 139, 144–146, 154–156, 158–159, 164, 166–167, 173, 177–178, 181, 183, 188, 190–192, 195, 199, 209–210, 227–228, 237, 256, 277–283
climate 126, 129–130, 141, 238, 242–247, 252; history 252
coast/coastline 18, 46–47, 62, 96, 122, 124, 133, 160, 209, 225–234, 261, 263
Colbert, Jean-Baptiste 43
Cold War 11, 59–61, 63–66, 68–69, 239
Cole, Tim 278–281
Colin, Francisco 248
Collins, Patricia Hill 77
colonialism 3, 39–40, 73, 79, 88, 150, 155, 159–167, 189, 192, 223, 239, 244, 246–247
colonies 150, 162, 188–189, 243, 245
communities 2, 39, 59–62, 65, 68, 88, 146, 155–156, 164, 168, 177, 208–209, 212–213, 219, 233, 238–239, 244, 258
concentration camps 276
Congress of Vienna 39, 44–45, 51
Constantinople 228
construction 3–4, 13, 17, 103, 115, 144, 155, 171–177, 179, 182–183, 188–190, 192, 195, 198–200, 208, 211, 228, 237, 240, 252, 257, 269, 275–277, 282
constructivism 189, 200
consumerism/consumption/consumption practices 56–57, 143, 146, 150, 158, 166, 255, 261, 263–264
'contact zones' *see* zones
Cook, James 4
Corner, Paul 149
corridors 106, 112, 176
Coryate, Thomas 211–215, 217
Cosgrove, Denis 8, 15, 208
country 3–4, 40, 43–45, 51, 60–61, 87, 92, 109, 111–112, 147–148, 154, 172, 175, 177–178, 183, 190–192, 196, 237, 239, 242, 244
countryside 1, 61, 69, 192, 210, 266
Cracow *see* Kraków
Crainic, Marina 68
Crang, Philip 8
Crawford, John 248
Cresswell, Tim 8
Crete 230, 255, 263
Crew Agreements 124–125, 130–131, 133

Crimean War 183
cultural geography *see* geography
cultural history 56, 173, 252
'cultural turn' 15, 17, 173, 252–253, 270
cyberspace 10
Cyprus 230, 253–256, 259–269
Czechoslovakia 60–61, 110–111

Damala 231
dance halls 79
Daniels, Stephen 15
D'Annunzio, Gabriele 144
Danube Delta 68, 154
Darby, H. C. 7
Davis, Mike 11
Davis Strait 122, 126
Dear, Michael 11
Debord, Guy 14
Decembrist Revolt 45, 51
Degano, Alois 105–106
Dennis, John 217–218
Derbyshire 218
Derrida, Jacques 12
deserts 156, 209
Deshima 192–193, 195–196, 198–199
'despatialization' 10
Dexterity Bay 133
diaries 77, 102, 110, 121, 125, 128, 135, 141–142, 145, 151, 192
dictator/dictatorship 61, 103, *107*, 110, 140–144, 146–150
Didyma 103
digital humanities 5, 90, 253, 256, 274–276; Digital Humanities Hub 2
digital mapping 2–3, 7, 261, 270, 274–284
districts 47, 78, 97, 145, 192–193, 218, 231, 255, 261, 278–280
Divall, Colin 173
Dixon, Deborah P. 8
domesticity 65, 74–76, 81–83, 103–106, 108, 112–116, 143, 210, 212, 218, 227
Driver, Felix 5
Duffy, Cian 216
Duncan, James 15
Duncan, Nancy 15
Dundee 122–124, 134
Dunlop, Catherine 39
Dutch East Indies 241
Dutch Empire 160, 191–198, 200, 241

'East', the 176, 242
ecology 88, 90, 93, 97, 154, 156, 232–233

economic history 172, 190, 252, 255, 258, 270; 'new' economic history 252
ecotones 232–233
Edney, Matthew H. 15, 39
Edo *see* Tokyo
'ego-documents' 141–142, 151
Egypt 223, 228–230, 238, 242
Elden, Stuart 5, 14
El Dorado 158–159, 163
empire 18–19, 40–41, 43–46, 51, 56, 95, 105–106, 124, 130, 147–148, 150, 155–156, 165, 172–177, 179–183, 189–191, 213, 222–223, 225–230, 232–234, 243, 245, 254
England 7, 13, 18, 73, 79, 87, 89–90, 191, 200, 211–212, 214, 217
Enlightenment 12, 179
environment 2–3, 5–6, 8, 18, 73–74, 77, 87, 94, 121–123, 126–127, 146, 154–156, 158, 166–167, 192, 196–197, 199–200, 212, 216, 222, 238, 244, 254, 270; determinism 16; environmental history/historians 2–3, 5–7, 19, 154, 252, 269–270
Ermacora, Chino 144
esercizi pubblici 143, 146
Ethiopian War 148
Euclidean geometry 14
Eurocentrism 8
Europe 4, 11, 15, 19, 41, 45, 47, 57, 59–61, 65, 67–69, 110, 114, 122, 155, 157–161, 163, 166–168, 172, 175–177, 180, 183, 188–193, 195–200, 212, 215, 219, 223, 225, 227–228, 237, 240, 242, 244–245, 248, 258, 278–279
Evans, Richard J. 14–15
expansionism 43, 109, 114, 150, 183, 243–244
expeditions 94, 149, 158, 163, 165–166
exploration 1, 3–4, 7–8, 12–13, 18–19, 40, 57, 59–60, 63, 65, 69, 88, 90, 93, 103, 128, 135–136, 139–142, 150–151, 154, 164–165, 193, 196, 211, 219, 226, 240, 243–246, 248, 254, 275, 277, 282–283

Fabritius-Dancu, Juliana 60, 63–67
Fabritius, Hermann 60, 63
Faith, Nicholas 173
farmers/farming/farms 74, 87, 89–90, 93, 95, 97, 167–168, 211–213, 256
fascism 63, 108, 140–151

Febvre, Lucien 15, 270
federal government 60, 64, 68, 75
Ferreira, Manuel 165, 167
Feuerbach, Anselm 113
fiction/fictional sources 74–79, 81, 83, 88, 105, 126, 151, 161, 240
Filipinos 244–248
films 14, *107*, 109, 151, 214
Finberg, H. P. R. 6–7
Finland 182
First World War 144, 173, 177, 239, 241
Fitzpatrick, Sheila 139
fleet 130, 229, 232–223
Flinders, Matthew 4
floor plan 102–103, 105–106, 109–110, 112–116
Florida 126
forests 39, 41, 88, 95, 157, 162–163, 168
fortifications 225, 229
Foster, Frank 80
Foucault, Michel 5, 9, 12–13, 103, 121–123, 125, 127, 130, 135, 223
France 17, 39, 43, 60, 212
Frank, Zephyr 4
Frederick the Great 112
Freeman, Michael 173
Fumimaro, Konoe 245

Galicia 44–47, 50
gardens 13, 90, 96
gates 90, 177, 192
Gdańsk 47
gender 6, 56, 65, 76–77, 82–83, 143, 188, 190–191, 195, 197–200, 258
Genoa 226
geographers 2–3, 5–9, 11–13, 15–18, 42, 59, 189, 208, 223, 228, 237, 243, 275, 282
geographic information systems/ geographical information science (GIS) 2–5, 7, 9–10, 19, 253, 259, 269, 274–278, 280–281, 283–284
geography 1–8, 10–11, 14–15, 17–19, 39, 41–42, 51, 63, 74–77, 79–83, 136, 147, 156, 160–161, 171–173, 175, 177–180, 182–183, 209–210, 223–224, 230–231, 237–238, 240–214, 252–270, 274–277, 284; 'end of geography' 10
geo-history 11, 15
GeoHumanities 6–9, 15
geo-locating 259–260, 262
geopolitics 6, 11, 66, 182–183
geovisualization 274, 276

Germany 7, 14–15, 17, 39, 49–50, 59–69, 104, 108–112, 139, 143, 155, 160, 173–176, 212, 214–215, 239, 241, 278
ghettoization/ghettos 199, 210, 274, 277–283
Giddens, Anthony 9
Gigliotti, Simone 283–284
Gilbert, Pamela K. 18
Gillis, John 232–233
global history 7
globalization 188
Global Positioning Systems (GPS) 10
'Global South', the 1, 237
'global village' 10
'Golden Ghetto' *see* Canton
Golden Horn 228
Goldsworthy, Vesna 57
Gottdiener, Mark 13
Graulich, Melody 76
'great confinement' 223
Great Trigonometric Survey of India 213
'Great War' *see* First World War
Greece 103, 214
Green Island 89
Greenland 122, 124, 128, 133
Gregory, Derek 11, 15
Gregory, Ian 2
Greytown 92
Gross, Lia 60, 63, 65
Grotius, Hugo 225
Guiana 160–161
guidebooks *see* travel guides

Habsburg monarchy 43, 45, 47, 227
Hadjikyriacou, Antonis 225
Hajime, Shimizu 241, 244
Hajj 179
Halifax, Lord 109
Hanson, Julienne 102
harbour 188, 190–191, 193, 196, 198, 200, 231, 245
Harley, (John) Brian 12, 43
Harvard University *see* universities
Harvey, David 6, 10, 13, 230
Haselberger, Lothar 103
Haus Wachenfeld 105–106
Hawai'i 241
Hawkins, Harriet 8
Heffernan, Michael 18
Hegel, Georg Wilhelm Friedrich 225
Heidegger, Martin 5, 14
Heimatkunde 62
Heine, Heinrich 10

Herschel, Sir John 103
heterotopias 13, 121–123, 125–127, 130, 135
hiking 18, 66, 68
Hillary, Edmund 214
Hillier, Bill 102
hills 91, 93–94, 96, 211, 213–216
Himalayas 242
Hindu 242
Hirado 191–192
Hirm, Ferdinand 63
historical geography 2, 4–7, 12, 18–19, 275
historical GIS *see* geographic information systems/geographical information science
historicism 9
historicization 1, 177, 254
history of science 6–7
Hitler, Adolf 103–105, 108–116
Hobsbawm, Eric 270
Hohenstein 63
Hohenzollern monarchy 45
Holocaust 247, 275, 277–278, 284
Holocaust Geographies Collaborative 277
Holy Land 216
Holy League 226
home 1, 3, 5, 8, 47, 60, 64, 68–69, 74, 76, 79–83, 90–92, 95–97, 104–105, 114, 116, 129, 140–141, 145, 148, 150, 156, 164, 166, 188, 193, 196–197, 209, 212, 214, 216, 243, 277, 279, 282
homeland 61–65, 68
Hopkins, Pauline 80
Horden, Peregrine 225
Hoskins, W. G. 7, 208
hospitals 13, 281, 283
houses/housing 18, 77–80, 94, 102, 111, 158, 166, 192–196, 256, 277–282
Howells, William Dean 82
Hromadka, Georg 60–61
hubs 2, 47, 156, 158, 160, 188, 199, 232
humanism 8, 15
humanities 2, 5, 8–9, 90, 253, 256, 274–276
Hungary 45–46, 49, 62, 144, 278
hydroelectric dams 154

imaginaries/imagination 1, 8, 10–11, 15–16, 19, 39–40, 44, 51, 58, 96, 144, 163, 171, 175, 177, 237–248, 254
'imaginative geography' 11, 238
Incas 155, 157

Index

independence 41, 43, 45, 50–51, 167, 227, 239, 245
India 195, 213, 223, 239, 241–212, 246
Indians 155, 159, 161, 164–165, 242, 244
indigenous people 133, 157–161, 163–164, 167–168, 243
Indochina 241, 244
Indonesia 39, 239
industrialization 172, 179
infrastructure 1, 19, 43, 51, 58, 171–183, 255, 266; 'critical' 173; technical 171–172
Institute of Pacific Relations 239
insularity 253–254, 269
intellectual landscape 8
intercultural encounters 156, 189–191
interdisciplinarity 2, 274, 282
internet 10, 88–89
Inverted Distance Weighted (IDW) Interpolation 259, 262
Iraq 223, 226
Ireland 96
Iron Curtain 63
irredentism 144
irrigation 155, 261
Islam 224
islands 87, 89, 92, 96, 122, 133, 192–193, 195–196, 199, 217, 226, 228–234, 241, 243, 253–255, 260–263, 266, 269
Istanbul 228, 230, 233
italianità (Italian-ness) 143, 146
Italian Renaissance 113
Italy 56, 78, 113, 140–144, 146–147, 149–151, 212, 215–216, 227, 277
Izmir 230

Jacobs, Jane M. 5
Jameson, Fredric 9, 11–12
Japan 181, 188–200, 237–248
Jaskot, Paul 275
Java 239
Jesuit missionaries 157, 161–165
Jews 78, 182, 192, 195, 274, 277–283
Johnson, Matthew 18
Jones, Thomas 108, 110
Jordan 223
journalists/journalism 10, 60, 80, 82, 105, 142–144
journals 3, 5–6, 8, 17, 42, 56, 110, 112, 125–126, 130, 133–135, 142, 146, 245
journeys 3, 19, 57, 63–64, 68, 82, 93, 114, 126, 130, 148, 158, 160–161, 172–173, 176–177, 191, 212, 215–216, 283
Judge, Joan 89

Kaizō 245
Kankrin, George von 179
Katkov, Mikhail 175
Keiji, Tanaka 243
Kenya 39
kernel density analysis 280
Kirchner, Hermann 214–215
Kirkpatrick, Ivone 109
kitchen 79, 90, 111, 210, 218
Klausenburg 67
knights 108
knowledge 1–2, 5–9, 12, 18, 43, 63, 75, 156, 163, 165–167, 179–180, 207, 256, 275; production 1, 5, 7, 9
Knowles, Anne Kelly 275, 284
Kolberg, Juliusz 41
Konstantin, Grand Duke 45
Konvitz, Josef 43
Korea 246
Koshar, Rudy 56
kraj see country
Kraków 45, 47
Krämer, Willi 68
Krasław 41
Krasnoiarsk 183
Kyoto School 245

laboratories 1
Labrador 122
Lake District, the 218
lakes 158–159, 161, 163, 218; Lake Wakatipu 95
Land, Isaac 232
land/landscapes 1, 3, 5–8, 15, 19, 39, 43, 46, 49, 57, 61–63, 65–66, 68–69, 75–78, 80, 83, 87, 89, 92–97, 115, 136, 154–155, 157, 160–161, 163, 166–167, 173, 176, 179–180, 192, 194, 198, 207–219, 222–234, 237, 239, 242, 244, 247–248, 254–256, 258, 262, 269; cultural construction of/cultural landscape 3, 208–210, 215, 219; history 7, 210; imagined landscapes 77; as 'way of seeing' 15, 208
Latvia 41, 49–50
League of Nations 241
Lefebvre, Henri 9, 13–17, 19, 171, 207–219, 221, 223, 284
Lelewel, Joachim 50–51
Le Roy Ladurie, Emmanuel 252

letters 3, 58–59, 63, 87, 89–90, 92, 94–95, 102, 134, 141–142, 147
Lewis, Martin 240
Libeskind, Daniel 103
libraries 40–41, 73–74, 87–88, 98, 109, 114–115, 131
Libya 223
lighthouses 92, 131
linearity 4, 10, 12, 63, 155, 163
linguistics 3, 12, 15, 44, 50, 197, 244
'linguistic turn' 12, 15
Lisbon 168
List, Friedrich 174–175
literary studies 6, 8, 18
literature 4, 12–13, 15, 17, 40, 76, 80, 89, 165, 189, 208, 240, 243, 246–247
Lithgow, William 216
Lithuania 40–42, 44, 47–51, 181
Lloyd George, David 108–110
local 6–7, 14, 18, 39, 50, 57, 63–68, 73, 75, 88, 97–98, *106*, 108, 140, 143–146, 149–150, 190–191, 193, 196–200, 211–213, 215, 219, 254–255, 269, 278; history 6–7
locale 16, 278
location 9, 13, 16, 66, 73, 148, 158–159, 164, 166, 191, 198, 207, 210, 214, 259
locomotive 172
lodging houses *see* boarding houses
logbook 128, 134
London 124, 177
London School of Economics 18
Lonely Planet 59
Lorraine 39
Los Angeles/Los Angeles School (of Urbanism) 11
Louis XIV 43
Lowenthal, David 8
Löw, Martina 16
Lüdtke, Alf 139
Luyia community 39
Lyons 215
Lyotard, Jean-François 12

MacArthur, Julie 39
magazines 105, 142, 214, 240, 245
Maier, Charles 175, 223, 225
Makart, Hans 108
Malta 226
Malthus, Thomas 18
Manchuria 183, 243, 245
Mander, Jane 87
Mania 227
Māori 96–97

Map of Cassini *see Carte Générale de la France*
Mappa (Mappy) Polski 40, 43–51
maps/mapping 1–3, 7, 9, 12, 14, 19, 39–51, 57–58, 62, 64–66, 68–69, 74, 77–78, 82–83, 89, 102–103, *107*, 108, *124*, 151, 160–166, 173–174, 176, 178, 188, 190, 197–198, 210, 223, 238, 240, 259, 261–263, 269–270, 274–284; *see also* digital mapping and GIS
Maramureş 60, 66, 68
Margariti, Roxani 230
maritoriality 19, 222–234
markets/market behaviour 10, 56, 74–76, 79, 83, 139–140, 150, 172–173, 176, 193, 195, 198, 212, 266, 277, 281
Marmara 228
Maruyama 197
Marxism 10, 13, 15, 225, 245, 258
Marx, Karl 10, 14
Massachusetts 73
Massey, Doreen 5, 17, 187, 189–190, 237–238
Masurovsky, Marc 283–284
materialism 252–253; 'new materialism' 252
materiality 1, 15–16, 18, 40, 42, 50, 252–253, 270; culture divide 252
Mayhew, Robert J. 18
Mays, Edward 128
McLuhan, Marshall 10
Mecca 179
Medina 179
Mediterranean Sea 19, 150, 192, 222, 226–233, 241, 254, 263, 266, 269
Meiji 241
Melanesia 241
Meléndez, Mariselle 18
memoirs 3, 75–78, 81, 83, 103, 105, 110, 141, 145, 283
mental maps 1, 57–58, 62, 64, 68–69, 77, 238, 283
Mercantile Marine Act 131
Mercantile Marine Office (MMO) 124, 131, 133–135
Merchant Shipping Act 131–132
Mesaoria 256–257, 263
Mesopotamia 103
Messina 216
Mexico 75
Micronesia 241, 243
'microregions' 18
middle classes 77, 79–82

Miki, Kiyoshi 245–248
mills 46–47, 91
Ming dynasty 190
Minh, Ho Chi 240
mining 46–47, 87, 95, 97, 164–167
Mitchell, Don 8
Mitford, Unity 108–109
Mittler, Barbara 89
mobility 10, 123, 125, 127, 130, 135, 171–173, 179–180, 182–183, 192, 196, 199
models/modelling 10, 87, 90, 97, 108, 113–134, 146, 156, 219, 224, 254, 260, 279, 281–283
modernism 9
modernity/modernization 14, 57, 61, 63, 66, 89, 143, 154, 172, 174, 177, 183, 223
monasteries 103, 256
Montemarciano 149
Montesquieu 244
Morea 229, 231
Moscow 139, 175–178, 192
Moscow News 175
Mosley, Oswald 108
mountains 9, 63, 66, 92, 95–96, 103–105, 108–109, 113, 155, 158, 208–219, 232, 255, 260, 263; Alps 69, 103, 110, 211–215, 217, 219; Andes 155, 163; Eiger 214; Făgăraş Mountains 63; Mont Blanc 214, 217; Mount Catherine 216; Mount Etna 216–217; Mount Everest 213–214; Mount Parnassus 214, 216; Mount Quarantine 216; Mount Untersberg 108; Rocciamelone 215, 217; Troodos Mountains 263
Mukerji, Chandra 223
Munich 105, 110, 112–113
Murray (John) 56, 59
museums 56, 67, 103, 110
Muslims 179, 195, 244, 257
Mussolini, Benito 141, 143–147, 149

Nabi 224, 227
Nagasaki 188, 190–197, 199–200
Nan'yō see 'South Seas'
Napoleon 51, 238
Napoleonic Wars 130
nationalism 2, 8, 14–15, 18, 51, 56–57, 63, 69, 87–88, 142–144, 150, 155, 168, 172–173, 175, 179, 181–182, 190–191, 198, 200, 237–238, 242, 247, 258, 278

national parks 210
nation/nation-state/nationhood 11, 40, 43, 47–48, 51, 58, 96, 146, 154–155, 160, 166–167, 174–175, 189, 239–242, 241–155, 258
native species 90, 92, 94–96
nature 3, 12, 63, 97, 108–109, 154, 156, 162, 230, 247, 252; culture divide 12, 162, 252
Nazi Germany 59, 139
Nazis/National Socialists 59, 63, 104, 109–110, 115, 139, 155, 278, 283
neighbourhoods 145–146, 237, 259
Neo-Kantian philosophy 245
networks/network analysis 2, 10, 12, 57–58, 60, 65, 68, 88, 90, 103, 146, 149, 158, 162–164, 167, 171–175, 177, 181–182, 195, 199, 244, 277–278, 281, 283
New Chancellery, Berlin 113
Newfoundland 122, 124, 126
New Guinea 239, 241
New Mexico 74
newspapers 87–98
Newton, Isaac 14
New York City 177
New Zealand 19, 87, 96–97, 241; South Island 87, 96; Victoria 88
Nicholas I, Tsar 175, 177
Nicholas II, Tsar 181, 183
Nicholson-Smith, Donald 13
Nietzsche, Friedrich 14
Noble, Charles Frederick 198
Norgay, Tenzing 214
North America 2, 60, 155
novels 19, 76–83, 87, 243

Obersalzberg 110–111
Oceania 241
Odessa 178
opera house 176
Opera Nazionale Dopolavoro 146
oral history 103, 135, 283
Ordnance Survey 210
Orellana 159
Orientalism 11, 238
'Orient', the 238–239, 241–242, 248
Orinoco 158–159
Orthodox Church 47–49, 182
Osbeck, Pehr 198
Osterhammel, Jürgen 15
osterie 143–144, 149, 151
Otago/*Otago Witness* 87, 89, 92, 96–97

Ottomans/Ottoman Empire 19, 222, 224, 226–234, 253, 255, 257–259, 266, 269

Pacific 156, 176, 239–240, 243
Pacific War 243
Palazzo Venezia 140
Palestine 228, 255, 263
Palgrave, William Gifford 248
panoramas 4, 87, 106, 109, 113, 173
Papua New Guinea 239, 241
Pará 155
paradise 63, 96–97
Paris 12, 14, 18, 41, 43, 51, 113, 159, 175–176
parish churches 18, 49, 96, 140
parks 80–81, 210
partitions of Poland-Lithuania 41, 44, 47–51
Parsi 195
Pasha, Hüseyin 226, 228, 233
passengers 125–126, 130, 133, 172–173, 176–180, 182
paths/pathways 60–61, 94–95, 102, 113, 156, 158, 166, 211–212, 216, 230
patriots/patriotism 39, 61, 64, 67, 145
Peacock, Andrew 227
Pearl Harbour 245
Pearson, Michael 232
Peking 176–177
Peloponnese 227
Peluso, Nancy Lee 39
Perthes publishing house 42
Pesne, Antoine 112
Petermann, August 42
Peukert, Detlev 139
Philippines 237, 239, 241, 245–248
Philo, Chris 5
philosophers/philosophy 5, 8, 12–13, 17, 207, 217, 245
photographs/photography 3, 58, 64, 74–75, 77, 102–103, 105, *107*, 109, 111–113, 151, 214
Piasecka, Janina 41
piazzas 148, 150
picturesque 61, 173
Piłsudski, Jósef 181
Piyade 131
place 1, 3–9, 11, 13, 16–18, 40, 43–44, 46, 56–58, 62, 68–69, 73, 75–81, 83, 87–92, 94–96, 103, 122, 125, 127–128, 132, 140–141, 143–144, 146–150, 156–161, *162*, 166–168, 172–173, 177, 182, 189–190, 199, 210, 212, 215–216, 218, 224–226, 229–230, 233, 240, 242, 255, 259, 263, 276–284; 'place-making' 3–4, 13; 'sense of place' 16
plans 102–103, 105–106, 109–110, 112–116, 151, 182, 217, 278, 282; Plan of Saint Gall 103
Plater, Ludwik 41, 50–51
Plater, Stanisław 40–42
plazas 73, 113, 157
poems 3, 61–63, 68, 222, 224, 234
Poland 39–51, 181–182, 278
polycentricity 156, 163
Polynesia 241
Porter-Szücs, Brian 45
Portman, John 11
Pomerania 46
Poniatowski, Stanisław August 49
Pontine Marshes 150
Poole, Joshua 214
population 18, 46, 49, 59, 78, 90, 96, 133, 156, 158, 161–162, 173, 177, 179, 183, 189, 208, 213, 223, 230, 233, 238–239, 244, 255–256, 258, 278
Porai-Koshits, Vladimir 180
ports 47, 123, 133, 177, 188, 191–193, 195–197, 199–200, 227, 230–231
Portugal 78, 165
Portuguese Empire 155, 158, 161–165, 167–168, 192, 197, 241
Posen, Grand Duchy of 46–47, 50
postcolonialism 189
postmodernism 9, 11
poststructuralism 12
Poznań 40, 44
Pratt, Mary-Louise 189
Presner, Todd 284
priests 157, 196, 198
principalities 174
prisons 13, 192, 276
projection 109, 140, 143, 225, 261
propaganda 61, 104, 149, 245, 247
prostitutes 77, 82, 197–198
Protestant church 47, 81, 96
Prussia 40–43, 45–47, 50
psychology/psychologists 58, 90, 102, 106, 111–112
public libraries 73, 88
public security 145, 147, 150
public transport 171, 178
public welfare 171
Purcell, Nicholas 225
Puritans 73

Qing dynasty 188–191
quantitative methods/sources 3, 78, 148, 254–255, 258, 261, 263, 269–270, 274–275
Qur'ān 224

race 75, 77, 80, 82, 197, 241–242, 244, 247
Raffestin, Claude 223
Ragıb, Koca 222, 234
railways/railroads 10, 66, 95, 127, 140, 171–183, 211; *Moscow-Vindava-Rybinsk Railway Company* (MVRRC) 177–178; Saint Petersburg-Warsaw Railroad 181; transcontinental railways 175–176, 179; Trans-Siberian Railroad 175–177, 182–183
Rajsko 283
Raleigh/Ralegh, Sir Walter 158–159
Rankin, William 4–5, 16
Ratzel, Friedrich 15
Rau, Susanne 17
region 7, 14, 19, 41, 45–46, 50, 58, 62–64, 66, 75–76, 83, 87, 89, 97, 143, 146, 154–155, 158–164, 166–168, 172, 175, 182, 189, 212, 237–248, 254–256, 259–260, 262–263, 266, 269
regional history 7, 14, 87
Registrar General of Shipping and Seamen (RGSS) 124, 131, 135
Reid, Anthony 240
Reis, Piri 231
religion 41, 44, 47, 96, 103–104, 110, 160, 190, 196–197, 215–216, 244
Relph, Edward 8
Renaissance humanism 8
Riis, Jacob 75, 78
Risorgimento 56
riverbanks 157–158, 162, 164, 166, 168
rivers 9, 19, 96, 154–168, 174, 182, 191, 196, 198, 200, 209, 213; Amazon River 154–164, 166–168; Amur River 182; Awamoko Stream 95; Caroni River 159; Columbia River 156; Danube River 68, 154; Madeira River 161, 163–165; Mississippi River 154, 158, 223; Pearl River 191, 200; River Nile 154–155; River Rhine 154; River Thames 154; Tapajós River 161, 163–168; Waiau River 96; Xingu River 163–164, 167
riverscape 163, 166, 168
riverways 161

Rizal, José 247–248
roads/road networks 3–4, 43, 87, 94, 156, 158, 167, 171, 175, 209, 211, 230, 278, 283
rocks 88, 92–94, 96–97, 124, 156, 163, 211, 214, 216, 231
Röhm, Ernst 112
Roman Empire 156
Romania 49, 59–68
Romanian Germans 52, 59–65, 67–68
Rome 47, 56, 103, 144–145, 147, 156
Rorty, Richard 12
routes 4, 66, 113, 159–160, 165, 198, 211–213, 216, 219, 227, 230, 233–234, 281–283
Royal Navy 130
Rüdiger, Jan 224, 226
ruins 217–218, 231, 259
Rupea 66
rural (areas) 7, 39, 58, 61, 64, 89, 93, 146, 255–256, 263, 269
Russia 19, 40–41, 43, 45–51, 139, 173–183, 227
Russian Revolution 181, 183
Russo-Japanese War 181
Ruthenian 49–50

Sachsenmaier, Dominic 8
Sack, Robert 223, 233
Said, Edward W. 11, 238
Saint Petersburg 177–178, 181
'Sakoku' 190
Santos, Epifanio de los 248
salt mines *see* mining
Salzburg 108, 115
Samuel, Raphael 5
San, Aung 240
Sauer, Carl 208
Saunders, Frances Stonor 11
Savoy 211
Sawyer, Frederic H. 248
scale 18, 60, 62, 65–66, 69, 74, 83, 102, 111, 113, 128, 132, 141, 150, 155–158, 161–164, 167–168, 176, 209, 237–238, 252, 254, 261, 263, 284
scene 75, 77, 83, 92, 94, 109, 129
Schama, Simon 155
Schivelbusch, Wolfgang 173
Schlögel, Karl 15, 18
Schmid, Christian 14
Schneider, Erika 60, 65
Schroer, Markus 16
Schuschnigg, Kurt 112

Scotland 87, 90, 121–122, 124, 128, 133, 135, 216, 218
Scudder, Vida 82–83
seas 9, 44, 62, 64, 94, 121–126, 128, 130–133, 156, 217, 222, 224–234, 238, 243, 245, 248, 254
Second World War 60, 103, 167, 171, 239–240
sedentary societies 224
Seiyō 242–243
Selim I 226
separatist movements 175, 181
sericulture 254, 266
servants 76–77, 108, 195–196, 198, 228–229
settlements 3, 15, 73, 77, 82–83, 134, 156–158, 162–164, 166–167, 192, 195, 198, 227, 231, 244, 247, 255–257, 259
Seven Years' War 227
sexuality 77, 79–82, 112
Shanghai 89
Shepperson, Mary 102–103
ships 19, 121–136, 159, 173, 191, 195–196, 199, 222, 227–230, 234; logs/records 124–130, 135–131, 133, 135; shipping routes 230, 233–234
shops 102, 140
Shōwa Research Association 245
Siam 241, 244
Siberia 175–177, 182–183
Sibiu/Hermannstadt 62–63, 67
Siebenbürgen/*Siebenbürgerlied* 62–63
Siebert, Loren 2
Silesia 46–47, 50
Simler, Josias 211–212
sistematizare 61
sites 1, 11–13, 18, 46–47, 56, 76, 79–80, 115, 122, 140, 145–146, 156–157, 164, 189, 208, 216, 225, 276–278, 281, 283; counter-sites 122
Situationism/Situationist International 13–14
sketches 66–67, 103, 105, 174, 178, 278, 284
Skinner, Barbara 49
Slawinski, Józef 41
Slovakia 49–50
Smith, Bernard 240
social formation 15, 207–208
social history/historians 5, 19, 121–122, 125, 136, 252
social integration 171, 175
socialism/socialists 59–63, 68–69, 144–145, 149, 181, 183

social reform 75, 78–81
social relations 1, 10, 17, 121, 283
social sciences 9, 253
social strata 178, 258
Soja, Edward 9–11, 13, 15
Sorbian 49–50
Sotira 256–257
South America 19, 154, 156–157, 160–161, 163, 166, 168
Southeast Asia 237–245
Southland 87–88, 96
'South Seas' 238, 241, 243, 248
Soviet Union 11, 59, 61, 139, 145
spaccio see bars
space/spaces: 'abolition'/'annihilation' of 10; absolute space 19, 189, 209, 230; abstract space 14, 229; administrative spaces 130, 133–136; architectural spaces 102; *see also* architects/architecture; blank/empty spaces 69, 283; border spaces 198–199; city spaces 73; *see also* cities; collective imagining of 57; colonial space/colonisation of space 39, 150; commercial spaces 196; communal spaces 108; conceived space 14; concept of 16–17, 188–189, 207; cultural constructedness of/social construction of 3–4, 208, 252, 269; differential space 14; diplomatic space 196; division/partitioning of 83, 180, 207–208, 210, 219, 223; domestic spaces 65, 74–76, 81–82, 104–106, 115; economic space 47, 253; everyday spaces 140–143, 145, 150–151; finite space 127; geometric space 14; harbour spaces 188, 190; heterotopic space 125–126, 130; historical contingency of 4, 11; 'hyperspace' 11; imagined space 223; imperial space 175; indigenous spaces 159–161; insular spaces 230, 269; intellectual space 8; interior spaces 103; itinerant spaces 127; layered space 122, 229; limitless space 109; linguistic spaces 3, 12, 15, 44, 50; liquid space 224–226, 231, 233; littoral space 232; lived space 14; 'living space' 18, 132; lost space 224; malleable spaces 127; mapping of *see* maps/mapping; materiality

of/physical space 1, 15–16, 18, 40, 132, 208, 223, 252–253, 270; metaphysical space 209, 219; mobile spaces 125, 127; 'monumental space' 150; moral/immoral spaces 79; multivalent spaces 148; 'natural' space 171; open space 76; organization of/ordering of 73–74, 83, 103; perceived spaces/perception of 121, 125, 127, 131; performance spaces/performative spaces 105, 116; as 'permanences' 230, 233; political space 143, 147, 149, 180; private/domestic space 65, 74–76, 81–82, 104–106, 115, 122; production/reproduction of 13, 171–173, 188, 209, 284; public space 73–74, 122, 150; regional spaces *see* region; remote spaces 167; representational spaces 14, 106, 207–210, 213–219, 284; representations of space/spatial representation 14, 121, 171, 195, 207–210, 217–219, 284; river/riverine spaces 154, 156, 158, 163–164, 166–167; *see also* rivers; sacred spaces 79; 'sense of space' 222, 224; shipboard spaces 121, 123, 125, 129, 131–132, 134, 136; *see also* ships; social space/spaces of sociability 14, 16, 142–143, 146, 150, 171–172, 189, 207, 209; spaces of encounter 156; spaces of intimacy 76; spaces of resistance 13; 'space-time' 17; 'spacing' 16; as a 'stage' 1, 3, 139, 225; symbolic space 1, 3, 12–13, 16, 103, 154, 208–209, 217; 'taming of' 189; trading spaces 193, 196, 198–199; tripartite division of 207–210, 219; urban space 77–78, 81, 209, 217; *see also* urban (areas); use of 79, 219; working-class spaces 75
Spain 73, 75, 157, 197, 226–227, 245–247
Spallek, Waldemar 41
spatial: agency 140; analysis 190, 192, 200, 259, 274, 278; attributes/characteristics 191, 253; categories 233, 241, 270; configurations 102; construction 195, 198; data 280; design 178; dimensions 1, 6–7, 10, 17, 62, 141, 256, 270; discourse 193; disintegration 175; dynamics 77, 84, 141, 151; expectations 76; experiences 4, 12, 104, 283; extension 163, 168; fiction 74–75; formations 73–75, 83; historians 8–9, 154, 284; imaginaries/imaginings 11, 16, 44, 51, 69, 175, 238–242, 248, 254; *see also* imaginaries/imagination; integration 173, 176–177; knowledge 2; logic 80; matters 6; order 179, 183, 257; patterns 163, 167–168, 274–275, 277, 281–282; perceptions 172–173; politics 59, 61, 64–65, 69; practices 14, 168, 207–219, 284; questions 284; relations 10, 135; restrictions 193, 199; rotation 163; routines 13; science 3, 8, 14; sensibilities 75; 'spatialization' 171; 'spatial turn' 9–18, 225, 252; Stanford *Spatial History Project* 4; statistics 274; theorists 5, 15, 17, 200; tools/methods for spatial analysis 2, 4–5, 7, 12, 19, 88, 90, 122, 188, 208, 210, 213, 219, 240, 253, 256, 259–262, 270, 274–278, 280–281, 283–284; transgression 81
'spatial history' 1–19, 57, 87, 97, 102, 121, 140–141, 150–151, 154–158, 166, 168, 171, 184, 209, 237, 240, 252–253, 269–270, 274, 277–279, 284; 'socio-spatial history' 102; as a 'way of seeing' 9, 141, 150, 281
'spatial turn' *see* spatial
Speer, Albert 108, 110–111, 113
spheres 5, 17, 65, 76, 79, 83, 165, 195, 231; domestic/public binary 76, 79, 83
squares 139–140, 145, 148
Sri Lanka 239
S.S. *Esquimaux* 121–123, 125–129, 131–135
Stadler, Matthew 109
Stalin, Joseph 62, 139
Stalintown (Oraşul Stalin) 62
Stanek, Łukasz 14
Stanford University *see* universities
state surveillance 141
stations 92, 128, 140, 171–173, 177–179, 225
statistics/statistical analysis 40–42, 44, 51, 173, 179, 253, 256, 260–261, 274

Stephani, Claus 66
Steiner, Erik 283–284
Stock, Paul 17–18
Stone, Lawrence 252, 270
Stork, Sophie 112
Storper, Michael 11
storytellers/storytelling 68, 75, 92, 95–96, 105, 154–155, 158–161, 163, 188–189, 210, 213–215, 219, 279
Stowe, William 56, 58
Strătescu, Stefan 68
streets/streetscapes 75, 77–82, 140, 145, 148, 193, 277–279, 281–282
structures 4, 19, 57, 73, 75, 89–90, 97, 102, 106, 109, 115–116, 127, 131, 134, 154, 158, 172, 174, 177, 190, 243, 252, 275
suburbs 279–280
subversion/'subversive acts' 42, 49, 62, 147–149
Sudan 223
Süleyman I 226
Sultan 226–228, 233–234, 258; Sultan Mustafa III 226, 233
Surrey 109
surveyors/surveying 3–4, 16, 43, 255, 260, 263
Swabians 60, 63, 68
Switzerland 60, 212
Syria 223, 227–228

Tabak, Faruk 254, 263, 269
tahrir 255, 257, 262
Taiwan 239
Tambiah, Stanley Jeyaraja 240
Tanaka, Keiji 243–244
Tapajós/Madeira complex 161
telecommunications 171
telegraphs 172, 181–182
terrain 4, 49–50, 81, 83
territory/territoriality/territorialization 1, 8, 11, 40, 43–49, 51, 144, 165, 167, 172–175, 177–183, 189, 191–192, 222–226, 229, 237, 239, 243, 245
Thailand 244
Thévenot, Jean de 216
'thick description' 89, 94
'Third World', the 11
'time-space compression' 10
Timor 239, 241
Titian 113
Titsingh, Isaac 193–194
Tōa 243, 245

Tokugawa 188–191, 196
Tokyo 2, 191
topography 7, 262
Tōru, Yano 244
tourism 56–57, 60–62, 70, 215
towns 4, 43, 46–47, 62, 82, 88, 139, 150, 156–158, 167, 217, 256, 276, 278; plans/planning 4, 217; townships 157
Tōyō see 'Orient', the
trains *see* railways
trattorie 143–144
transatlantic voyages 122, 126, 163
transnational history 7, 252
Transylvania 59–62, 66–68
travel guides/travelogues 1, 19, 56–69, 142, 151, 173, 180, 198; *osterie* guides 151
travellers 56–58, 173, 178, 211–212, 214–218, 248
Treaty of Carlowitz 227
Treaty of Madrid 162
trigonometrical surveys 213, 259
Troost, Gerdy 105, 109
tropical/the tropics 164, 244–245, 247–248
Tsarist Empire 43, 45, 175, 177, 179–183
Tuan, Yi-Fu 8
Turin 149, 215
Turkey 103
Turkish 227, 259
Turkestan 179, 183

Udine 144
Ukraine 44, 49, 182, 226
Ülgün 227
United States, the 2–3, 5–6, 8–9, 13, 15, 60, 74–77, 80, 89, 124, 133, 155–156, 172, 177, 196, 199, 223, 227, 239, 242, 245–246, 248
universities 1–2, 7, 14, 41, 77, 79, 243, 277; Harvard University 79; Stanford University 2, 4, 277; University of Hiroshima 243; University of Lancaster 2; University of Leicester 7; University of Nanterre 14; University of Warsaw 41; Wellesley College 79, 82
urban (areas) 3, 7, 9, 11, 13–14, 19, 57, 73–74, 77–83, 103, 111, 177, 179, 192, 209, 217–218; geography 79, 82–83; history 7, 19, 73, 83; planning 14, 218; urbanization 179
Urry, John 56

Venice 226, 231, 255, 259
Verne, Jules 176
Vicentini, Raffaele 145
Victorian society 80; Victorian cities 82
Vidal de la Blache, Paul 15
Vienna 39, 44–45, 51
Vietnam 239–240
villages 10, 73–74, 92, 94, 149–150, 162, 164–167, 211, 232, 256–257, 259–262
Vilna 181
Virilio, Paul 10
Viscri 62, 66
Vistula 47
visualization 2, 40–41, 51, 103, 210, 261, 269, 277, 280, 283–284
Vitebsk 44, 178
Vitte, Sergei 176, 182
volcanoes 216
voyages 121–124, 126–128, 132–133, 164–165

Wagner, Richard 114
Wahnfried 114
Walcott, Derek 224
Walker, Andrew Barclay 125–130, 133–135
Wallachia 49
walls 4, 11, 83, 93, 102–103, 109, 150, 191–193, 281–283
warehouses 82, 193
Warf, Barney 18
Warmia 47
wars: War of American Independence 227; War of the Austrian Succession 227; War of Cyprus 263
Warsaw 41, 45, 51, 181
waterscape 164

Westin Bonaventure Hotel 11
'West', the 1, 11, 60–61, 64–65, 69, 76, 156, 176, 183, 237, 242
whalers/whaling 121–123, 125, 127–130, 133–135
Whampoa 196
White, Hayden 12, 75
White, Richard 2, 156
Wick, Alexis 224–225
Wigen, Kären 240
'wilderness' 214, 237
Williams, Raymond 89
Withers, Charles 5
Wittfogel, Karl 155–156
Wittstock-Reich, Rohtraut 67
Wodzisław 283
Wolch, Jennifer 11
Wolfe, Albert 79
Wolters, Oliver William 240
woodland 94
workhouses 13
workplace 14, 16, 76, 80
World Exhibition 175–176
World Wide Web 10
Worster, Donald 156

Yemen 226
yūjo 197–198

Zipser, the 66
zones 77, 116, 147, 156–157, 165, 223; border 18–19, 188–200; 'contact zones' 189, 191–192, 195–196, 200, 227
Zorzi, Elio 144
Zmeu, Costache 61
Zweier, Ewalt 66

CPSIA information can be obtained
at www.ICGtesting.com
Printed in the USA
LVHW081044060822
725339LV00023B/514